GEOPHYSICS & GEODESICS

GEOPHYSICS & GEODESICS

Editor

Julius Schult

scitus academics

Geophysics & Geodesics

Edited by **Julius Schult**

This Book is Printed on Acid-Free Paper. ♾

ISBN: 9781681178981

Library of Congress Control Number: 2017948927

SCITUS Academics LLC,

616, Corporate Way, Suite 2, 4766,

Valley Cottage, NY 10989

www.scitusacademics.com

Notice

Printed in United States of America.

Preface

Geophysics is broadly the study of the physics of planetary bodies and atmospheres, including the flow of energy through these systems. We have seen a remarkable quarter-century of technological expansion in geophysical methods and this growth is continuing. Geophysical observations are fundamental to our understanding of the Earth and how it works for and against us. For example, motion in the Earth's deep core gives rise to our magnetic field, which protects us from harmful solar radiation. Earthquakes and volcanic eruptions, on the other hand, have the ability destroy on a terrible scale. Geophysics is also fundamental to the needs of society – it is essential for exploring for energy, water, and mineral resources, monitoring environmental impact and change and assessing natural and manmade hazards. The use of geophysical methods in petroleum exploration continues to expand and is an integral discipline in the discovery and production of oil and gas deposits. Similarly, the geodetic infrastructure provides the foundation for numerous applications with broad societal and commercial impact, from early warning systems for hazards to intelligent transportation systems. Over time, the precision and timeliness of these applications has improved, with operational applications now routinely working at accuracies only recently achievable, while the scientific applications are approaching the part-per-billion accuracy level and near-real-time operations.

This book encompasses the topics of the major sciences – physics, astronomy, planetary science, geology, environmental science, oceanography, meteorology, and scientific problems of geodesy. It presents on theoretical or modeling studies, and on results of experiments and interpretations. This will be of immense interest for students, researchers as well as scientists from different disciplines as geophysicists, including seismologists, meteorologists, glaciologists, geomagnetists and geodesists.

Table of Contents

CHAPTER 1

Planetary Sciences, Geodynamics, Impacts, Mass Extinctions, and Evolution: Developments and Interconnections

Jaime Urrutia-Fucugauchi and Ligia Pérez-Cruz

Programa Universitario de Perforaciones en Océanos y Continentes, Departamento de Geomagnetismo y Exploración Geofísica, Instituto de Geofísica, Universidad Nacional Autónoma de México, Delegación Coyoacán, 04510 Mexico City, DF, Mexico

ABSTRACT

Research frontiers in geophysics are being expanded, with development of new fields resulting from technological advances such as the Earth observation satellite network, global positioning system, high pressure-temperature physics, tomographic methods, and big data computing. Planetary missions and enhanced exoplanets detection capabilities, with discovery of a wide range of exoplanets and multiple systems, have renewed attention to models of planetary system formation and planet's characteristics, Earth's interior, and geodynamics, highlighting the need to better understand the Earth system, processes, and spatio-temporal scales. Here we review the emerging interconnections resulting from advances in planetary sciences, geodynamics, high pressure-temperature physics, meteorite impacts, and mass extinctions.

1. INTRODUCTION

In the 16th and 17th centuries, physics encompassed a wide field of inquiry with significant advances coming from many widely separated endeavors in what are now astronomy, optics, mechanics, gas chemistry, thermodynamics, and so forth. They included development of the heliocentric model for the solar system, formulation of the laws of planetary motion, experimental and mathematical descriptions of pendular and parabolic motion, the law of universal gravitation, inertial reference frame and laws of motion, the pressure-volume Boyle law, and the ideal gas law, among many other discoveries. Modern research in physics continues to encompass a wide field of inquiry, which is reflected into the different disciplines and emerging frontiers.

Increased awareness on the role of interactions among the Earth's components of the atmosphere, hydrosphere, lithosphere, ionosphere, and biosphere (Figure 1) leads to integrative approaches in Earth system science. This has led to better understanding of interactions, component flow, and feedback mechanisms acting with distinct spatial-temporal scales and manifested in the geochemical cycles, surface processes, and Earth's climate. Recent advances in the study of the Earth's interior and external processes in the near Earth environment and the solar system are resulting in broad integrative approaches.

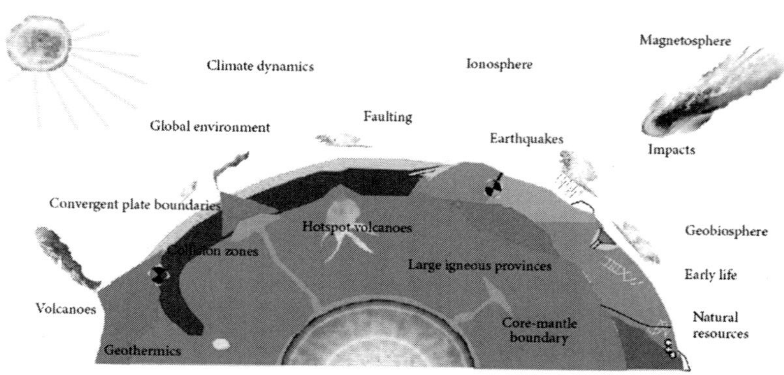

Figure 1: Earth system components connecting the Earth's deep interior, core, mantle, lithosphere, atmosphere, oceans, magnetosphere, and ionosphere with external processes like asteroid impacts, cosmic radiation, and solar winds (credits: International Scientific Continental Drilling Program (ICDP) website: http://www.icdp-online.org).

Recent and long standing questions are being investigated using technological and theoretical developments, which include high performance computing, big data analysis, satellite observation system, instrumental networks, and planetary missions. Planetary missions to the solar system and the discovery of exoplanets and multiple systems provide a broad context for Earth's studies, integrating studies and challenging models and theories. Here, we review developments in geodynamics, high pressure mineral physics, meteorite impacts, mass extinctions, and planetary sciences and in the emerging interconnections as fields develop.

2. GEODYNAMICS AND EARTH'S DEEP INTERIOR

In the 1960s and early 1970s, development of plate tectonics provided a new paradigm for the Earth sciences, with Earth's upper layer divided into several plates undergoing large-scale plate motions [1]. Plate tectonics integrated surface tectonic processes with the Earth's interior and deep energy sources, linking magmatism, seismicity, mountain building, and metallogeny in a unified way. The theory integrated the rich long-held archive of geological and geophysical data on the continents with the more recently acquired information on the oceans, particularly on the mid-ocean ridges, fracture zones, and trenches [2–4]. Plate tectonics provides a kinematic framework building on a global synthesis of geologic mapping, structural geology, stratigraphy, paleontology, petrology, geochemistry, seismology, paleomagnetism, geodesy and marine geology, and geophysics.

In the past three decades, plate tectonics has proved highly successful, prompting multi- and interdisciplinary studies. Understanding how the planet works has however remained a challenge, with the dynamics of deep and surface processes, mechanisms, and energy sources only partly investigated. Key aspects of plate dynamics, mantle convection, hotspot magmatism, mantle layered structure, convection, core-mantle, intraplate deformation, vertical motions, polar wandering, and plate driving forces still remain only partly understood.

Plate tectonics provide a global model for the lithosphere, which is broken into several plates undergoing relative motion at plate boundaries (Figure 2(a)). Oceanic lithosphere is created at ridges and recycled back into the mantle at subduction zones. The advent of international and regional broadband seismological and GPS networks has provided instantaneous plate motion data. This has opened new ways to study plate kinematics, with improved spatial and temporal resolution. Plate models integrating geological and geodetic data are being constructed, which permit analyzing plate reorganizations for the past few million years and evaluating plate deformation and diffuse plate boundaries. The recent synthesis by DeMets et al. [5] incorporates 27 plates, including six small plates not directly linked to the ridge system, and gives a high resolution plate kinematic model (Figure 2(b)). Their results confirm the rigid plate assumption and provide constraints on plate deformation resulting from thermal contraction and wide plate boundaries.

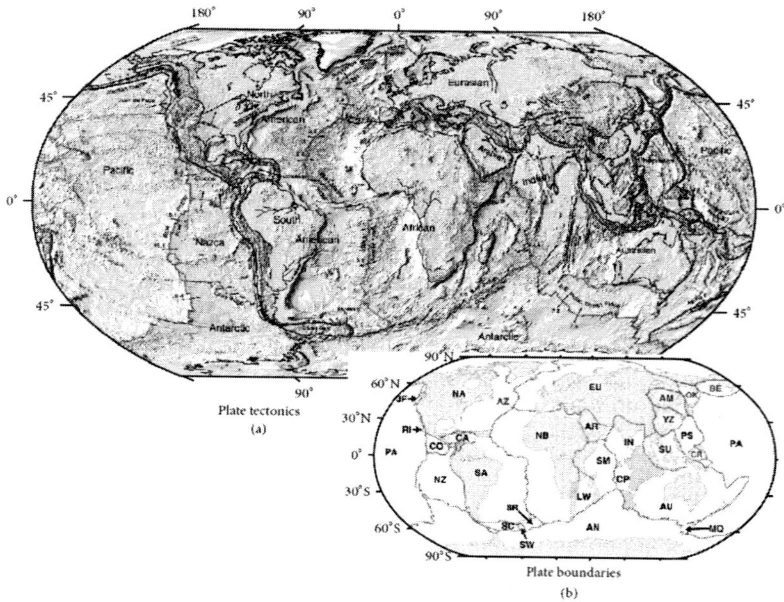

Figure 2: The Earth's lithosphere is divided into several tectonic plates that undergo relative motion. Plate boundaries are divergent boundaries (seafloor spreading ridges), convergent boundaries (subduction zones), and transform boundaries (transform faults). (a) Plate tectonic boundaries. Structural information on normal and reverse faults and volcanic centers is added (credits: NASA Earth Observatory and Goddard Space Flight Center website: http://earthobservatory.nasa.gov). (b) Global plate model incorporating 27 plates in a high resolution plate kinematic model (adapted from DeMets et al. [5]).

Over long time scales, plate motions have undergone major changes and plate reorganizations with ocean basins closing and opening, which relate to deep processes and mantle convection [6]. Geological estimates of plate motion using the marine magnetic anomalies, fracture zones, hotspot tracks, and paleomagnetic directions have been used to reconstruct plate kinematics for the past 200 Ma. Studies of oceanic plateaus, igneous provinces, orogenic belts, and volcanic arcs provide tight constraints on plate motions and mantle convection for the Phanerozoic and Precambrian, with formation of supercontinent assemblies and continental breakup [7]. The role of deep mantle structures in plate tectonics can be observed on the residual geoid long wavelength characteristics and shear wave velocity zones in the deep mantle and core-mantle boundary.

The challenge is how to use the improved resolution on plate kinematics for modeling plate dynamics [8]. The forces that control plate motion and the relation to deep processes in the mantle, the nature of hotspots, fate of subducted lithosphere, and processes at the core-mantle D″ zone are in general poorly constrained. Earth's deep structure, mineral composition, convection, high pressure/temperature physics, and energy sources remain as a major frontier [9, 10].

Advances are being made from seismological analyses imaging velocity anomalies, wave polarization, and seismic anisotropy features in the mantle and core [11]. Seismic wave attenuation anomalies have been documented with depth, which correlate with estimates from mantle viscosity from geodynamic modeling. Measurements of attenuation and other anelastic properties have been linked to rheological properties, which are investigated in theoretical models and laboratory experiments. The layered structure of Earth's interior is characterized by increase of pressure and temperature, with variation of physical properties and mineralogy and phase changes. Pressure increases from about 24 GPa in the crust to 364 GPa in the inner core (Figure 3). Recently, physical and compositional structural mineral properties are being determined at increasing pressure and temperature using diamond-anvil cells, laser beams, noble gas graphite furnaces, and synchrotron sources. MgSiO-rich perovskite is the main constituent of the lower mantle down to 2900 km. This mantle mineral undergoes a phase transformation to denser postperovskite at core-mantle conditions, characterizing the physical properties at the D″ layer. Iron and iron-silica alloys are investigated at simulated outer and inner core conditions, with pressures and temperatures up to 257 GPa and 2400 K [12] and 364 GPa and 5500 K [13]. Experiments on high pressure mineral physics are providing novel data on the mineralogy and physical properties like anelasticity and plasticity, which are coupled from first principles calculations in constraining phase transformations and depth variations [14–16].

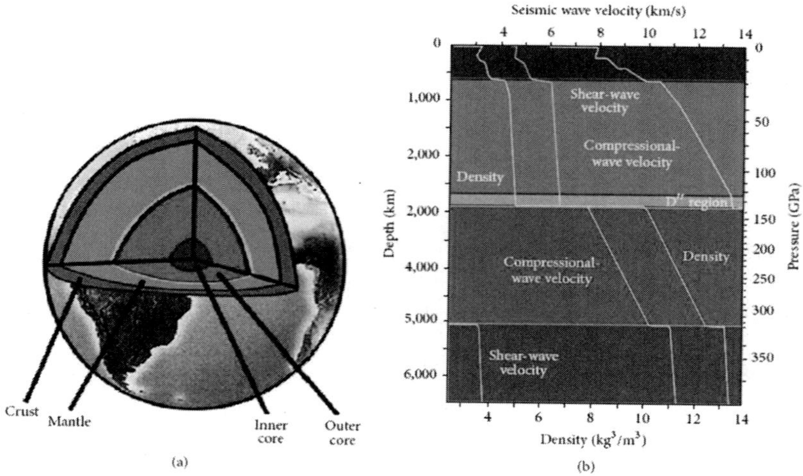

Figure 3: Density and seismic velocity variation with depth in Earth's interior (adapted from Romanowicz [9]).

Computer modeling of convection permits testing different boundary conditions, property contrasts, and geometries, including those long explored of whole-mantle and double-layer convection. Dynamo modeling for geomagnetic field generation simulates short- and long-term variations observed at the surface in secular variation and regional anomalies, including polarity reversals. Thermal boundary conditions play major roles in dynamo behavior. Increasing computational power permits simulating fine mesh geometries with higher resolution. The field of geodynamic modeling, coupled to deep interior models for layered convection, mantle viscosity, and physical property contrast regional anomalies, has greatly expended in recent years showing large potential for further developments [17].

Plate boundaries are locations for active exchange interactions from the deep mantle to the surface, which manifest in seismicity, heat flow, and magmatic activity (Figure 4). Regional instrumental networks, geophysical surveys, and modeling on zones like the San Andreas transform fault in western United States, the Dead Sea and Anatolian faults in the Middle East, or the Honshu subduction zone in Japan are providing fresh high resolution data. Studies also address the economic implications, where significant mineral and energy resources concentrate at plate boundaries and related hazards associated with earthquakes and volcanic eruptions [18, 19].

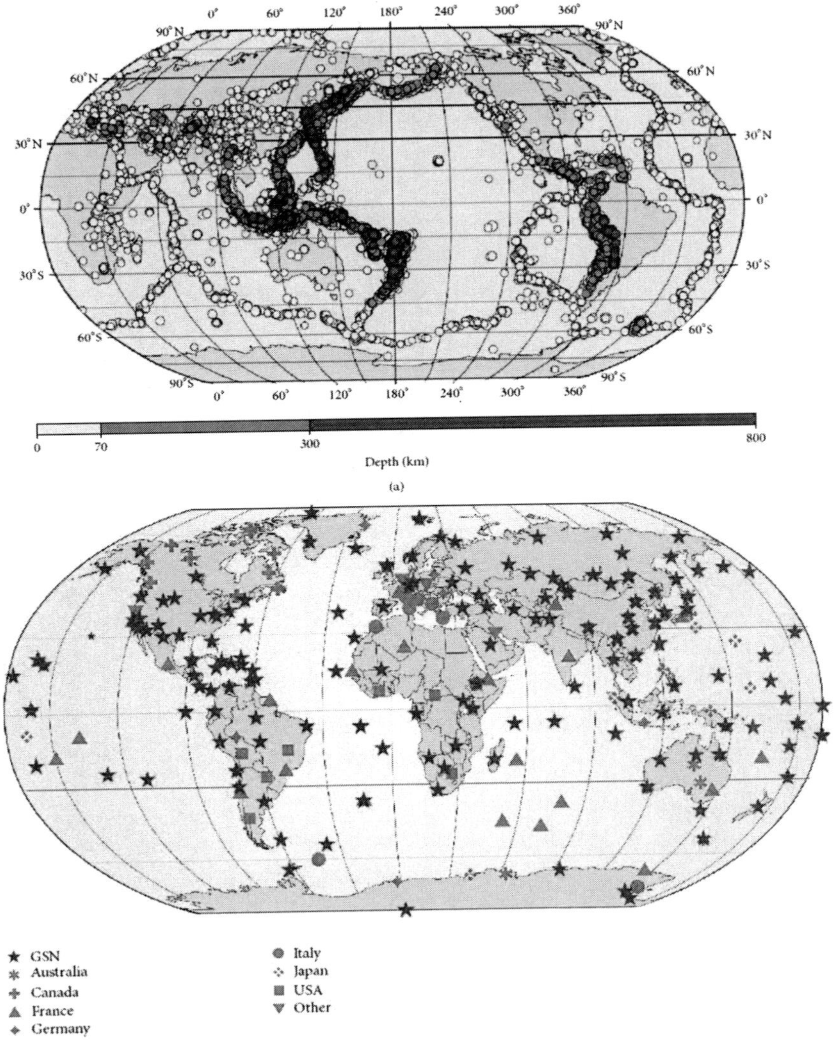

Figure 4: (a) Seismicity and plate boundaries, with focal depth distribution. (b) Global seismic networks (adapted from Romanowicz [9]).

Research on earthquakes, slow slip events, and volcanic eruptions provides enhanced understanding of mechanisms and developing new monitoring tools. Studies are addressing megathrust earthquakes like the Tokai-Oki magnitude 9.0 earthquake and the plate subduction process [20]. Active volcanoes present special challenges, particularly to model magma inside the conduits and deep connections in the mantle, which has prompted development of a range of methods of remote sensing, GPS, tiltmeters, broadband seismic networks, and integrated potential field and electromagnetic surveys. New tools being added include muons tomography exploiting secondary cosmic rays produced in the upper atmosphere with enhanced capabilities for imaging deep volcano structures [21, 22].

3. IMPACTS, MASS EXTINCTIONS, AND EVOLUTION

The evolution of life had been mostly studied from the fossil record, which provides evidence on past living organisms preserved along Earth's history. Paleontological studies have built a broad picture of life evolution from the single-celled organisms in the Precambrian to the multicellular organism in the Phanerozoic, providing a spatial-temporal reference system incorporated into the geological time scale. The field moved from stratigraphic, fossilization, and taxonomic based studies to exploring the ecosystems, physiology, reproductive traits, organism diseases, climate and environmental interactions, and feedbacks. With the introduction of isotope geochemistry and molecular studies the paleobiology field is being expanded, becoming increasingly multi- and interdisciplinary.

The extinction rates have been climbing as a result of the effects of climate and environmental changes and anthropogenic activity. The global warming, ocean acidification, deforestation, and pollution are affecting the ecosystems, with the extinction of species in the land and marine realms. Over a longer time span from the last deglaciation at the Late Pleistocene and Holocene transition, a large number of species including many land and marine vertebrates has disappeared. The extinction rates and magnitude has increased interest in studying past extinction events, particularly those associated with the five mass extinctions in the Phanerozoic (Figure 5). Mass extinctions are characterized by being above the rates of background extinction levels, occurring over a relatively short time [23, 24]. Barnosky et al. [25] have analyzed the recent extinctions in a geological context and compared them with the past five events. Most of the species that have ever developed are extinct, so studies of extinction rates and mechanisms are critical for understanding the evolution processes.

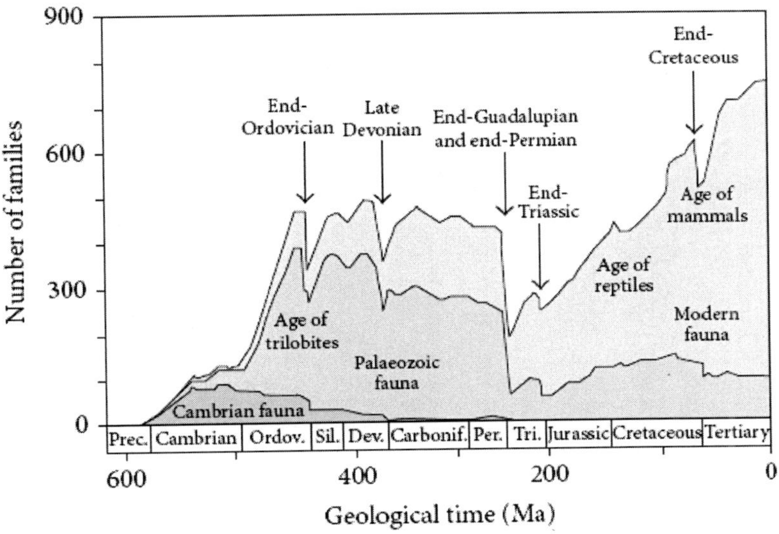

Percentage of extinct families: late Ordovician 12%, late Devonian 14%, late Permian 52%, late Triassic 12%, late Cretaceous 11%

Figure 5: Number of families as a function of geologic time, showing the five major extinction events marked by sharp biodiversity decrease (adapted from Raup and Sepkoski [23]).

The end-Cretaceous mass extinction, the second in severity in the Phanerozoic and most recent one, is being intensely studied. It affected significant numbers of species and genera, with extinction of the dinosaurs, pterosaurs, ammonites and numerous marine microorganisms, causing the disappearance of about 75% of the species. The mass extinction marks the end of the Mesozoic Era. The Cretaceous/Paleogene (K/Pg) boundary is recognized by a globally distributed thin clay layer (Figure 6), which represents the fine-grain-sized fraction of the ejecta from the Chicxulub impact [26–28]. The K/Pg boundary layer is a global stratigraphic marker, which permits unprecedented temporal resolution and lateral correlation of events.

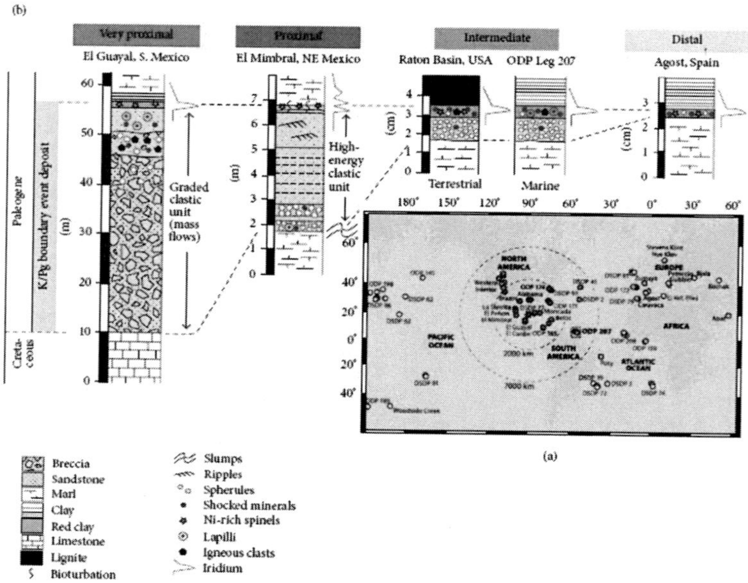

Figure 6: Cretaceous/Paleogene (K/Pg) boundary sections for distal, intermediate, proximal, and very proximal sites. Schematic K/Pg boundary sections (b). (a) Distribution of K/Pg boundary sites (Schulte et al. [26]).

The K/Pg layer is a few millimeter-to-centimeters thick, formed by a basal spherulitic layer, representing parabolic-emplaced melted droplets or condensates from a high temperature ejecta cloud, and the clay representing the fine-grained ejecta emplaced in the upper stratosphere (Figure 6). In the Gulf of Mexico-Caribbean Sea region, it has a more complex structure with high-energy tsunami deposit and a high temperature layer. Analyses of the layer distribution, composition, and physical properties permit reconstructing the dynamics of the impact event. Studies of K/Pg boundary sections provide data on the climatic and environmental changes and effects on the biota. Studies include analyses on the extinct species, ecosystem disruption, surviving species, short- and long-term postimpact effects, recovery patterns, and diversification. The problem in interpreting the mechanisms of extinction and effects on the biota has been the precision needed in dating and correlation. Separating events on the scale of seconds to months involved in the impact event in the geologic record are a major challenge, which has sparked attempts in refining the dating methods and stratigraphy. The most recent analysis by Renne et al. [29] has reduced the uncertainties in dating the K/Pg boundary to within ~30 ka, which represents a sharp improvement in dating capabilities.

Studies on the K/Pg boundary, impact event, and mass extinction are expanding, addressing life evolution at short and long time scales. One of the processes investigated addresses the evolution on maximum body size of terrestrial mammals, which coexisted with dinosaurs during most of the Mesozoic. For about 140 Ma mammals coexisted with the dinosaurs, restricted to small body sizes and ecosystems. Following the extinction of dinosaurs, first the birds increased their size, including some large predators. Later, mammals started to diversify and increase their maximum body size during the Paleocene and early Eocene. Smith et al. [30] have analyzed the evolution of maximum body size for terrestrial mammals showing that the groups increase their body mass by the late Eocene, irrespective of the landmass.

The fossil record provides a punctuated view of life evolution, biased to certain geological settings, environments, and life forms that are more easily preserved. Dating and lateral correlation of rock strata present a further complication, with less resolution as we go back in time. High resolution stratigraphic methods, making use of multiproxy methods integrating statistical, spectral, and numerical simulation analyses, are being developed. Radiometric dating has improved, which is being applied combined with astronomical, magnetic polarity, and cyclostratigraphy, resulting in high resolution chronologies. The developments are applied to calibrating the geological time scale with increased precision.

Studies of the fossil record and evolution are closely related to the climatic and environmental factors, which are linked from the early beginnings in the Precambrian with the oxygenation of the atmosphere and oceans, the advent of the eukaryotes and evolution of life, and climate and environment during the Phanerozoic. Studies are focusing on early life forms, formation of the iron banded formations, global glaciations, and the construction of the life tree. New tools for climate reconstruction with increased high resolution are being developed using a wide range of biological, chemical, isotopic, and physical proxies. In Mexico and North and Central America, studies assess the effects, mechanisms, and interconnections of the Inter-tropical Convergence Zone latitudinal migration, North American monsoon, El Niño-Southern Oscillation, Pacific Decadal Oscillation, solar irradiance, and teleconnections [31, 32]. The studies are addressing climate evolution at different spatial and temporal scales, which are coupled with computational simulations and theoretical models for millennial, centennial to decadal resolution. Recent studies explore the links and influence of climatic and environmental factors on evolutionary patterns and the interconnections of the biosphere with climate [33].

A major development has come from the molecular clocks, which have significantly impacted methods to calibrate evolutionary time [34, 35]. Modeling tools for molecular tree analysis have rapidly evolved, providing estimates for branching events that are calibrated against the minimum ages from the fossil record. Improved understanding of the different genomes and rates of change has remained a major challenge in using molecular clocks to provide absolute dates. Given the advances in instrumentation and methods that are capable of providing vast amounts of data and processing power, the molecular clock will provide higher resolution in investigating evolutionary time. Multigene clocks applied to multitaxa are already giving unprecedented details in branching points, integrating phylogenetic reconstructions, the fossil record, and constraints on genome evolutionary rates [36].

Molecular analysis is well suited for studying macroevolutionary evolution, for instance, the appearance of eukaryotes, which in the fossil record appear at about 800 Ma when global changes in the oceans and climate were occurring. The molecular estimates for the early eukaryotic diversification are younger at around 1866 to 1679 Ma [36]. This older date is consistent with reports on eukaryotic microfossils, indicating a long time span in the diversification of the major eukaryotic lineages [33, 36]. Studies are addressing evolutionary traits at genomic level, investigating eukaryotic evolution over million-year periods across species. Organism complexity is related to genomic features such as cell type number, gene contents, protein length, proteome disorder, and protein interactivity, which are being quantified [37, 38]. In the 1.4 Ga evolution of eukaryotes, alternative splicing has steadily increased with organism complexity [38].

4. PLANETARY SCIENCES

Exploration of the solar system using Earth based multispectral remote sensing and space probes has opened new research frontiers. Planetary missions to the terrestrial planets and moons of the gas giant planets have provided data on the structure, surface morphology, magmatic activity, tectonic styles, and deep interiors.

Observations of the surfaces of the inner planets and moons show that they are characterized by craters of different sizes and morphologies. They have been formed by collision of asteroid and cometary fragments over time, from small sized impacts to the large peak ring and multiring basin impacts. Large impacts produce deep transient excavation cavities in the curst, fragmenting and removing large volumes of rock and redistributing crustal material. On Earth, the active tectonic environment and erosion have effectively erased the record of impacts, with a relatively small number of craters documented and only three large multiring basins [39]. The Chicxulub crater, with a ~200 km rim diameter formed at the K/Pg boundary, is the youngest of the multiring basins and the only one with the ejecta preserved [26, 27, 29]. The other two structures formed in Precambrian times: Sudbury at about 1.8 and Vredefort at about 2 Ga ago. Chicxulub crater is located in the Yucatan platform in the southern Gulf of Mexico. The structure is covered by carbonate sediments and is being investigated by geophysical methods and deep drilling (Figure 7) [40, 41].

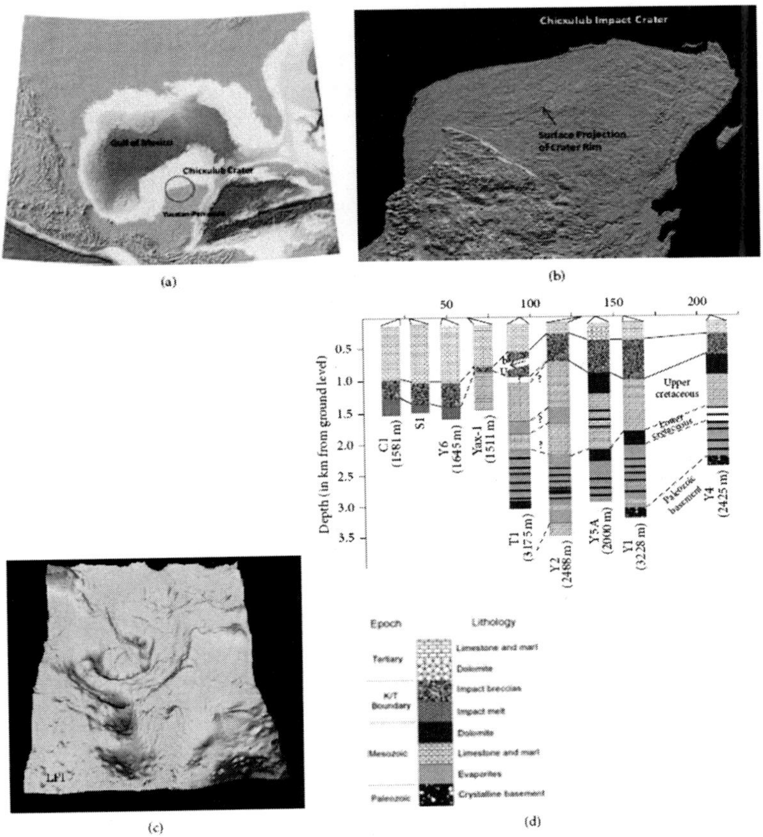

Figure 7: Chicxulub impact crater. (a) Gulf of Mexico and location of Chicxulub crater in the Yucatan platform. (b) Satellite interferometric radar image of Yucatan peninsula (credits: JPL-Caltech NASA), showing surface features associated with the buried crater structure. (c) Bouguer gravity anomaly of the Chicxulub crater (Sharpton et al. [40]). (d) Schematic lithological columns and lateral correlation for deep boreholes in the Chicxulub crater area, plotted as a function of relative distance to crater center (Urrutia-Fucugauchi et al. [27, 41]).

Impacts produce deformation at various depths, generating thermal anomalies and forming long-lived hydrothermal systems. The craters showing hydrothermal alteration are being investigated for manifestations of life forms, forming part of the exobiology programs. Studies of impact craters in the terrestrial record and elsewhere are enhancing understanding of these highly energetic phenomena in shaping planetary surfaces, including those in the asteroid belt.

Analysis of frequency, density, and size distribution of craters permits estimating the age of the planetary surfaces, with ancient surfaces marked by high density of craters, often including the large multiring basins [39]. The size-frequency crater relationships are also related to the geodynamics and deep structure. Plate tectonics appears restricted to Earth [10, 42]. Magmatic activity is observed in other bodies, including Mars, Venus, and Io. Mars lithosphere appears not being fragmented and under relative motion. Venus shows intense deformation and experienced a catastrophic resurfacing event about 500 Ma ago.

Evidence on the deep structure, thermal state, and convection comes from studies of meteorites, magnetic fields, and core dynamos. Meteorites have long been used for studying the origin and early stages of evolution of the planetary system (Figure 8). Analyses of chondrites and other primitive meteorites have documented the age of the first solids represented by refractory inclusions and chondrules, chemical composition of the solar nebula, and formation of planetesimals [43]. Studies are providing increasing resolution on the evolutionary stages (e.g., [43, 44]). Studies on chondrites and iron and stony-iron meteorites support that their planetesimals had differentiated iron cores capable of sustaining dynamo action for ~10 Ma periods [45–49]. The paleomagnetic record of main group pallasites supports the fact that they come from near the core-mantle boundary of differentiated planetesimals that sustained internal magnetic fields [47]. Partly differentiated planetesimals might have been relatively abundant in the early stages of the solar system [48]. Many were destroyed by energetic collisions, and a fraction of them are preserved in the asteroid belt. Recent analyses show that asteroid Vesta had a convecting iron core in the early stages [49].

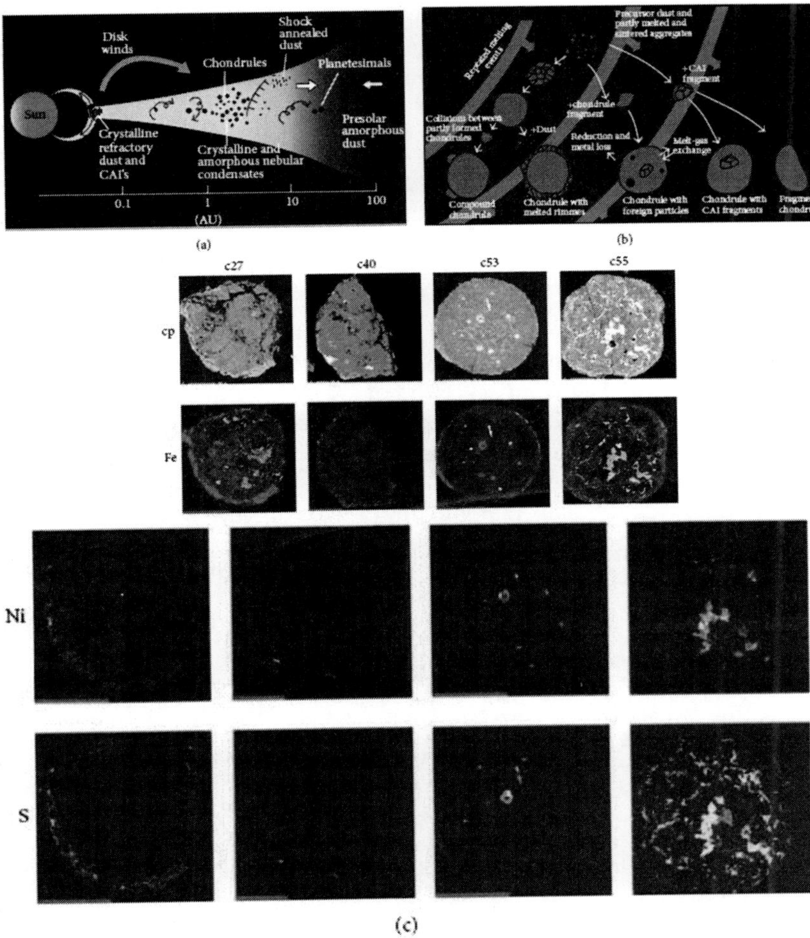

Figure 8: Schematic model of formation of chondrules and calcium-aluminium inclusions CAIs. (a) Protoplanetary disk. (b) Chondrule types with different morphologies and internal structures (adapted from Scott [43]). (c) Scanning electron microscopy images of individual chondrules from the Allende meteorite, showing the different morphologies, internal structures, and Fe, Ni, and S compositions. Numbers refer to laboratory sample identifications (Urrutia-Fucugauchi et al. [45]).

Planetary exploration is one of the most rapidly expanding frontiers in geophysics, with new data coming from the solar system missions and new exciting findings of exoplanets and planetary systems. The recent discoveries of exoplanets and multiple systems challenge the models for formation and early evolution of planetary systems based on observations of our solar system [50]. The large number of exoplanets discovered revives interest in planetary models with distinct formation zones for gas-icy giants and rocky planets within given regions of the accretion disk and models involving large-scale planet migration.

With increasing resolution and detection capacity, smaller Earth-sized planets are being detected. The Kepler space-based telescope mission is currently analyzing thousands of candidates, including several small mass planets. Recently, Quintana et al. [51] reported the finding of Kepler-186f, a 1.11 Earth-radius exoplanet in an orbit within the habitable zone around a M1-type dwarf star of the main sequence (Figure 9). Kepler-186f is the outermost planet of a five-planet system characterized by coplanar orbits. The multiplanet system is compatible with formation in a protoplanetary disk, with planets formed from accretion of local material and/or collisional growth of planetesimals. Numerical simulations conducted by Quintana and coauthors [51] for the Kepler-186 system show that too steep density configurations, with dense accretion disk close to the star, are required. These results suggest that planets underwent inward migration while forming or a late stage perturbation.

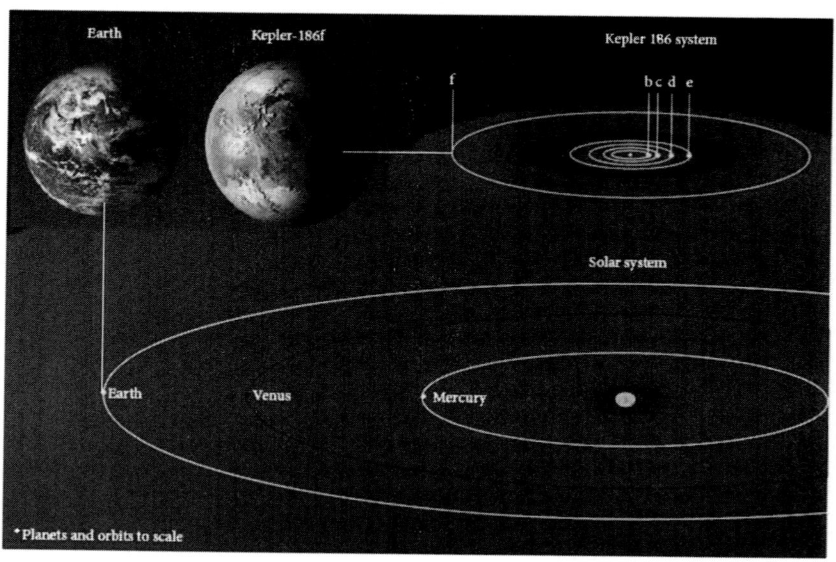

Figure 9: Schematic artistic representation of Kepler-186 multiple system compared with the inner solar system. Kepler-186 is a five-planet system

located ~500 light-years away orbiting an M star half the Sun mass (Quintana et al. [51]) (credits: NASA Ames/SETI Institute/JPL-Caltech).

Detection methods focus mainly on large planets close to the star, so most discoveries are large gas planets in orbits close to their stars. Detecting small Earth-like planets remains a challenge. Robertson et al. [52] analyzed the system around the M dwarf Gliese 581 star, showing that stellar activity might cause interference resulting in false exoplanet detection. Their results show that the signal for GJ 581 g, one of the four exoplanets in the system, depends on the eccentricity assumed for the companion GJ 581 d.

A major challenge in studying exoplanets lies in constraining the mass, density, composition, and orbital parameters. Recent developments start to provide new tools and data. Rocky planets are expected to have smaller sizes than gas and icy planets, but additional observations are required, which can be explored from the star metallicity. Buchhave et al. [53] analyzed the abundance of elements heavier than hydrogen and helium for 405 exoplanet host stars, finding that the exoplanet sizes separate into three metallicity regions. The three populations are interpreted in terms of rocky, gas dwarf, and gas-icy giant exoplanets. Another field of intense scrutiny is the detection of atmospheres for the super-Earths, gas dwarfs, and icy-gas giants [54]. Recent studies using transmission spectroscopy data report absorption features giving details on the atmosphere properties, confirming clouds in a super-Earth [55]. Considering that a significant fraction of exoplanets so far detected range in size between Earth and Neptune, the new studies open an interesting research field.

Determining the orbital parameters and spin provides important constraints on the planet ambient characteristics. Many exoplanets detected show orbits close to the stars, which are easier to detect with current methods. Spectroscopy observations can provide data on the spin velocity, which has been recently reported for gas giant planet β Pictoris b [56]. The exoplanet is located far from the star, about twice the distance of Jupiter in our system, and is quite bright. The spin determination comes from (blue) shifted carbon monoxide spectral signals from the planet, which gives an estimate of 25 km/s. In the solar system, spin correlates with the mass, showing a broad trend with the exception of Mercury and Venus. The fast rotation velocity, about 2 and 50 times greater than Jupiter's and Earth's, fits well with the planet mass. The study adds an interesting tool for characterizing multiplanet systems, which can provide constraints for models of planetary formation.

Interest in extraterrestrial life, which for a long time remained limited to theoretical analyses, has led to studies of organisms in extreme environments. Studies of extremophile communities from the deep crust, ocean thermal vents, hyperarid deserts, or polar caps have expanded understanding on food webs, energy sources, reproductive strategies, and metabolic states. Planetary missions are being directed to extraterrestrial life searches. Several missions have been directed to Mars, since the Viking missions experiments have tested the properties of the soils and atmosphere looking for evidence on liquid water and organic compounds. Recent missions are expanding the characterization of surface liquid water, hydrothermal activity, organic compounds, and fossil clues. New missions and spectroscopy observations use remote sensing clues of life activity in the planetary atmospheres.

Until the mid-1990s, the only planetary system known was our own. Models for evolution of planetary nebula predicted the formation of planets from planetary disks, but no observational evidence was available. The recent reports of hundreds of exoplanets and multiple planet systems and the observations on their sizes, orbits, and star characteristics are drastically changing and expanding theories and models for formation of planets and planetary systems [57–60].

5. CONCLUSIONS

New tools like the Earth observation satellite network, the global positioning system, planetary missions, high pressure/temperature experiments, high resolution tomography, and high performance computing play a major role in expanding research frontiers in geophysics. Increased interest in understanding Earth processes and new developments in instrumentation, modeling, and observation capabilities also comes from population growth and demographic changes, which increase global demand for minerals, water, and energy resources, resulting in pollution, land use changes, deforestation, environmental degradation, organism extinction, changes in atmospheric gas composition, and global warming. In this context, understanding Earth's subsystems of the atmosphere, oceans, continents, ionosphere, magnetosphere, biosphere, and deep interior, their interconnections, cycles, spatio-temporal scales, and feedback mechanisms has become a major priority. The anthropogenic induced changes are comparable to those caused by geologic forces on the planet, highlighting the importance of integrated research. This has prompted global approaches in Earth system science and development of research fields, many of them at cross-disciplinary borders like biogeosciences, environmental geophysics, exobiology, and planetary sciences.

In a broad general context, the developments in high performance computing power, personal computers, telecommunications, electronics, and advent of the internet are profoundly changing the scientific research enterprise. The developments touch practically every area related to research with electronic databases, publications, electronic archives, search engines, software, and personal and group interactions. The capacity for analyzing massive data sets using supercomputers and computer networks facilitates using numerical methods and complex simulations. High performance computing allows modeling of the complex climate system, core and mantle tomography, Earth observational satellite multispectral data, or exoplanet detection systems with the massive data sets from the space-telescope Kepler and other search missions.

Studies in widely different fields are interconnected with the recent developments, opening bridges across previously separated endeavors. Studies on the origin and evolution of the solar system are linked to the new areas of planetary sciences, which challenge current models opening new questions. Most of the exoplanets discovered are in the size range between Earth and Neptune, for which there are no analogs in the solar system. Studies of the structure and properties, orbital characteristics, and formation mechanisms for the super-Earths and gas giants are giving fresh insights on planetary evolution [58]. Studies are addressing finer details in the characteristics of exoplanets in addition to size, orbit, and mass, such as the spin, surface temperature, and presence and composition of atmospheres and clouds [59, 60]. The mass-spin relation in the solar system is related to the breakup velocity and impacts added angular momentum. The estimation of the fast spin for βPictoris b, which fits with the trend for fast spin and large mass, opens the link of impacts in the formation of planets [57, 59]. β Pictoris b is a young planet still contracting and cooling, towards a size comparable to Jupiter. Determination of the spin characteristics for a larger group of exoplanets will allow investigating how planets form and evolve in different protoplanetary disks environments.

Exoplanet research and planetary missions connect with investigation of the cratering record on Earth and in other bodies of the solar system, including the large impacts during the early stages of planet formation. Satellites in the solar system show different characteristics of the rocky and gas-icy planets, with small satellites in large planets and larger satellites in small planets. Studies on the tectonics and deep structure on Earth are now related to planetary research on the planet interiors, planet formation models, and thermal states [42]. Results from high pressure and temperature mineral physics [11–15] relate and constrain models of formation of super-Earth and giant icy-gas exoplanets [51–60], as well as the planets in the solar system [50]. We have similar links between studies of life on extreme terrestrial environments, origin, and evolution of life in the young Earth and studies of exobiology [61]. Studies are uncovering relationships and exploring new questions and interconnections.

COMPETING INTERESTS

The authors declare that there is no conflict of interests regarding the publication of this paper.

ACKNOWLEDGMENTS

The authors thank Ana Escalante and Miguel Angel Diaz for assistance with the figures. This study forms part of National University of Mexico Programs on the Chicxulub Impact, the Cretaceous/Paleogene Boundary, and MeteorPlan. Partial support comes from Papiit IG-101115 and Conacyt grants.

REFERENCES

1. X. LePichon, J. Francheatau, and J. Bonin, Plate Tectonics, Elsevier, Amsterdam, The Netherlands, 1973.
2. J. T. Wilson, "A new class of faults and their bearing on continental drift," Nature, vol. 207, no. 4995, pp. 343–347, 1965. r
3. W. J. Morgan, "Rises, trenches, great faults, and crustal blocks," Journal of Geophysical Research, vol. 73, no. 6, pp. 1959–1982, 1968. r ·
4. D. P. McKenzie and R. L. Parker, "The North Pacific: an example of tectonics on a sphere," Nature, vol. 216, no. 5122, pp. 1276–1280, 1967. r ··
5. C. DeMets, R. G. Gordon, and D. F. Argus, "Geologically current plate motions," Geophysical Journal International, vol. 181, no. 1, pp. 1–80, 2010. r · ·
6. K. Burke, "Plate tectonics, the wilson cycle, and mantle plumes: geodynamics from the top," Annual Review of Earth and Planetary Sciences, vol. 39, pp. 1–29, 2011.
7. R. N. Mitchell, T. M. Kilian, and D. A. D. Evans, "Supercontinent cycles and the calculation of absolute palaeolongitude in deep time," Nature, vol. 482, no. 7384, pp. 208–211, 2012. r · ·
8. D. L. Turcotte and G. Schubert, Geodynamics: Applications of Continuum Physics to Geological Problems, John Wiley & Sons, New York, NY, USA, 1982.
9. B. Romanowicz, "Using seismic waves to image Earth's internal structure," Nature, vol. 451, no. 7176, pp. 266–268, 2008. r ·
10. G. Schubert, D. Turcotte, and P. Olson, Mantle Convection in the Earth and Planets, Cambridge University Press, Cambridge, UK, 2001.
11. S. A. Karato, A. M. Forte, R. C. Liebermann, G. Masters, and L. Stixrude, Eds., Earth's Deep Interior: Mineral Physics and Tomography from the Atomic to the Global Scale, vol. 117 of AGU Geophysical Monograph, American Geophysical Union, 2000.
12. H. Asanuma, E. Ohtani, T. Sakai et al., "Phase relations of Fe-Si alloy up to core conditions: implications for the Earth inner core," Geophysical Research Letters, vol. 35, no. 12, Article ID L12307, 2008. r · ·
13. S. Tateno, K. Hirose, Y. Ohishi, and Y. Tatsumi, "The structure of iron in Earth's inner core," Science, vol. 330, no. 6002, pp. 359–361, 2010. r · ·
14. M. Murakami, K. Hirose, K. Kawamura, N. Sata, and Y. Ohishi, "Post-perovskite phase transition in $MgSiO_3$," Science, vol. 304, no. 5672, pp. 855–858, 2004. r · ·

15. D. C. Rubie, T. Duffy, and E. Ohtani, "New developments in high pressure mineral physics and applications to the Earth's interior," Physics of the Earth and Planetary Interiors, vol. 143-144, pp. 1–3, 2004. r ·
16. J.-F. Lin, W. Sturhahn, J. Zhao, G. Shen, H.-K. Mao, and R. J. Hemley, "Sound velocities of hot dense iron: Birch's Law revisited," Science, vol. 308, no. 5730, pp. 1892–1894, 2005. r · ·
17. L. Hwang, T. Jordan, L. Kellog, J. Tromp, and R. Wiellemann, Advancing Solid Earth System Science Through High-Performance Computing, Computational Infrastructure for Geodynamics, University of California, Davis, Calif, USA, 2014.
18. ICSU, Earth System Science for Global Sustainability: The Grand Challenges, International Council for Science, Paris, France, 2010.
19. A. Ismail-Zadeh, J. Urrutia-Fucugauchi, A. Kijko, K. Takeuchi, and I. Zialapin, Eds., Extreme Natural Hazards, Disaster Risks and Societal Implications, Cambridge University Press, Cambridge, UK, 2014.
20. M. Simons, S. E. Minson, A. Sladen et al., "The 2011 magnitude 9.0 Tohoku-Oki earthquake: mosaicking the megathrust from seconds to centuries," Science, vol. 332, no. 6036, pp. 1421–1425, 2011. r · ·
21. H. K. M. Tanaka, T. Uchida, M. Tanaka, H. Shinohara, and H. Taira, "Cosmic-ray muon imaging of magma in a conduit: degassing process of Satsuma-Iwojima Volcano, Japan," Geophysical Research Letters, vol. 36, no. 1, Article ID L01304, 2009. r · ·
22. V. Grabski, R. Nuñez, S. Aguilar et al., "Use of horizontal cosmic muons to study density distribution variations in the Popocatepetl volcano," in Proceedings of the 33rd International Cosmic Ray Conference (ICRC '13), vol. 33, pp. 1–4, Rio de Janeiro, Brazil, July 2013.
23. D. M. Raup and J. J. Sepkoski Jr., "Mass extinctions in the marine fossil record," Science, vol. 215, no. 4539, pp. 1501–1503, 1982. r · ·
24. J. J. Sepkoski Jr., "Patterns of phanerozoic extinction: a perspective from global data bases," in Global Events and Event Stratigraphy in the Phanerozoic, O. H. Walliser, Ed., pp. 35–51, Springer, New York, NY, USA, 1996. r ·
25. A. D. Barnosky, N. Matzke, S. Tomiya et al., "Has the Earth's sixth mass extinction already arrived?" Nature, vol. 471, no. 7336, pp. 51–57, 2011. r · ·
26. P. Schulte, L. Alegret, I. Arenillas et al., "The Chicxulub asteroid impact and mass extinction at the Cretaceous-paleogene boundary," Science, vol. 327, no. 5970, pp. 1214–1218, 2010. r · ·
27. J. Urrutia-Fucugauchi, A. Camargo-Zanoguera, and L. Pérez-Cruz, "Discovery and focused study of the Chicxulub impact crater," Eos, vol. 92, no. 25, pp. 209–210, 2011. r · ·
28. L. W. Alvarez, W. Alvarez, F. Asaro, and H. V. Michel, "Extraterrestrial cause for the Cretaceous-Tertiary extinction," Science, vol. 208, no. 4448, pp. 1095–1108, 1980. r · ·
29. P. R. Renne, A. L. Deino, F. J. Hilgen et al., "Time scales of critical events around the cretaceous-paleogene boundary," Science, vol. 339, no. 6120, pp. 684–687, 2013. r · ·
30. F. A. Smith, A. G. Boyer, J. H. Brown et al., "The evolution of maximum body size of terrestrial mammals," Science, vol. 330, no. 6008, pp. 1216–1219, 2010. r · ·
31. G. H. Haug, K. A. Hughen, D. M. Sigman, L. C. Peterson, and U. Röhl, "Southward migration of the intertropical convergence zone through the holocene," Science, vol. 293, no. 5533, pp. 1304–1308, 2001.r · ·
32. L. Pérez-Cruz, "Hydrological changes and paleoproductivity in the Gulf of California during middle and late Holocene and their relationship with ITCZ and North American Monsoon variability," Quaternary Research, vol. 79, no. 2, pp. 138–151, 2013. r · ·
33. J. L. Blois and E. A. Hadly, "Mammalian response to cenozoic climatic change," Annual Review of Earth and Planetary Sciences, vol. 37, pp. 181–208, 2009. r · ·

34. S. Kumar, "Molecular clocks: four decades of evolution," Nature Reviews Genetics, vol. 6, no. 8, pp. 654–662, 2005. r · ·
35. S. Kumar and S. B. Hedges, "A molecular timescale for vertebrate evolution," Nature, vol. 392, no. 6679, pp. 917–920, 1998. r · ·
36. L. W. Parfrey, D. J. G. Lahr, A. H. Knoll, and L. A. Katz, "Estimating the timing of early eukaryotic diversification with multigene molecular clocks," Proceedings of the National Academy of Sciences of the United States of America, vol. 108, no. 33, pp. 13624–13629, 2011. r · ·
37. E. Schad, P. Tompa, and H. Hegyi, "The relationship between proteome size, structural disorder and organism complexity," Genome Biology, vol. 12, article R120, 2011. r · ·
38. L. Chen, S. J. Bush, J. M. Tovar-Corona, A. Castillo-Morales, and A. O. Urrutia, "Correcting for differential transcript coverage reveals a strong relationship between alternative splicing and organism complexity," Molecular Biology and Evolution, vol. 31, no. 6, pp. 1402–1413, 2014. r · ·
39. J. Urrutia-Fucugauchi and L. Perez-Cruz, "Multiring-forming large bolide impacts and evolution of planetary surfaces," International Geology Review, vol. 51, no. 12, pp. 1079–1102, 2009. r· ·
40. V. L. Sharpton, K. Burke, A. Camargo-Zanoguera et al., "Chicxulub multiring impact basin: size and other characteristics derived from gravity analysis," Science, vol. 261, no. 5128, pp. 1564–1567, 1993.r · ·
41. J. Urrutia-Fucugauchi, A. Camargo-Zanoguera, L. Pérez-Cruz, and G. Pérez-Cruz, "The Chicxulub multi-ring impact crater, yucatan carbonate platform, Gulf of Mexico," Geofisica Internacional, vol. 50, no. 1, pp. 99–127, 2011. ·
42. C. O'Neill, A. M. Jellinek, and A. Lenardic, "Conditions for the onset of plate tectonics on terrestrial planets and moons," Earth and Planetary Science Letters, vol. 261, no. 1-2, pp. 20–32, 2007. r · ·
43. E. R. D. Scott, "Chondrites and the protoplanetary disk," Annual Review of Earth and Planetary Sciences, vol. 35, pp. 577–620, 2007. r · ·
44. J. N. Connelly, M. Bizzarro, A. N. Krot, Å. Nordlund, D. Wielandt, and M. A. Ivanova, "The absolute chronology and thermal processing of solids in the solar protoplanetary disk," Science, vol. 338, no. 6107, pp. 651–655, 2012. r · ·
45. J. Urrutia-Fucugauchi, L. Pérez-Cruz, and D. Flores-Gutiérrez, "Meteorite paleomagnetism—from magnetic domains to planetary fields and core dynamos," Geofisica Internacional, vol. 53, no. 3, pp. 343–363, 2014. r · ·
46. L. T. Elkins-Tanton, B. P. Weiss, and M. T. Zuber, "Chondrites as samples of differentiated planetesimals," Earth and Planetary Science Letters, vol. 305, no. 1-2, pp. 1–10, 2011. r · ·
47. J. A. Tarduno, R. D. Cottrell, F. Nimmo et al., "Evidence for a dynamo in the main group pallasite parent body," Science, vol. 338, no. 6109, pp. 939–942, 2012. r · ·
48. B. P. Weiss and L. T. Elkins-Tanton, "Differentiated planetesimals and the parent bodies of chondrites," Annual Review of Earth and Planetary Sciences, vol. 41, pp. 529–560, 2013. r · ·
49. R. R. Fu, B. P. Weiss, D. L. Shuster et al., "An ancient core dynamo in asteroid Vesta," Science, vol. 338, no. 6104, pp. 238–241, 2012. r · ·
50. A. Morbidelli, J. I. Lunine, D. P. O'Brien, S. N. Raymond, and K. J. Walsh, "Building terrestrial planets," Annual Review of Earth and Planetary Sciences, vol. 40, pp. 251–275, 2012. r · ·
51. E. V. Quintana, T. Barclay, S. N. Raymond et al., "An Earth-sized planet in the habitable zone of a cool star," Science, vol. 344, no. 6181, pp. 277–280, 2014. r · ·
52. P. Robertson, S. Mahadevan, M. Endl, and A. Roy, "Stellar activity masquerading as planets in the habitable zone of the M dwarf Gliese 581," Science, vol. 345, no. 6195, pp. 440–444, 2014. r · ·
53. L. A. Buchhave, M. Bizzarro, D. W. Latham et al., "Three regimes of extrasolar planet radius inferred from host star metallicities," Nature, vol. 509, no. 7502, pp. 593–595, 2014. r · ·

54. H. A. Knutson, B. Benneke, D. Deming, and D. Homeier, "A featureless transmission spectrum for the Neptune-mass exoplanet GJ436b," Nature, vol. 505, no. 7481, pp. 66–68, 2014. r · ·
55. L. Kreidberg, J. L. Bean, J.-M. Désert et al., "Clouds in the atmosphere of the super-Earth exoplanet GJ 1214b," Nature, vol. 505, no. 7481, pp. 69–72, 2014. r · ·
56. I. A. G. Snellen, B. R. Brandl, R. J. De Kok, M. Brogi, J. Birkby, and H. Schwarz, "Fast spin of the young extrasolar planet β Pictoris b," Nature, vol. 508, no. 7498, pp. 63–65, 2014. r · ·
57. A. W. Howard, "Observed properties of extrasolar planets," Science, vol. 340, no. 6132, pp. 572–576, 2013.r · ·
58. T. Barman, "Astronomy: a new spin on exoplanets," Nature, vol. 508, no. 7498, pp. 41–42, 2014. r · ·
59. X. Dumusque, F. Pepe, C. Lovis et al., "An Earth-mass planet orbiting α Centauri B," Nature, vol. 491, no. 7423, pp. 207–211, 2012. r · ·
60. R. M. Canup and W. R. Ward, "A common mass scaling for satellite systems of gaseous planets," Nature, vol. 441, no. 7095, pp. 834–839, 2006. r · ·
61. C. S. Cockell, Astrobiology: Understanding Life in the Universe, Wiley-Blackwell, 2015.

CHAPTER 2

Enhancing The Resolution of Airborne Gamma-Ray Data using Horizontal Gradients

David Beamish

British Geological Survey, Keyworth, Nottingham NG12 5GG, UK

ABSTRACT

The spatial resolution characteristics of airborne gamma-ray data are largely controlled by survey elevation and line separation. In the UK, although low nominal surveys altitudes may be permitted, regulatory zones with elevations in excess of 180 m are required above conurbations. Since the data, typically in the form of grids, are evaluated alongside many other detailed geoscientific spatial datasets their absolute resolution limits, together with their spatial characteristics, become relevant. Here, using published software, we study the theoretical resolution characteristics of this form of survey data obtained with a line separation of 200 m. Of particular interest is the airborne response behaviour when non-uniform distributions of radioactivity are encountered. Although ultimately a function of the radioelement-concentration contrast encountered, the calculations reveal that such zones are most difficult to identify when their scale length decreases below the scale of the line separation. This limited resolution then further decreases with elevation. In order to increase our ability to resolve the edges of non-uniform source regions we calculate the horizontal gradient magnitude (HGM) of the observed data. While the data used can be the estimated radioelement concentrations (potassium, thorium and uranium) or their ratios, we demonstrate that the total count is particularly suited to this type of analysis. The theoretical calculations are supported by an examination of survey data across a series of isolated bodies (offshore islands). This empirical study indicates the practical limits to resolution when using the horizontal gradient and these are governed by the survey line separation. The HGM response provides an enhanced mapping of the edges of zones associated with a contrast in flux behaviour. The edges are detected using the maxima in the response and these can be additionally examined using grid curvature analysis. The technique is assessed using recent survey data containing geological, soil and environmental influences. The results demonstrate the spatially pervasive nature of flux contrasts associated with soil and environmental contributions which potentially mask, or perturb, the underlying bedrock geological response.

Keywords: Airborne gamma-ray data,Data processing, Horizontal gradient, Soil, Geology

1. INTRODUCTION

Radiometric, or gamma-ray spectrometric surveys are used to calculate concentrations of the radioelements potassium, thorium and uranium by measuring the gamma-rays which the radioactive isotopes of these elements emit during radioactive decay. The fundamentals of airborne radiometric surveys are described by Minty (1967). The geochemical interpretation of such data, in the regolith context of Australia, is discussed by Dickson and Scott (1997). Their application to soil/regolith mapping is described by Wilford et al. (1997). In the UK, airborne total gamma count surveys were first flown over SW England in 1957–9 as part of the UK Atomic Energy Authority uranium exploration programme (Kimbell et al., 2003). Higher resolution, regional scale, multi-parameter, airborne geophysical surveys commenced in 1998 (Peart et al., 2003). Since then a number of further surveys, incorporating radiometric spectrometric data and using flight-lines spaced at 200 m have been conducted at relatively low nominal altitudes (56 to 90 m). A feature of these surveys is a regulatory requirement to increase survey elevation (to > 180 m) over conurbations.

The modern surveys have been conducted for resource, soil and environmental assessments. Soils in the UK generally formed after the last ice age (< 12,000 years ago) and are therefore distinct from the ancient soils of Pangea (e.g. those of Africa and Australia). Two particular aspects of gamma-ray flux behaviour arise in the case of the UK (i.e. a temperate, mid-latitude zone) and these are (a) the prevalence of a 'lower-bound' on soil moisture and (b) the occurrence of low density, organo-mineral and organic soils (Beamish, 2015). The soil, when sufficiently thick, attenuates the flux.

The degree of attenuation with depth is governed by the density and wetness of the soil profile as described by Beamish (2013, 2015). The conceptual model of soil-superficial-bedrock behaviour is that the bedrock acts as a parent material to the soil so that the particle size, mineralogy and radioelement concentrations of the soil derive from the bedrock material. Significant exceptions may occur in the case of material transport (e.g. weathering, erosion, groundwater flow and leaching).

Since the radiometric data are evaluated alongside many other detailed geoscientific datasets the absolute resolution limits of the airborne data, together with their spatial characteristics, become relevant. In order to assess the volume of material contributing to the gamma-ray flux measured by an airborne detector, the concept of a circle of investigation was presented by Duval et al. (1971). It was found that altitude is the most important parameter in determining the volume of material that produces a given percentage of the total observed flux and the results discussed by Duval et al., (1971) are often quoted. The calculations however only relate to an infinitely large, homogenous material.

Further calculations relating to the design parameters for airborne surveys were presented by Pitkin and Duval (1980) building on the earlier work. The study included the effects of finite 'anomalous' source regions (e.g. regions of enhanced radioactivity) and therefore the calculations can be used to assess the spatial mapping capabilities for any specified airborne survey. Here we revisit the calculations of Pitkin and Duval (1980) using software made available by the U.S. Geological Survey: USGS (Duval, 1997). The calculations performed here provide detector flux responses due to finite source concentrations at survey elevations of 60 and 120 m. Here we primarily consider circular source regions (e.g. a localised soil or environmental response) and an infinite linear strip (e.g. a geological response). Linear dimensions of the source region range from 50 to 400 m and are considered in relation to a flight line spacing of 200 m. For more extensive sources, the detector response becomes uniform across the source region before the source edge modifies the response.

The behaviour of the anomalous response characteristics indicate that the lateral scale of the source response is governed by the scale length of the anomalous region. The edges of the source region are identified by the response decrease to a background value. This type of behaviour is also found in potential field data. Magnetic and gravity spatial derivatives have a well-established role in the interpretation of such data. Many methods have been developed to estimate the edge-location of singular bodies (Nabighian et al., 2005). In the same context of edge detection we here consider the behaviour of the horizontal gradient magnitude (HGM) of the anomalous radiometric response. The horizontal gradients are calculated from the set of theoretical anomaly response profiles and it is demonstrated that maxima are associated with the source edge locations except for the smallest bodies. A uniform HGM response (no horizontal gradient) is necessarily associated with extended uniform source regions. Based on the these results, it is suggested that a horizontal gradient analysis of airborne radiometric data offers enhanced resolution of regions characterised by changes in concentration and hence flux behaviour.

The majority of airborne radiometric data are assessed using gridded data sets that comprise the three main radioelements (and their summary ternary image) together with their associated ratios. Additionally total count (TC) data are used to summarise the total flux observed across a broader energy range. In theory horizontal gradient analysis can be applied to any of the three individual radioelement distributions, their ratios and the total count. The radioelement concentration data and total count are one-sided, positive response measurements with a noise floor that permeates the low count behaviour. The ratio information is subject to error propagation from the constituent data and is correspondingly noisier. The horizontal gradient calculation may amplify the noise content of a particular data set and it is relevant to consider the signal to noise characteristics of the data used. In previous UK studies of the attenuation characteristics of radiometric data (e.g. Beamish, 2013) the total count, being a spectral summation, has been used to provide a high signal-to-noise ratio and is also used here in assessing the HGM response.

Since the theoretical calculations are idealised (no noise), we consider the HGM response obtained from survey data across a control study area. The data used are total count obtained across a series of small islands in which the background material (the sea) provides a null response. The degree to which the outline of each island may be detected in the HGM edge response is demonstrated. We also assess the application of curvature analysis (Blakely and Simpson, 1986; Phillips et al., 2007) to map peak (ridge) locations determined by the HGM analysis. The control study uses high-contrast materials so that it also necessary to consider the general case of survey data which encounter more spatially-complex soil, environmental and bedrock variations. Recent UK survey data are used to study the behaviour of the HGM response. A 22 × 20 km study area from SW England provides examples of geological, soil and environmental responses and their characterisation using the HGM total count response. The study is then extended to the local scale to examine the detailed characteristics of some of the soil and environmental responses detected. Finally the HGM response obtained using total count is compared with the ternary response obtained using the HGM of the three individual radioelements.

2. THEORY

2.1. Field of view

In order to assess the volume of material contributing to the gamma-ray flux measured by an airborne detector, Duval et al. (1971) introduced the concept of a circle of investigation. The authors performed theoretical calculations based on a flat homogenous material that is infinitely large and infinitely thick (the so-called infinite source). The calculations are valid for gamma-ray energies from 1.5 to 3.0 MeV and therefore for energies associated with the 3 main naturally-occurring radioelements. The effects of altitude, air and source material density together with their attenuation coefficients were investigated. It was found that altitude is the most important parameter in determining the volume of material that produces a given percentage of the total observed flux. In a more wide-ranging set of theoretical calculations, Pitkin and Duval (1980) considered the broader design requirements of airborne surveys in terms of their resolution capabilities. The parameters considered included aircraft speed, flight line spacing and detector volume. The calculations were based on the same conceptual model of the infinite source yield and circle of investigation and relate the airborne detector response to the ground area viewed by the detector. Importantly, the study included the effects of finite 'anomalous' source regions (e.g. a region of enhanced radioactivity) and therefore the calculations can be used to assess the spatial mapping characteristics for any specified airborne survey.

Duval (1997) produced a U.S. Geological Survey (USGS) report, providing and describing, a set of gamma-ray modelling programmes that are based on the previous publications and allow the user to perform their own evaluations. This software is used here to provide an understanding of UK airborne data with an elevation range from 60 to 120 m and a flight line spacing that is typically 200 m. Initially we consider a so-called strip model which extends the circle of investigation analysis to a moving detector (here we use a velocity of 64 $m \cdot s^{-1}$). Flux contributions from strips parallel to the flight path are summed in order to provide a spatial description of the areal contribution from an infinitely thick source. Secondly we consider the effects of anomalous finite gamma-ray sources. These are finite areas defined by a relative increase in their source concentration. The areas considered can be circular, rectangular or an infinite linear feature of constant width (any orientation to the flight line). The latter is probably more relevant to a geological context, although all three display similar edge behaviour in relation to the intrinsic scale of the anomalous feature. Here, for brevity, we consider circular features. The calculations use a set of defined parameters. The parameters used here are (i) a material attenuation coefficient of 0.0444 $cm^2 \cdot g^{-1}$ (e.g. at energies associated with potassium), (ii) a detector area of 0.02 m^2, (iii) a detector photo-peak efficiency of 0.15, (iv) an air density of 0.0012 $g \cdot cm^{-3}$ and (v) a material density of 1.1 $g \cdot cm^{-3}$ (i.e. soil).

The strip calculations for a moving detector across an infinite and uniform source region are presented in a number of useful ways by Pitkin and Duval (1980). They are only partially repeated here for the sake of completeness. The cumulative percentage of total flux arising from strip widths (2 × the symmetrical radius perpendicular to the flight line) is shown in Figure 1a for flying heights of 60, 120 and 180 m. The behaviour observed is a series of bell-type curves that allow the circles of investigation to be compared at different flying heights. Thus 90% of the total flux is obtained across scale lengths of 250 m (60 m elevation), 360 m (120 m elevation) and 440 mm (180 m elevation). The same information can also be presented in a different way by extracting a set of values of the cumulative response and plotting the strip width associated with flying height as shown in Figure 1b. The results demonstrate the degree to which airborne measurements, measured at intervals from 60 to 70 m (using typical aircraft speeds of 60 to 70 m/s) along flight lines, constitute a moving-average spatial measurement thus limiting precise resolution of flux behaviour.

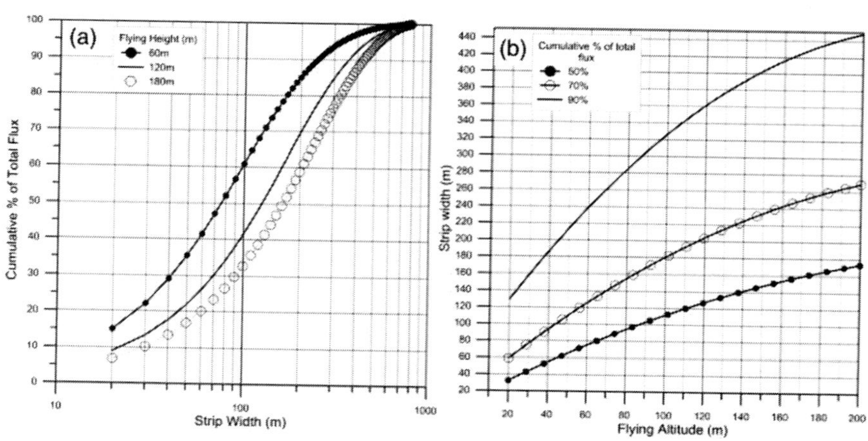

Figure 1. Results of the strip calculation for a moving detector across a uniform half-space. (a) Cumulative % of total flux as a function of strip width for 3 survey heights. (b) Variation of fixed cumulative % of total flux as a function of strip width and survey height.

Of more relevance to most survey data is the response behaviour when non-uniform distributions of radioactivity are encountered. The model considered is a finite area (a circle) defined by a relative increase in source concentration. The flight line bisects the anomalous region and the centre of the anomalous region is here set at 500 m. The source and background regions are assumed to be infinitely thick and a large relative concentration increase of 50 is used for the anomalous area. The source contributions for circles with radii from 25 to 200 m (lateral distances of 50 to 400 m) are shown in Figure 2 for two survey heights of 60 and 120 m. The decrease in count rates with survey altitude is most evident together with a reduced response with decreasing source dimension. Obviously when considering real data and their associated errors the detectability of small anomalous areas with lateral dimensions of < 100 m is likely to be low. The half-widths (the distance across which the peak amplitude is reduced by 50%) for the data shown in Figure 2 are summarised in Table 1.

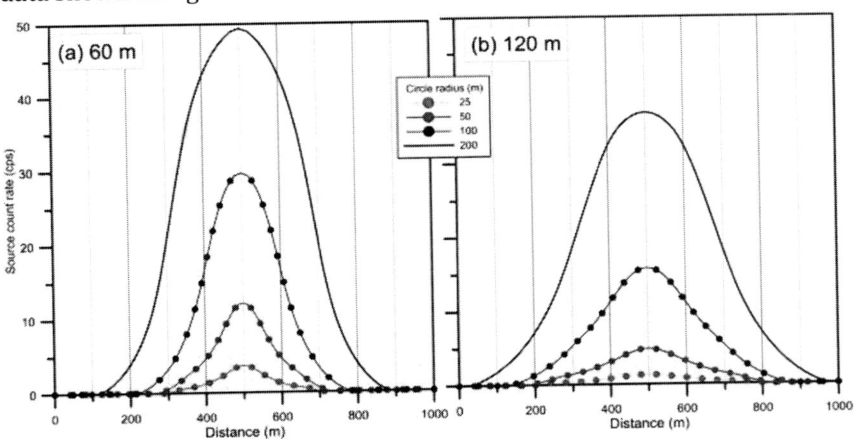

Figure 2. Variation of detector count rate for circular source regions with radii from 25 to 200 m and centred at 500 m. (a) 60 m survey height. (b) 120 m survey height.

Table 1. The half-widths (the distance across which the peak amplitude is reduced by 50%) for the data shown in Figure 2 for elevations of 60 and 120 m.

Body radius (m)	60 m half-width (m)	120 m half-width (m)
25	140	250
50	160	260
100	216	300
200	370	392

The resolution of each feature decreases with the source dimension. The largest region (a lateral dimension of 400 m) is found to provide a half-width > 350 m. Assuming a survey with 200 m flight lines that are symmetrical about the centre of the anomalous region, then the larger area will produce a significant response across 3 flight lines (e.g. at lateral distances from 300, 500 and 700 m in Figure 2). As the dimension of the circular body increases beyond a radius of 200 m, the source contribution becomes uniform at the peak response observed in Figure 2. This uniform response then decays to zero towards the edges of the region and follows the spatial decay behaviour shown in Figure 2.

We have also compared the behaviour of the infinite linear source (perpendicular to flight line) with that of the circular response discussed above. In order to compare like-with-like, we consider that a linear dimension (a width) of a linear source should be equated to the diameter of the circular source. When this comparison is undertaken it is found that the behaviour of the infinite source follows the general behaviour of the circular source (i.e. the behaviour shown in Figure 2 and summarised in Table 1). One distinct difference is that the infinite linear source produces a larger amplitude response for the same source concentration contrast (when the survey line is perpendicular to the source strike). This implies that, for the same source concentration contrast, an extensive linear anomaly (i.e. a geological anomaly) is more detectable than an equivalent circular feature (i.e. a localised environmental anomaly).

2.2. Horizontal derivatives

The behaviour of the anomalous response characteristics shown in Figure 2 indicates that the lateral scale of the source response is governed by the scale length of the anomalous region. The edges of the source region are then identified by the response decay to a background value (a value of zero in this theoretical case study). This type of behaviour is also found in potential field data. Magnetic and gravity spatial derivatives have a well-established role in the interpretation of such data (Grauch et al., 2001) but can also be usefully applied to non-potential field data (Beamish, 2012). The source parameters that are typically estimated are the boundaries or edges of singular source bodies (Blakely, 1995; Blakely and Simpson, 1986). In the same context of edge detection we here consider the behaviour of the horizontal gradient magnitude of the anomalous response.

The horizontal gradients calculated from the set of anomaly response profiles of Figure 2are shown in Figure 3. The symmetrical behaviour displays a minimum at the origin of the anomalous region (located at 500 m). Maxima are then observed towards the edges of the anomalous region. At a survey elevation of 60 m (Figure 3a), the maxima are observed at the edge locations of each circular region for radii of 200 m, 100 m and 50 m. The maxima for the 25 m circle are close to those of the 50 m model and probably indicate a resolution limit has been reached in the calculations employed. The behaviour is repeated at a survey elevation of 120 m for radii of 200 m and 100 m, but the spatial gradients are less distinct for radii of 50 and 25 m due to the low amplitudes of the response used in the calculation. Shoulders appear in the gradient response beyond the position of maximum amplitude. The shoulders become increasingly significant both with decreasing body size and increasing elevation. The results indicate the likely practical limits of small body detection when using 200 m flight lines obtained with an envelope of survey elevations (rural and conurbation).

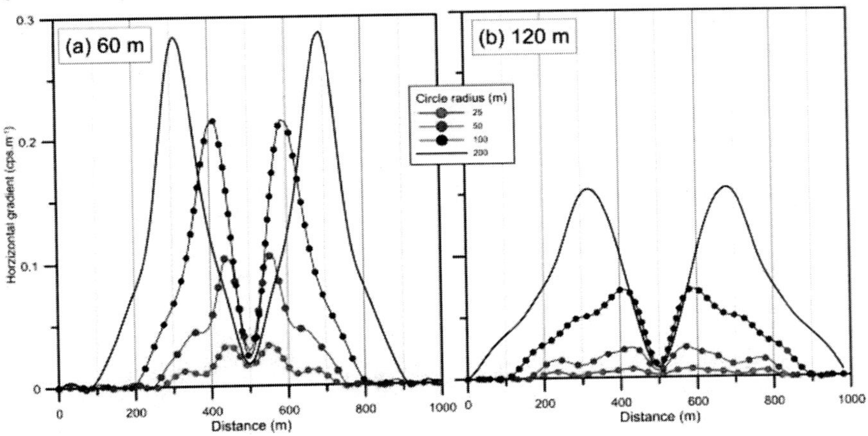

Figure 3. Horizontal gradients calculated from the detector response functions of Figure 2. Circular source radii range from 25 to 200 m and are centred at 500 m.

For larger bodies the horizontal gradient response is a minimum over the source region (the response is uniform over the source region and produces no horizontal gradient). Horizontal gradients are then produced across the edges in accord with those observed in Figure 3. It is suggested that a horizontal gradient analysis of airborne radiometric data offers enhanced resolution of regions characterised by a change in concentration and hence flux behaviour. The analysis is summarised in Figure 4 which considers a circular anomaly with a radius of 100 m and plots the response behaviour, observed at 60 m, as a 3D perspective view. The source detector response (Figure 2a) is shown as a wireframe and the high values of the horizontal gradient response (Figure 3a) are shown as a colour contour plot. The peak in the horizontal gradient is a circle which is associated with the body edges.

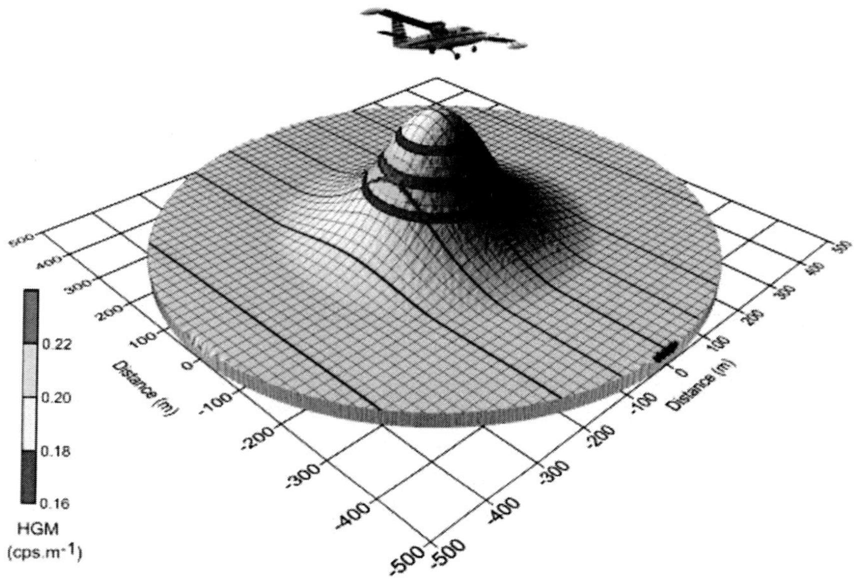

Figure 4. 3D perspective view of the detector count rate response (greyscale and wireframe) and high values of the horizontal gradient magnitude (HGM), values > 0.16 cps·m⁻¹. Survey elevation is 60 m. The source circle has a diameter of 200 m and is centred at the origin. The aircraft and altitude, representing the detector, are not to scale.

Thus far we have considered only idealised (no noise) data. In contrast to potential field data, the gamma-ray data are positive response measurements with a noise floor that permeates the low count behaviour and limits the idealised resolution characteristics. In theory horizontal gradient analysis can be applied to any of the three individual radioelement distributions, or their ratios. The horizontal gradient calculation may amplify the noise content of a particular data set and it is relevant to consider the signal to noise characteristics of the data used. A horizontal gradient analysis may be applied to line-based (profile) data or to gridded data sets. When a grid analysis is undertaken, the total horizontal gradient magnitude (HGM) is based on the x-derivative (dx) and y-derivative (dy) responses and is defined as the positive-only response:

The HGM is commonly referred to as an edge-detection filter and is widely used in geophysical interpretation (e.g. Grauch et al., 2001).

3. MATERIALS AND METHODS

3.1. The airborne radiometric data

Here we use UK airborne radiometric data sets obtained with a line-separation of 200 m which were obtained at low elevation, subject to regulatory increases in elevation above conurbations. A small (3 × 3 km) control data set is taken from the Tellus survey of Northern Ireland (2005–2006) described by Beamish and Young (2009). A larger study area is extracted from the TellusSW survey flown in the latter half of 2013 and described by Beamish and White (2014). In both surveys, a radiometric data sampling of 1 Hz provides an along-line sampling largely in the range from 60 to 70 m.

The radiometric data were acquired with a 256 channel gamma spectrometer system (GeoExploranium GR-820/3) comprising 32 l of downward-looking NaI(Tl) detectors and 8 l of upward-looking detectors. Uranium (^{238}U) is estimated through the radon daughter ^{214}Bi in its decay chain, while thorium (^{232}Th) is estimated through ^{208}Tl in its decay chain. Potassium (^{40}K) is measured directly at 1.461 MeV. Conventionally secular equilibrium in the decay chains of uranium and thorium is assumed (Minty, 1967) and the ground concentration results are reported as equivalent uranium (eU, ppm) and equivalent thorium (eTh, ppm). Potassium is reported as %K. Total count (TC) data (energy range from 0.41 to 2.81 MeV) is reported in cps. A vertically uniform activity concentration is assumed.

The gamma radiation registered by the detector is composed of contributions from soil/rock, the atmosphere, the aircraft and cosmic radiation. In order to calibrate airborne radiometric data, the commonly adopted standard is to follow the recommendations made in a series of technical documents and publications from the International Atomic Energy Agency (IAEA). The set of procedures applied here are based on protocols described in IAEA (1991, 2003, 2010) and by Grasty and Minty (1995).

The zero-level characteristics of the processed radiometric data were obtained from survey data over deep water. In the case of Northern Ireland data, the standard deviation of the total count null data was reported to be 40 cps. In the case of the TellusSW data, the figure increases to 76 cps.

3.2. Data processing

Guidelines for the mapping of radiometric data have been published (IAEA, 2003) and these indicate the most frequently used algorithms are the bidirectional gridding and minimum curvature techniques. Kriging has not been widely used as it is difficult to account for the anisotropy in the sample density when modelling the semivariogram. Here the grids are obtained using a minimum curvature algorithm which produces a smooth grid by iteratively solving a set of difference equations minimising the total second horizontal derivative while attempting to honour the input data.

For the analysis conducted here we use minimum-curvature grids with a cell-resolution of 40 m. Having obtained a gridded data set of the HGM response, a common processing procedure is to extract the maxima of the HGM response as described by Blakely and Simpson (1986). The methodology was further considered by Phillips et al. (2007). These authors employed a 3 × 3 (grid cell) moving-window to assess the local curvature and define turning points. Here we use a ridge analysis to isolate maxima. The application of these procedures to the HGM estimates from TC data, and the individual radioelements, to improve edge detection of anomalous source regions, is also considered.

4. RESULTS

4.1. A control example

Ideally the detection and resolution limits of TC data and their associated horizontal gradients could be demonstrated using data obtained across isolated zones of varying sizes. In the UK, many airborne surveys contain an offshore extension to accommodate the acquisition of magnetic data. Offshore islands then offer isolated zones with a given TC (soil-geological) response with a surrounding sea providing a null (theoretically zero) response. Survey data across a 3 × 3 km area in Northern Ireland, comprising a number of islands, typical of drowned drumlin topography, were used by Beamish (2012) to study the resolution characteristics of airborne electromagnetic data. Here the same area is studied in relation to the radiometric TC response. The islands lie within the tidal Strangford Lough on the eastern seaboard. The area is shown on a base topographic map in Figure 5a with the onshore area appearing in the west. Figure 5 also shows the sampling locations of the airborne survey which used 200 m flight line separations at an azimuth of 345°. One advantage of this type of area is that there are only small survey elevation adjustments rather than the sometimes large adjustments that may occur across coastal cliffs. The mean survey elevation across the area was 56 m with a standard deviation of 10 m. The average distance between survey points is 52 m. The noise level of the TC data was estimated using a portion of the data entirely confined to a larger area of the sea. These data provided a typical close-to-normal distribution with a mean value of 169 cps. The full range of the null data was from 34 to 229 cps. The along-line TC data are shown as four range colour values in Figure 5b. The 2 lowest count ranges define a potentially noise (or null) level below and above the mean value of 169 cps. The two higher ranges should then represent increasing (above noise) response behaviour.

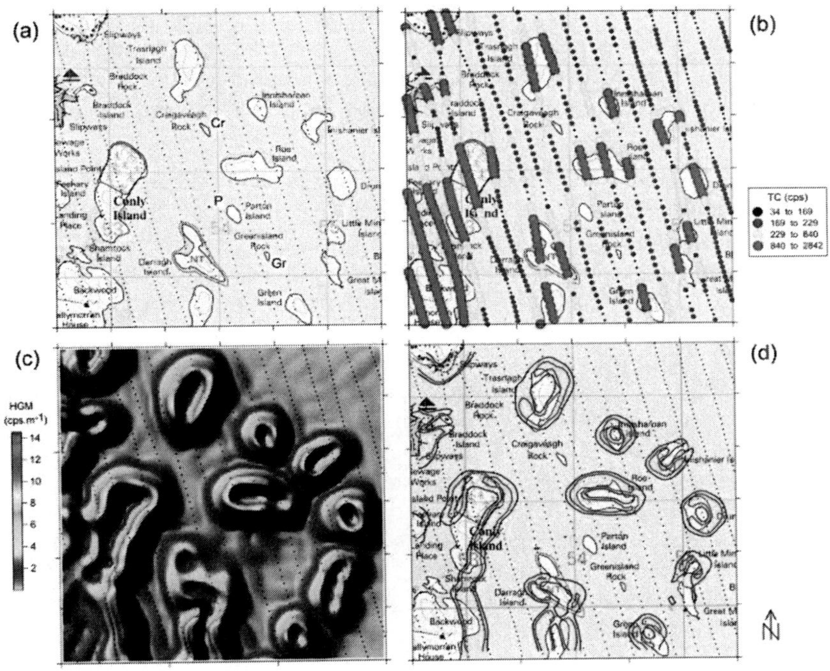

Figure 5. The 3 × 3 km control study area in which the survey sample points (200 m line spacing) are shown as dots. (a) Topographic base map with the 3 smallest islands/rocks identified as Cr (Craigaveegh rock) Gr (Greenisland rock) and P (Parton Island). (b) Total count data shown as 4 colour-range posted values. (c) Continuous colour image, with shade, of the horizontal gradient magnitude (TC data). (d) Background grey-scale topographic map and two contours of the horizontal gradient magnitude, 6 cps·m^{-1} in blue and 8 cps·m^{-1} in red.

The mainland and larger islands are all associated with high TC responses. We are not particularly interested in the precise amplitude levels but rather their decay to a zero response as indicated by the third class interval (yellow). These decays can be observed both along-line and on adjacent over-sea lines reflecting the spatial sensitivity of the response behaviour.

The smallest island is a rock (Gr, Figure 5a) and is located midway between flight lines and no response is observed. The larger Parton Island (P) is also situated midway between flight lines and this produces a small muted response as does Craigaveegh rock (Cr) to the north. The response behaviour across the set of islands can be further studied manually to arrive at an empirical estimate of a circle of investigation in relation to the edge of these isolated bodies. In detail the edge of each body is not precisely defined due to the occurrence of sand beaches. Despite this the analysis suggests that, for these TC data, the lateral distance across which the response is reduced by 90% ranges from 42 to 182 m with a mean of 101 m.

The HGM response of these data is shown as a continuous colour (linear) image in Figure 5c. The high amplitudes clearly outline the flux contrasts provided by the land-sea edges. The 3 smaller rock-island features, previously noted, do not provide a significant HGM response. In order to clarify the behaviour of the HGM, Figure 5d shows two high contour levels (6 and 8 $cps\cdot m^{-1}$) of the HGM response and shows more clearly the extent to which the edges of individual islands are resolved. The application of the curvature-ridge analysis of Phillips et al. (2007) applied to the HGM response is summarised in Figure 6. This form of moving-window analysis is known to produce a level of spurious behaviour in the ridge solutions obtained. Here we retain only the highest amplitudes (HGM > 3.5 $cps\cdot m^{-1}$) for evaluation. For these data we note that the ridge solutions (and indeed the HGM) can only partially replicate the complex spatial form of the islands and the associated changes in radiometric contributions. Comparing Figure 5c, d and Figure 6 we note that the continuous image of the HGM is potentially the most useful in providing the largest dynamic range of the behaviour generated by the lateral contrasts.

The coastal example obviously provides high levels of flux contrast which are unlikely to be representative of the behaviour of the majority of soil and geological contrasts which control the behaviour of the majority of survey data. We now consider further soil, geological and environmental examples taken from a recent airborne geophysical survey of SW England. We note that when examined at the detailed scale (as above), the HGM response is unlikely to provide an exact edge, or boundary, description of the flux contrast when compact and spatially complex areas are encountered.

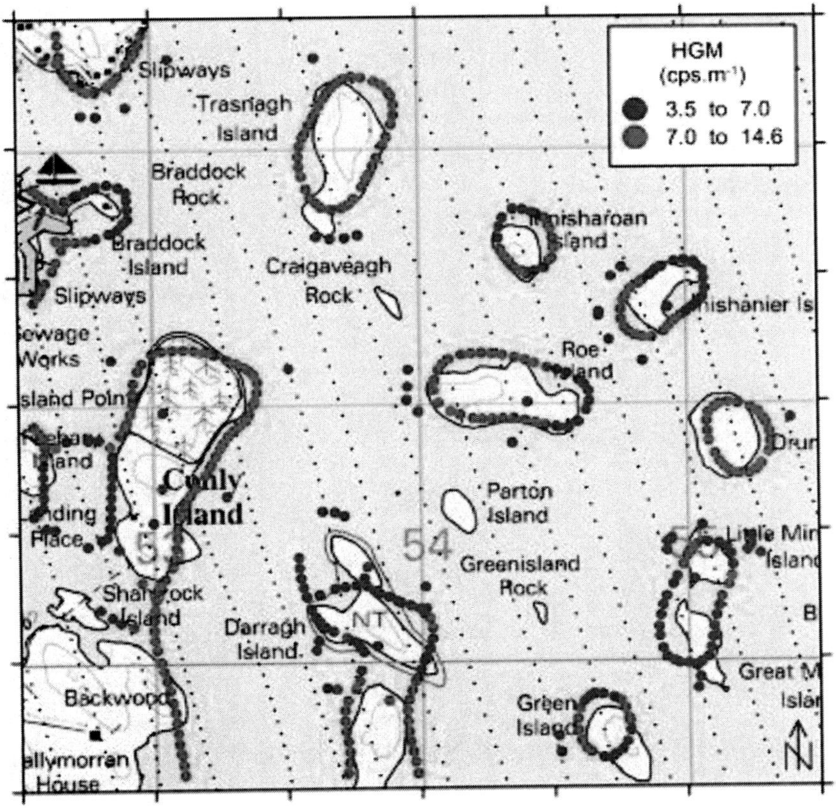

Figure 6. The 3 × 3 km control study area in which the survey sample points (200 m line spacing) are shown as dots. Posted values of the largest amplitudes (> 3.5 cps·m⁻¹) of the curvature-ridge-analysis of the horizontal gradient magnitude (TC data).

4.2. Whole survey

The radiometric data from the TellusSW survey is displayed as a ternary image alongside a simplified version of the bedrock geology in Figure 7. A granite batholith intruded the existing country rock during the Variscan Orogeny producing metamorphic zones and establishing fluid circulation zones responsible for the extensive polymetallic mineralisation of the region (Edmonds et al., 1975). Figure 7a shows the granite outcrops in white. The majority of the metasediments are pelites or graywackes and are of Devonian age although a large extensional, structure preserved

Carboniferous sediments as the Culm Basin (Figure 7a). Following Leveridge and Hartley (2006) a sequence of basins are identified in Figure 7a to allow comparison with the radiometric data. A 22 × 20 km rectangle identifies the study area considered here which covers part of a granite (Bodmin, B in Figure 7a) and the Devonian-Carboniferous contact area.

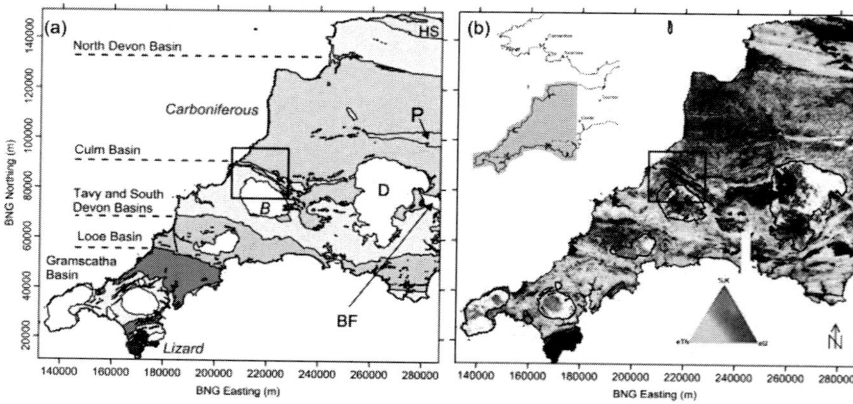

Figure 7. TellusSW survey area and location map (inset). Central rectangle identifies the main study area. BNG refers to British National Grid. (a) Simplified geology based on basins identified by Leveridge and Hartley (2006). HS: Hangman Sandstone formation, P: Permian formation, BF: Palaeogene Bovey Formation. Five outcropping granites are identified in white, B (Bodmin) and D (Dartmoor). (b) The radiometric ternary colour image obtained using the 3 radioelement ground concentrations.

The ternary image (Figure 7b) is a 3-way colour stretch formed from the distributions of potassium (red), thorium (green) and uranium (blue). The image is cut-to-coast to remove the null responses over the offshore area. White responses indicate areas in which all three radioelements have high concentrations while black responses represent lows. The zones with the highest radiometric responses are associated with, and largely confined to, the outcropping highly radiogenic granite zones. Within the granite outcrops, areas of preferential potassium enrichment (i.e. biotite granite) are observed. It is worth noting that the radiometric data across the Bodmin granite have identified a previously unknown internal edge, clearly defined by high values, within the outcrop (see also Figure 8). Low value responses occur in association with the ultramafic Lizard ophiolite complex and, in the north, with a Devonian lithology named the Hangman Sandstone (HS) formation which has a specific deltaic origin. Other contrasting response characteristics are observed with the Permian (P) lithologies associated with the Crediton Trough and the Palaeogene Bovey Formation (BF) in the east.

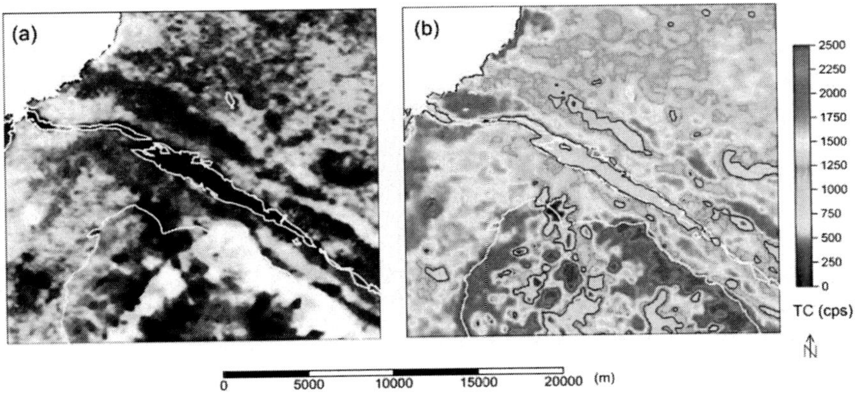

Figure 8. The 22 × 20 km main study area with outlines of the Bodmin granite outcrop and the Tintagel volcanic formation in white. (a) Ternary image, legend as in Figure 7. (b) Total count image using a linear scale. A single black contour identifies a value of 1000 cps.

Comparing Figure 7a and b it is evident that, at the large scale shown, the bedrock formations control the spectral response (colour) of the radiometric data. Within this framework many other detailed variations are observed. Within the Bodmin (B) and Dartmoor (D) granites low responses are caused by significant areas of peat formed on the high ground (moors) above the granites. The largest area of attenuation is due to blanket bog covering the majority of the western portion of the Dartmoor granite. In order to fully understand the radiometric response observed the role of both soils and bedrock must be considered. The selected 22 × 20 km sub-area contains a complex mix of geological, soil and environmental radiometric responses.

4.3. Main study area

The ternary image and total count response across the study area are shown in Figure 8. Although the area is relatively rural, survey altitudes range from 55 to 257 m, with a median of 85 m. The higher values are associated with conurbations. Both images reference the Bodmin granite outcrop in the south (see also Figure 7) and a detailed (1:50 k) bedrock mapping of a single extensive Carboniferous volcanic unit (the Tintagel formation) which influences the radiometric response. The 1:50 k bedrock mapping across the area is particularly complex and fragmented so a single unit is used as an example of a geological response. The Tintagel volcanic formation (identified in Figure 8) comprising lava, tuffs and agglomerates, crosses the entire area and is identified by a low count response

(< 1000 cps). The ternary image of Figure 8a clearly provides a variety of information on the radioelement concentrations of the area both at a regional and local scale. The Bodmin granite response appears spatially complex as are all the highly radiogenic granites of the survey region (Figure 7b). A major internal edge is apparent within the outcrop but a large number of small scale features appear to perturb the response. A clear and continuous granitic edge is only defined along the eastern margin of the granite.

An image of the HGM response is shown in Figure 9a alongside a classification of the largest HGM ridge amplitudes (Figure 9b). The grey-scale image of Figure 9a attempts to preserve the dynamic range of the response with black representing a value of zero and pure white representing a maximum value. The colour-coded classification of the curvature-ridge analysis values associated with the data (Figure 9b) provides a useful simplification of the information in which the lower amplitude responses are omitted. While the classified ridge amplitude data are of immediate interest (the edges that are most clearly defined) we note the lower amplitude characterisation of the HGM (Figure 9a) also warrants assessment.

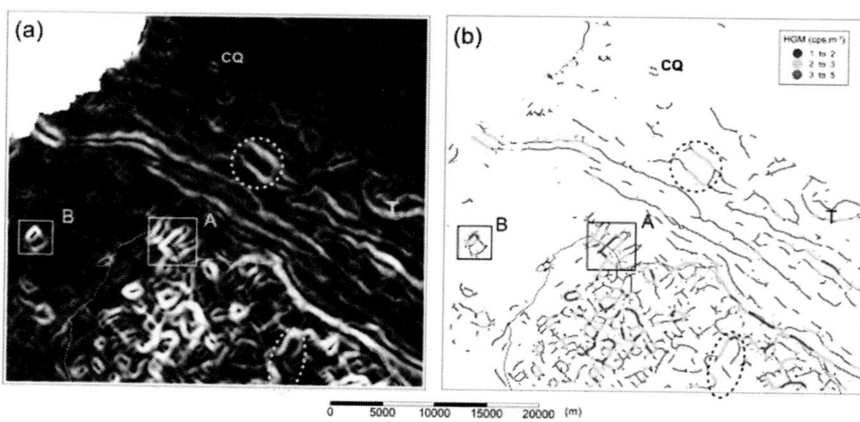

Figure 9. The 22 × 20 km main study area with outline of the Bodmin granite outcrop. Two regions with dotted outlines are extensive zones of woodland. T refers to a response associated with one outcrop of the Teign chert formation. CQ refers to Cansford quarry. Two rectangles labelled A and B define 2 further study areas. (a) Continuous grey-scale image of the horizontal gradient magnitude (TC data) with black representing zero and white representing a maximum value. (b) Curvature ridge analysis of the horizontal gradient magnitude showing values above 1 $cps \cdot m^{-1}$.

Studies of UK radiometric data (e.g. Beamish and White, 2011; Beamish, 2015) have highlighted the spatially-complex nature of their radioelement and attenuation characteristics. The HGM, as calculated here, should provide edge detection of anomalous areas defined as having higher or lower relative flux characteristics (a flux contrast). The amplitude of the TC data must therefore be used in conjunction with the HGM information. The radiometric data carry information on both bedrock geology and soils and these may also be perturbed by near-surface environmental influences (e.g. water bodies and mineral extraction sites). When providing an assessment of bedrock radioelements, all non-bedrock responses can be considered noise. The pervasive nature of the many localised bodies identified by the HGM suggests the scale of the problem that will be encountered when radioelement information (e.g. Figure 8a) is used in a bedrock assessment.

The HGM information, as show in Figure 9 may be used, alongside the conventional radiometric measurements, in conjunction with a range of additional geoscientific information (e.g. soil parent material, land-use and vegetation). Since the edge information for the area is extensive and complex only some comments and two detailed examples are provided here. Edges associated with one of the more extensive geological responses (the Tintagel volcanic zone shown in Figure 8) are mapped across the area and some low amplitude internal boundaries within the formation are detected. A series of sub-parallel geological edges appearing to the NE of the granite are also resolved. There are other bedrock HGM responses and a closed boundary defining one area of the Teign chert formation is identified at T. Two areas, shown with broken lines in Figure 9, show an association with 2 of the larger areas of mixed conifer and broadleaf woodland. The attenuated response across certain woodland zones remains enigmatic since not all such features give rise to the effect (Beamish, 2015).

We have chosen 2 areas, identified by rectangles in Figure 9, to illustrate the detail contained in the HGM response. The high spatial density of small scale effects across the granite outcrop is partially related to zones of wet peat. Rectangle A defines a 2.5 × 2.5 area across the granite margin used as an example of an attenuation effect. Zones of enhanced concentration are also observed in the HGM response. Rectangle B defines a 1.75 × 1.75 km area where an absence of soil cover across a quarry site generates an enhanced response. A second quarry enhancement location (Cansford quarry, CQ) is also identified in Figure 9.

4.4. Sub-area A

A topographic base map of the 2.5 × 2.5 km sub-area is shown in Figure 10a. The survey elevations range from 66 to 98 m with a median value of 81 m. Measurement locations are identified with symbols in Figure 10a,b. The presence of the central water reservoir complicates the data assessment due to the absence of a subsurface response. Two woodland areas can also be noted and these appear not to perturb the response characteristics. The main influences on the radiometric data are compact areas of peat. The 4 peat areas, as mapped by a 1:50 k BGS (British Geological Survey) superficial cover database, are shown in Figure 10b with a background soil texture map identifying light soils (L) associated with the granite and medium to light (silty) to heavy soils (M_L_H) elsewhere. The change in soil type also identifies the granite margin. Any potential change in bedrock response across the margin is effectively masked by the soil response. The boundaries of the peat areas are retained for reference in the following two images. Figure 10c shows the TC data which approach, but do not reach a value of zero over the central water body. The 3 larger peat areas provide attenuation zones while the smaller peat area in the SW produces only a small reduction in amplitude. The water body edges inevitably produce the largest HGM response shown in Figure 10d. Here the maximum HGM amplitude is seen to track the peat outlines in a reasonable manner. In detail the results suggest that the mapped peat zone in the NW extends to the NE below the conifer forest. The ability of the UK radiometric data to provide mapping of peat zones was previously described by Beamish (2014).

4.5. Sub-area B

Delabole slate quarry near Pengelly has operated continuously since the 15th century. A topographic base map of the area is shown in Figure 11a in which a small water body can be observed at the quarry centre. Soils and bedrock across the 1.75 × 1.75 km area are uniform. The total count response is shown in Figure 12b and amplitudes are enhanced across the main pit and along the spoil zone to the south (see also Figure 12d) where soil is absent or thin. A localised larger amplitude feature is observed on the edge of the spoil zone. The HGM response shown in Figure 12c and the HGM ridge-analysis of Figure 12d, define a closed body around the main quarry and a subsidiary, lower amplitude feature beyond the edge of the spoil zone. It should be noted that although the depth of this quarry is exceptional (> 140 m), similar total count enhancements are routinely observed at most larger quarry sites in the UK (see also CQ in Figure 9) and are interpreted as being due to the absence of soil cover.

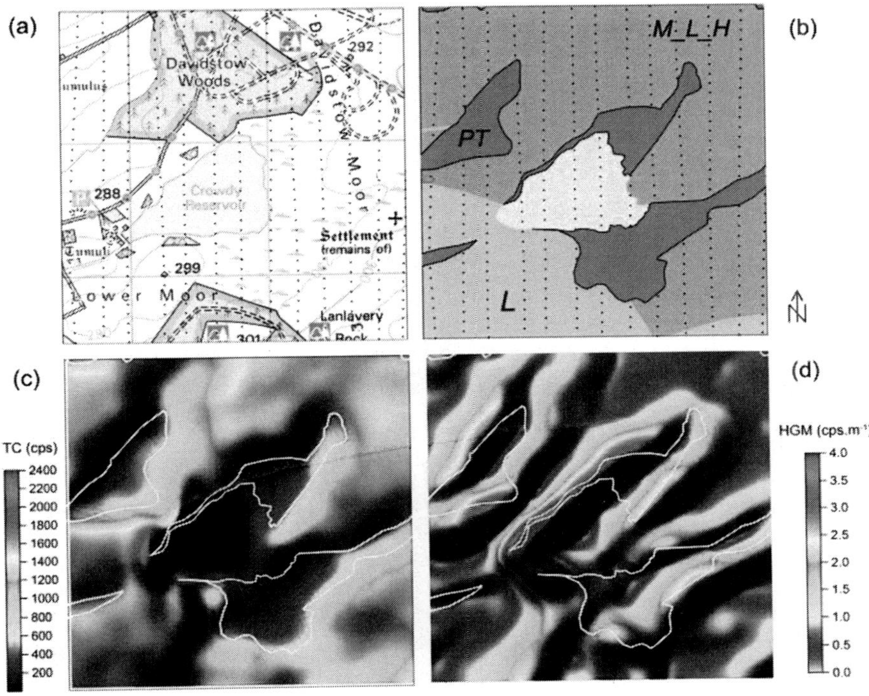

Figure 10. The 2.5 × 2.5 km sub-area A. (a) Topographic base map with survey sample points as dots. (b) Soil map showing water body, zones of peat (PT), light soils (L) and medium to light (silty) to heavy soils (M_L_H). (c) Continuous colour image, with shade, of the total count (TC). (d) Continuous colour image, with shade, of the horizontal gradient magnitude (HGM) obtained from the TC data.

5. DISCUSSION

Although the application of HGM analysis has been demonstrated using total count data, it was previously noted that the technique can be applied to individual radioelement distributions. When undertaking such an analysis the lower signal-to-noise of these distributions, particularly that of uranium, needs to be acknowledged. Despite this, the spectral response of the HGM applied to the individual data may be considered useful. We return to the main study area for this analysis. The TC-HGM response is shown in Figure 12a as a continuous colour image and provides a reference for Figure 12b. The ternary image obtained from HGM-potassium, HGM-thorium and HGM-uranium for the main study area is shown in Figure 12b.

The ternary response RGB colour scale is the same as that used in Figure 7b. In Figure 12b there is a degree of 'colour bleeding' mainly in the blue (uranium) channel due to the higher comparative noise level in the uranium channel. Despite this, edges in the ternary-HGM response clearly track those in the TC-HGM response and display a range of spectral characteristics. In the case of the regionally-extensive geological responses, a centre-line of the zone(s) is defined in black with edges defined by a spectral response (towards yellow) that indicates the absence of a uranium (blue) component edge. The white responses in Figure 12b result from relative highs in all 3 radioelements although subtle variations in spectral character are observed. Thus a slight thorium enhancement is seen in association with the northern and eastern granite edge. The spectral behaviour assists with the identification of potential continuity of edge effects.

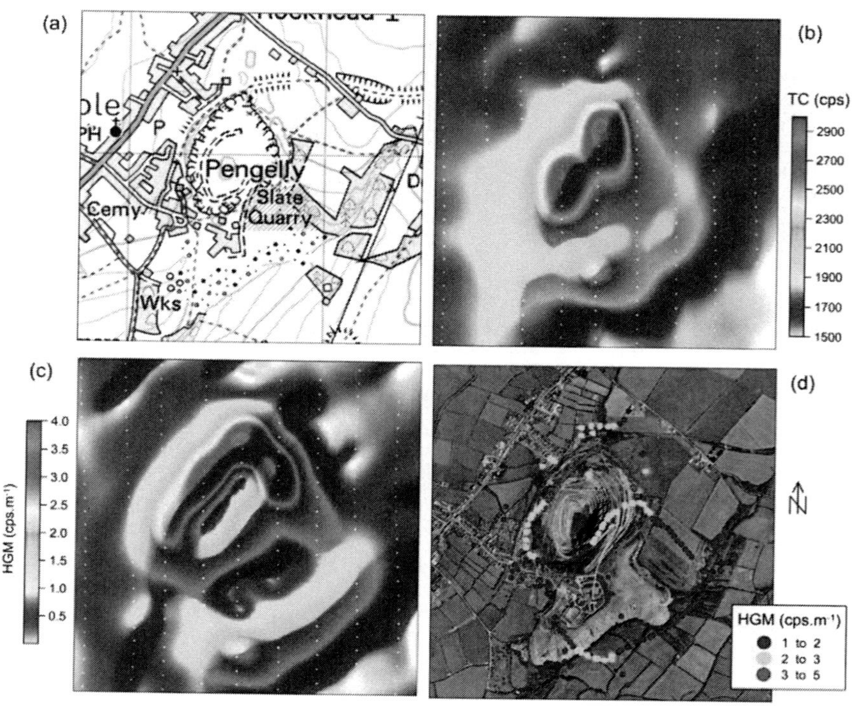

Figure 11. The 1.75 × 1.75 km sub-area B. (a) Topographic base map. (b) Continuous colour image, with shade, of total count (TC), white dots denote survey sample points (c) Continuous colour image, with shade, of the horizontal gradient magnitude (HGM) obtained from the TC data, white dots denote survey sample points. (d) Curvature ridge analysis of the horizontal gradient magnitude showing values above 1 cps·m^{-1}.

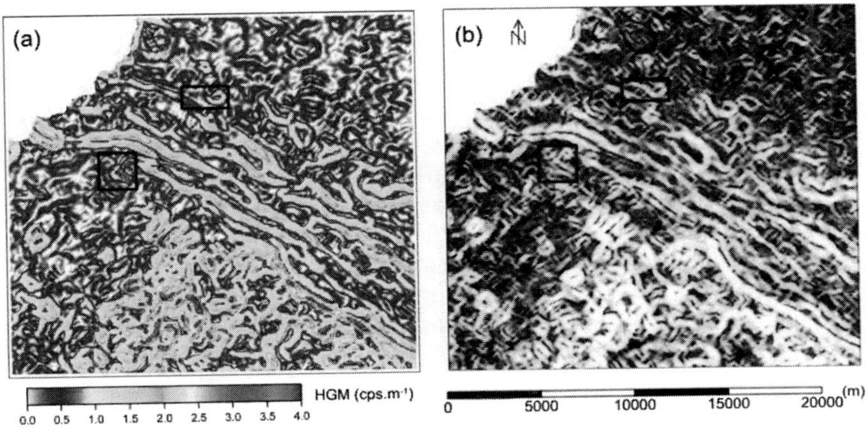

Figure 12. The 22 × 20 km main study area. The 2 rectangles are referred to in the text. (a) Continuous colour image, with shade, of the horizontal gradient magnitude (HGM) obtained from the TC data. (b) Ternary image of the HGM of potassium (red), HGM of thorium (green) and HGM of uranium (blue).

The TC-HGM response of Figure 12a contains a noise level that can be associated with the lowest amplitudes. This information is not immediately apparent in the ternary-HGM of Figure 12b and so it may be prudent to assess their joint behaviour. In order to introduce a noise assessment into the images of Figure 12 it is possible to blank regions (grid cells) falling below a selected amplitude threshold. In the case of the Figure 12a this would be based on the TC-HGM amplitude. In the case of Figure 12b, the same procedure can be transferred to the ternary image or a different threshold can be developed using the amplitudes, or amplitude combinations, of the 3 radioelement HGM amplitudes. Much of the detailed behaviour in Figure 12 does not have a random (noise) appearance. The character of the edges detected will be governed by the limits to spatial resolution of finite areas that were discussed, both theoretically and empirically, previously. Many of the localised responses are likely to be soil related. Two rectangles, displaying such effects, allow the slightly different character of the two responses to be compared. Much soil mapping is based on texture whereas the flux changes assessed here largely relate to changes in soil properties (in the absence of bedrock effects). According to Beamish (2015), the main soil properties involved are density and wetness.

6. CONCLUSIONS

The theoretical results of the study indicate that 90% of the total flux is obtained across ground scale lengths of 250 m (60 m elevation), 360 m (120 m elevation) and 440 m (180 m elevation). The airborne measurements therefore constitute a moving-average spatial measurement thus limiting precise resolution of flux behaviour. A further study of anomalies of finite lateral extent indicated the decrease in flux count with survey elevation and a reduced response with decreasing source dimension. The detectability of small anomalous areas (< 100 m, or half the flight line spacing) is likely to be low or non-existent. The results indicate that, for lateral dimensions greater than the flight line spacing, the lateral scale of the anomalous source response is governed by the scale length of the anomalous region. A study of the horizontal gradient magnitude (HGM) of the anomalous response indicated that the edges of the anomalous region could be identified as maxima in the HGM response. The technique therefore offers increased resolution of the lateral contributions to changes in flux behaviour.

A control study area containing isolated island responses of varying scale-lengths was used to further assess the HGM resolution characteristics using survey data. Small islands less than the flight-line spacing are not detected unless a survey line directly overflies the island. The intrinsic shape of larger features is only partially replicated and again demonstrates the limited detailed resolution characteristics of airborne survey data. Although the application of moving-window curvature analysis to the HGM response was demonstrated, a continuous image of the HGM response is potentially the most useful in providing the largest dynamic range of the behaviour generated by the lateral contrasts.

HGM analysis was applied to recent survey data to allow an assessment of geological, soil and environmental effects. The amplitudes of the data used must also be used to distinguish the form of the relative flux behaviour (i.e. an attenuation or enhancement). While at the large (survey-wide) scale, there is a clear association between major bedrock types and the radiometric spectral response, the high resolution HGM analysis reveals the extent of small scale contributions to the response. Such effects are pervasive and include contributions from bedrock lithologies, changes in soil properties, man-made soil removal (quarrying), water bodies and certain types of vegetation. These can be studied using the high signal total count data and/or the spectral response formed by the HGM response of individual radioelements.

ACKNOWLEDGEMENTS

My thanks go to Jim White for an internal review and to Joe Duval and the USGS for providing the software used in the theoretical study. Topographic map data used here are based upon Ordnance Survey data with the permission of the Controller of Her Majesty's Stationery Office, © Crown copyright. The Northern Island data used in the study come from the Tellus Project which was funded by DETI and by the Building Sustainable Prosperity scheme of the Rural Development Programme (Department of Agriculture and Rural Development of Northern Ireland). This paper is published with the permission of the Executive Director, British Geological Survey (NERC).

REFERENCES

1. Beamish, D., 2012. The application of spatial derivatives to non-potential field data. Geophys. Prospect. 60, 337–360.
2. Beamish, D., 2013. Gamma ray attenuation in the soils of Northern Ireland, with special reference to peat. J. Environ. Radioact. 115, 13–27.
3. Beamish, D., 2014. Peat mapping associations of airborne radiometric survey data. Remote Sens. 6, 521–539.
4. Beamish, D., 2015. Relationships between gamma-ray attenuation and soils in SW England. Geoderma 259-260, 174–186.
5. Beamish, D., White, J.C., 2011. A radiometric airborne geophysical survey of the Isle of Wight. Proc. Geol. Assoc. 122, 787–799.
6. Beamish, D., White, J., 2014. TellusSW: airborne geophysical data and processing report. British Geological Survey Open Report, OR/14/014. Beamish, D., Young, M., 2009. Geophysics of Northern Ireland: the Tellus effect. First Break 27, 43–49.
7. Blakely, R.J., 1995. Potential Theory in Gravity and Magnetic Applications. Cambridge University Press, Cambridge, U.K. Blakely, R.J., Simpson, R.W., 1986. Approximating edges of source bodies from magnetic or gravity anomalies. Geophysics 51, 1494–1498.
8. Dickson, B.L., Scott, K.M., 1997. Interpretation of aerial gamma-ray surveys – adding the geochemical factors. AGSO J. Aust. Geol. Geophys. 17, 187–200.
9. Duval, J.S., 1997. Gamma-ray modelling programs. U.S. Geol. Surv. Open File Rep. 97–96.
10. Duval, J.S., Cook, B., Adams, J.A.S., 1971. Circle of investigation of air-borne gamma-ray spectrometer. J. Geophys. Res. 76, 8466–8470.
11. Edmonds, E.A., McKeown, M.C., Williams, M., 1975. British Regional Geology: South-West England. fourth ed. HMSO, London. Grasty, R.L., Minty, B.R.S., 1995. A guide to the technical specifications for airborne gammaray surveys. AGSO Australian Geological Survey Organisation, Record 1996/60.
12. Grauch, V.J.S., Hudson, M.R., Minor, S.A., 2001. Aeromagnetic expression of faults that offset basin fill, Albuquerque basin, New Mexico. Geophysics 66, 707–720.
13. IAEA, 1991. Airborne Gamma Ray Spectrometer Surveying. Technical Report SeriesInternational Atomic Energy Agency, Vienna (No. 323). IAEA, 2003. Guidelines for Radioelement Mapping Using Gamma Ray Spectrometry. Technical Report SeriesInternational Atomic Energy Agency, Vienna (No. 136).

14. IAEA, 2010. Radioelement Mapping. Nuclear Energy SeriesInternational Atomic Energy Agency, Vienna (No. NF-T-1.3).
15. Kimbell, S.F., Cooper, D.C., Scrivener, R.C., 2003. The digitisation and geological interpretation of data from the 1957-9 airborne radiometric survey of south-west England. Geoscience in South-West England. 10, pp. 410–416.
16. Leveridge, B.E., Hartley, A.J., 2006. The Variscan orogeny: the development and deformation of Devonian/Carboniferous basins in SW England and South Wales. In: Brenchley, P.J., Rawson, P.F. (Eds.), The Geology of England and Wales, second ed. Geological Society, London, pp. 225–255.
17. Minty, B.R.S., 1967. Fundamentals of airborne gamma-ray spectrometry. AGSO J. Aust. Geol. Geophys. 17, 39–50.
18. Nabighian, M.N., Grauch, V.J.S., Hansen, R.O., LaFehr, T.R., Li, Y., Peirce, J.W., Phillips, J.D., Ruder, M.E., 2005. The historical development of the magnetic method in exploration. Geophysics 70, 33–61.
19. Peart, R.J., Cuss, R.J., Beamish, D., Jones, D.G., 2003. The High Resolution Airborne Resource and Environmental Survey-Phase 1 (HiRES-1): Background, Data Processing and Dissemination and Future Prospects. British Geological Survey Internal Report IR/03/112.
20. Phillips, J.D., Hansen, R.O., Blakely, R.J., 2007. The use of curvature in potential-field interpretation. Explor. Geophys. 38, 111–119.
21. Pitkin, J.A., Duval, J.S., 1980. Design parameters for aerial gamma ray surveys. Geophysics 45, 1427–1439.
22. Wilford, J.R., Bierworth, P.N., Craig, M.A., 1997. Application of airborne gamma-ray spectrometry in soil/regolith mapping and applied geomorphology. AGSO J. Aust. Geol. Geophys. 17, 201–216.

CHAPTER 3

Geophysical and Hydrological Investigations of the Northern Wadis Area of Azraq Basin for Groundwater Artificial Recharge Purposes

Hani Al-Amoush[1*], Abdel Rahman Al-Shabeeb[1], Saad Al-Ayyash[2], Rida Al-Adamat[1], Majed Ibrahim[1], A'kif Al-Fugara[2], Jaafar Abu Rajab[3]

[1]*Institute of Earth and Environmental Sciences, Al al-Bayt University, Mafraq, Jordan*

[2]*Faculty of Engineering, Al al-Bayt University, Mafraq, Jordan*

[3]*Natural Resources Faculty, Hashemite University, Zarqa, Jordan*

ABSTRACT

Geophysical study and watershed hydrological delineation have been integrated at downstream of Alasra dam site Norh Azraq area to investigate their potential for artificial groundwater recharge. The total surface area of the watershed was found to be about 195 square kilometers. The estimated annual runoff volumes for the Alasra watershed ranged between 1.2 and 1.8 MCM. Moreover, the interpretation of Ten Time Domain Electromagnetic (TDEM) soundings suggested three principal subsurface layers. The top surface layer has an intermediate resistivity (90 - 110 Ohm · m) with a thickness ranging from a few meters to around 50 m. This layer was interpreted as superficial deposits. The second subsurface layer with variably high resistivity values is composed of unsaturated massive basalt layer and probably belongs to Madhala Olivine Phyric Basalt Formation (MOB). The large variations in resistivity could be ascribed to the degree of water saturation (as a result of groundwater recharge from the nearby harvested water dam), or lithological variations (clay content) and/or due to structural control. The third subsurface layer has low resistivity values (<10 Ω·m to 40 Ω·m) and was found at a depth ranging from 120 to 150 m. This layer could represent a saturated basalt layer with high clay contents. The subsurface structures and major faults have been identified. Based on the results of this study, a combination of surface and subsurface artificial groundwater recharge techniques is highly recommended.

Keywords: TDEM, Groundwater Recharge, Alasra, Hydrogeophysics, Azraq, Jordan

1. INTRODUCTION

Jordan is located in an arid to semi-arid region with variable topography features. The far western part of the country is mountainous terrains, while the eastern and southeastern parts are dominated by flat-desert terrains. Due to these topographic variations, rainfall distribution varies considerably. The annual rainfall intensity ranges from 600 mm in the northwest to less than 50 mm in eastern and southern parts of the country [1] . Jordan is suffering from scarcity and shortage of water resources. The available per capita water resources are projected to decline from about 150 m^3/year to only 90 m^3/year by the year 2020, putting Jordan in the category of absolute water shortage [2] . The political situations in the region, especially the recent waves of Syrian refugees have added an extra pressure on the natural resources and in particular the water resources. During winter times, the eastern and desert areas received extensive rain storms and monsoons, by which, a large quantity of surface water are drained through wadis and spread over low lands, surface and flat areas and eventually lost to evaporation. In recent years, several studies have been conducted on water harvesting and capturing surface water in dams in the eastern and northeastern Jordan (e.g. [3] - [6]). During the last four decades, the groundwater table has declined dramatically due to the extensive pumping (either for domestic and agricultural activities). Groundwater artificial recharge is an effective technique to fill aquifers using different types of water, and to augment ground- water level. It has many advantages over water harvesting, as it is less affected by the climatic conditions and evaporation. In addition, the recharged water is subjected to self-purification (natural attenuation) which will be enhanced naturally through percolation. Groundwater artificial recharge studies have been done using different geophysical and GIS techniques in several areas worldwide as well as in Jordan (e.g. [4] [7] - [13]).

TDEM technique has been previously used in numerous hydrogeological applications (e.g. [14] - [21]). In this study, the TDEM technique and hydrological study were integrated to investigate the potentials for artificial groundwater recharge at the dam downstream area of at Alasra Dam site as a tool for conservation of the harvested and flooding water to be used at times of dry and hot summer seasons.

2. DESCRIPTION OF STUDY AREA

The investigated area- which is part of Jordan's eastern plateau - is located in the northwestern part of Azraq basin between [480000 - 492000] E, [550000 - 575000] N according to Jordan Transverse Mercator (JTM) (Figure 1). Relief topography of Alasra watershed area shows that the elevations are ranging from 740 - 1000 m above sea level (masl). The average annual rainfall varies from 70 mm in the far eastern and southern part of the study area to about 150 mm in the northwestern part. The temperature may reaches more than 40°C during summer days and drops to a few degrees below zero in winter, especially during the night [1] . The relative humidity in eastern plateau is low. In winter, it is generally around 50% - 60% and drops to 15% in summer [1] . The potential evaporation rates increase east and south-wards directions and may exceed 4000 mm/y in the center of plateau [1] .

3. GEOLOGY OF THE STUDY AREA

The geological setting of Neogene continental basalts exposed at NE-Jordan have been investigated by several researchers (e.g. [22] - [35]). The study area is part of Harrat ash-sham Basalts super Group and forms the northern extension of Arabian Volcanic province [35] . A brief description of the major geological formations found in the study area (Figure 2) can be summarized as follows [35] [36] .

3.1. Abed Olivine Phyric Basalt Formation (AOB) (Safawi Group)

It comprises thick and massive flow units up to 10 m with a total thickness of up to 100 m. It contains several basaltic flood lava and feeder dike systems and is characterized by a massive, blocky and, columnar joint with polygon upper surfaces. There is a strong vesicularity in the upper part of the flow edifices.

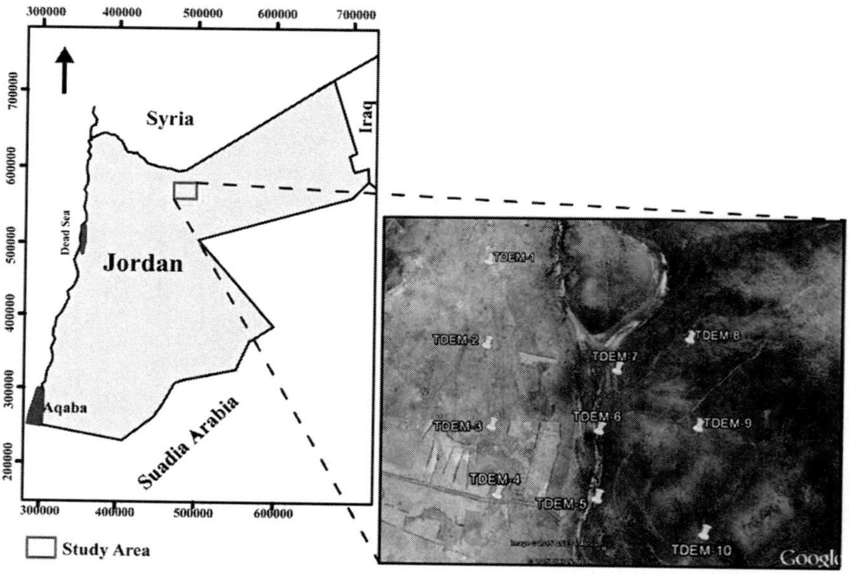

Figure 1. Location map of the study area and TDEM soundings sites on Google earth image. Coordinate system: Jordan Transverse Mercator (JTM).

3.2. Ufayhim Xenolith Basalt Formation (UM) (Asfar Group)

This formation was formed by eruption from the Jebel al Aritin volcano which flooded west- and south-wards along an old wadi. It typically displays well developed columnar jointing and non-systematic closely spaced horizontal jointing. It has host mantle xenolith.

3.3. Ushayhib Olivine Pyroxene Phyric Basalt Formation (UB) (Asfar Group)

Lava of the formation is extruded from the central vent of the Jabal Ushayhib volcano and the associated parasitic volcanoes. It forms an upstanding semi-vertical cliff above the plain of the Abed Olivine Phyric Basalt Formation (AOB) with up to 15 m thick.

3.4. Hashimyya Aphanitic Basalt (HAB) (Asfar Group)

Similar to Ufayhim Formation, the Hashimyya Formation typically displays columnar jointing, and less developed non-systematic closely spaced horizontal jointing. It is characterized by the presence of volcaniclastic deposits underneath. The thickness of Hashimyya Formation is less than 15 m and comprises thin flow units mostly ranges between 3 and 5 m. it is characterized by the presence of amygdaloidal texture, where vesicles are filled with calcite.

3.5. Madhala Olivine Phyric Basalt Formation (MOB) (Asfar Group)

Lava of Madhala Olivine Phyric Basalt Formation erupted from volcanic vent centers, including shield volcanoes, composite volcano vents and cinder scoria cones in the area. Its thickness varies greatly and reaches its maximum 100 m around the volcanic vents. It is mainly consists of massive and bedded basalt with polygonal joints, vesicles are present. There are large similarities between its lithology and the Abed Olivine Phyric Basal Formation.

3.6. Mahadda Basalt Formation (M) (Asfar Group)

It differs from the other units of the Asfar group in its hummocky and rough upper surface which comprises weathered wadi-fill lava with poorly preserved pressure ridges. The thickness of the formation varies from 10m to 25 m. It is lithologically similar to Madhala Olivine Phyric Basalt Formation.

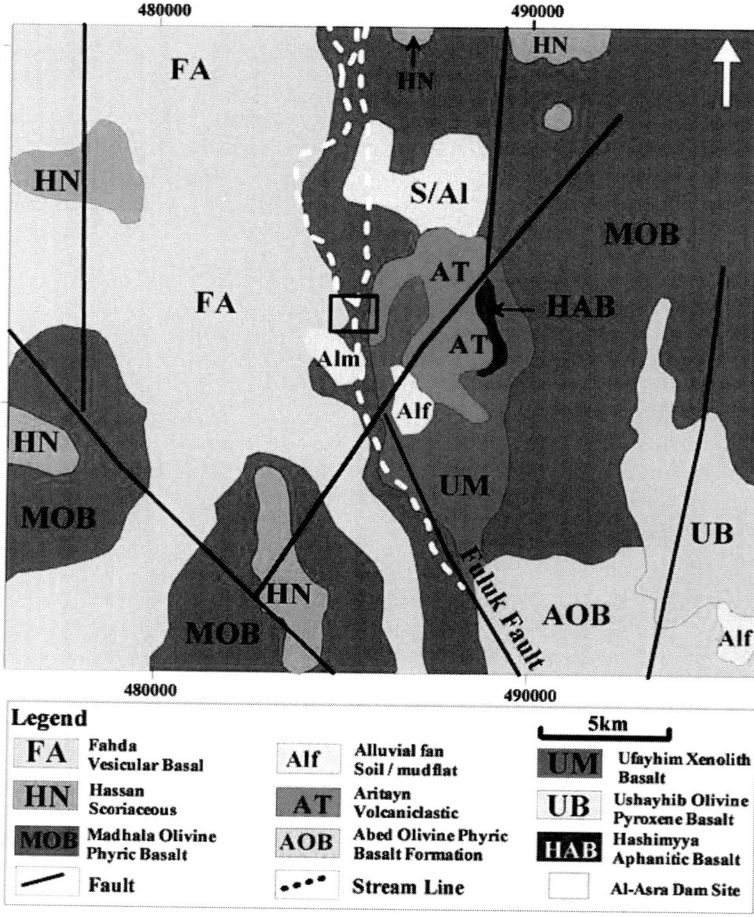

Figure 2. Geological map of the study area (modified after [36]).

3.7. Hassan Scoriaceous Formation (HN) (Rimah Group)

The 15 - 40 m deposits of Hassan Formation include bombs of lava spatter with various shapes along with coarse scoriaceous blocks. It is composed of poorly bedded, friable and very coarse-grained scoriaceous pyroclasts.

3.8. Aritayn Volcaniclastic Formation (AT) (Rimah Group)

The Aritayn Volcaniclastic Formation is developed from stratified cinder cones, occasionally composite. These pyroclastic deposits interbedded with lava flows. The formation consists of bedded, poorly cemented air-fall tephra, which is typified by its stratified form and typically cavernous. Similar to the Hassan formation, it is characterized by a smooth ground surface color. The boulder cover is mainly pyroclasts mixed with basaltic boulders.

3.9. Bishriyya Group (BY)

This group includes the youngest volcanism recorded in the study area. It is subdivided into two formations based on truncation flow lines which is a reflection of different ages, otherwise, the area is lithologically and morphological similar. These are Fahda Vesicular Basal Formation (FA) and Wadi Manasif Basalt Formation (WMF) which are not shown in the investigated area. The surface of the flows is characterized by a rubbly and very vesicular ground cover. Boulders are characterized by a purplish-black weathering color. They show a columnar jointing and exhibit typical polygonal lobes of pahoehoe lava types. This group has a thickness ranging from 25 - 60 m.

3.10. Superfacial Deposits

3.10.1. Alluvium and Wadi Sediments (Al)

Alluvium and wadi sediments consist of gravels and, sand deposits.

3.10.2. Alluvial Fans (Alf)

Alluvial fans occur in two places, to the southeast of Jabal Al Aritin and south of Wadi Al Aritin. The fans are composed of stony, or sometimes boulder, deposits with arcuate low ridges of gravels, sand and silt.

3.10.3. Alluvial Mudflats and Silt Flats (Alm)

Several mudflats are surrounding the investigated area. They were probably formed as a result of regional and local faults which produced small depressions, where mud accumulated from standing water fed by ephemeral wades discharged into the mudflat. They consist of soft and silty clay and rock fragments.

4. DATA ACQUISITION AND MODELING

4.1. Hydrology and Watershed Modeling

The watershed model of Alasra site (Figure 3) was developed using the geographic information system (ArcGIS10.3), where the SRTM digital elevation data was used to delineate the boundaries of the watershed (its sub-basins and the wadi network within the watershed). The total area of the watershed was found to be about 195 square kilometers, extending northward (inside the Syrian border) to the south with a total length of about 40 kilometers.

A preliminary runoff investigation for part of Alasra site was conducted by [3] . The study showed that the average annual surface water which can be collected at the outlet of the watershed was 0.2 MCM. This quantity constitutes about 8% of the annual rainfall volume received by the watershed. In another study for the region, where the study area is located [37] , the annual runoff coefficient varied between 5% and 8% with an average annual rainfall in the area of 120 mm; given these figures, the annual runoff volume for the Alasra watershed could be estimated between 1.2 and 1.8 MCM.

4.2. Time Domain Electromagnetic Method (TDEM)

In the downstream of Alasra harvested dam site (Figure 1), TDEM soundings were conducted to characterize the subsurface layers and structures and to investigate their potentials for artificial groundwater recharge. The field geometry of the TDEM's soundings was gridded with 200 - 300 m spacing. This allows for obtaining different 2D resistivity models representing the subsurface resistivity variations with different directions. Table 1 lists the coordinates, elevations and root mean square error (RMSE) for each modeled TDEM soundings. The square transmitter loop was 50 m × 50 m with estimated depth of about 130 m. In TDEM theory, a direct current (DC) is transmitting through loop wire. This produces a static primary magnetic field. The current is shut off abruptly after a while, which due to Faraday's law induces an electrical field in the surroundings. In the ground, this will again produce a magnetic field; the secondary field. Since the secondary magnetic field is generated during the period when the primary magnetic field is off, it can be measured relatively easily. The decaying of the secondary magnetic field in conductive bodies is slower than that in poor conductors. Measurements of the rate of change (decay) of the secondary field thus provide a means of detecting subsurface conductive bodies and estimating their conductivities [38] [39] . The decaying transient signal is sampled at various time gates to yield a sounding apparent resistivity curve. In this study, ATEM FAST 48 HPC system has been used for data acquisition. The system was set to transmit current up to 4 Ampere using 12 Voltage source batteries with 48 active time gates (15,360 μs—t center). The stacking time was set to about 7 minutes with 50 Hz noise filter in order to avoid aliasing effects of

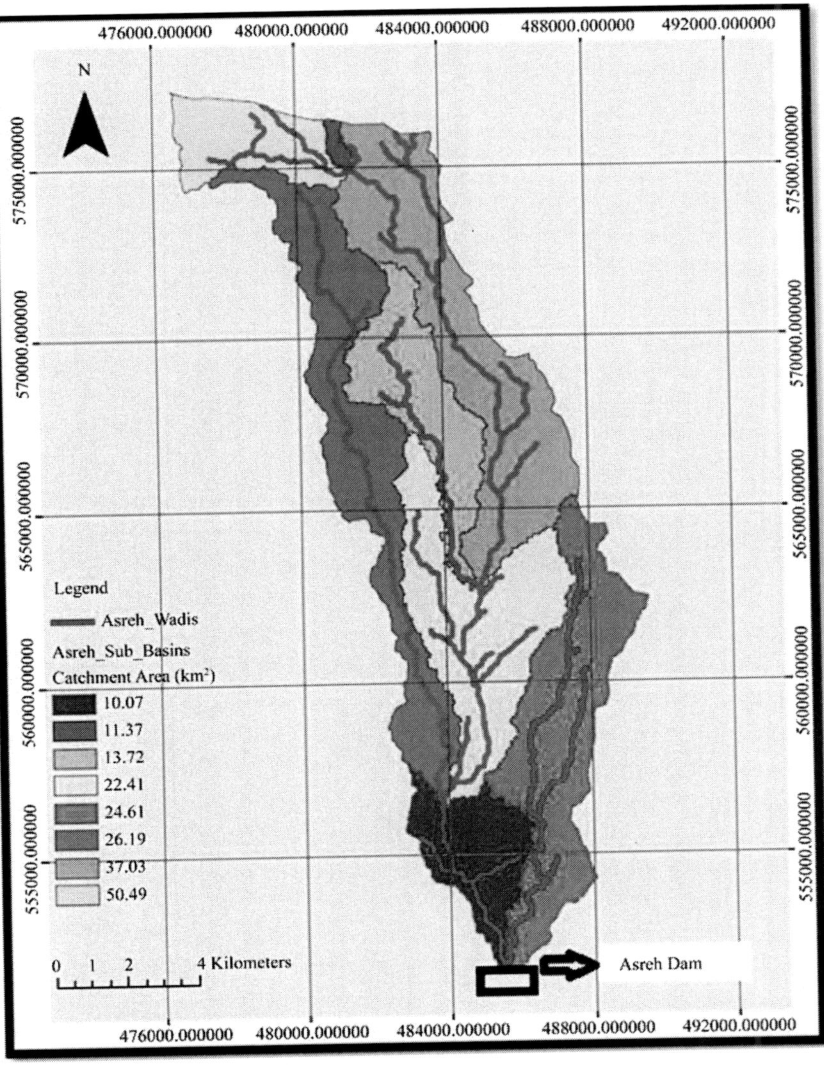

Figure 3. Watershed and sub-watershed areas of Alasra dam site.

Table 1. Coordinates, elevations and Root Mean Square Error (RMSE) of TDEM data.

Sounding	Longitude	Latitude	Elevation	RMSE (%)
TDEM-1	485153	552067	724	1.85
TDEM-2	485169	551802	715	3.09
TDEM-3	485198	551576	710	1.31
TDEM-4	485224	551410	718	5.62
TDEM-5	485448	551407	705	3.88
TDEM-6	485453	551573	703	6.74
TDEM-7	485498	551740	703	1.58
TDEM-8	485685	551830	735	2.75
TDEM-9	485686	551584	740	4.61
TDEM-10	485681	551331	735	5.46

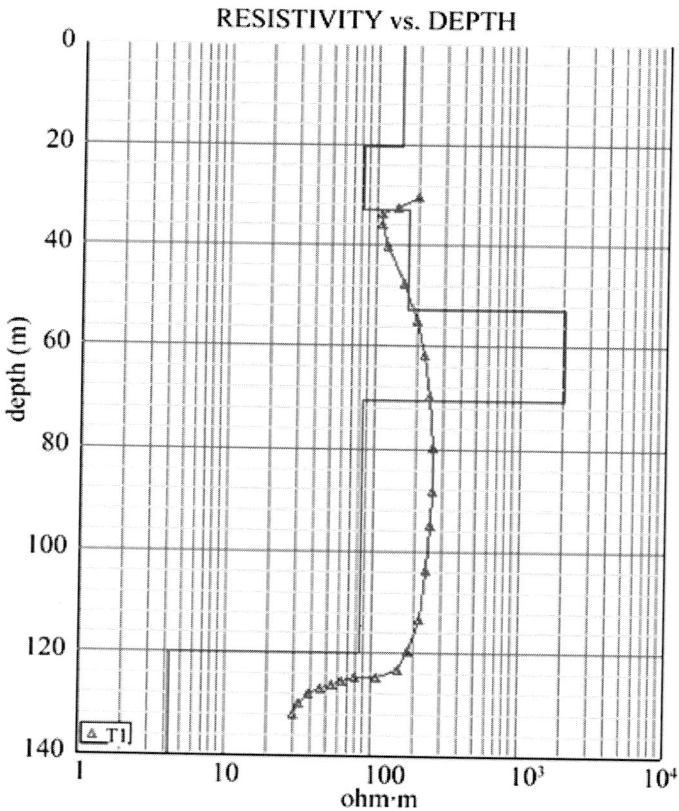

(a)

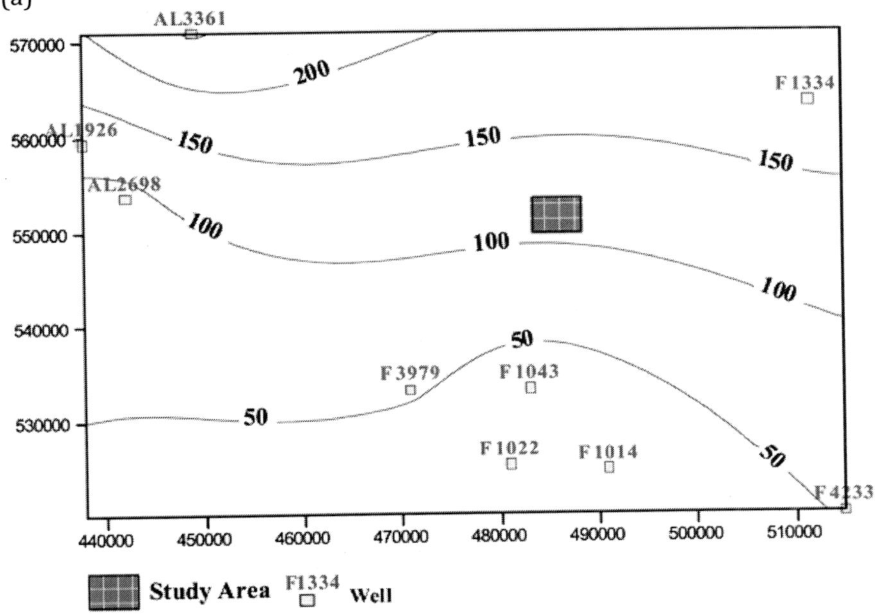

(b)

Figure 4. (a) An example of a 1-D layering model and measured curve of TDEM-1 (the figure shows the variations of resistivity with depth and inverted model of TDEM-1) (RMES = 1.8%, see Table 2 for interpretation). At depth 120 m below ground surface, the resistivity is abruptly declining to reach 30 ohm·m which could be ascribed to groundwater aquifer (see **Figure 4**(b)); (b) Static water level map of the basaltic aquifer in the study area and its surrounding. The static water level in the study area is located at depth ranging between countour lines 125 m and 150 m below ground surface. This is highly correlated with the results of the inverted model of TDEM's (**Figure 4**(a)). Data obtained from Ministriy of Water and Irrigation and interpolated using Kriging method.

possible galvanic interference. The system is comprising a Transmitter-Receive control and managed by HP-IPAQ Pocket system. TDEM measurements were analyzed using TEM RESEAR- CHER Software, a component of TEM FAST 48 HPC system. Data were processed by fitting the theoretical model with field data. TEM RESEARCHER Software enables to combine multi TDEM soundings in a 2-D cross section of resistivity distribution. It allows for interpreting TDEM data in a single curve (Figure 4) or in a profile of multi TDEM soundings (Figures 5-10).

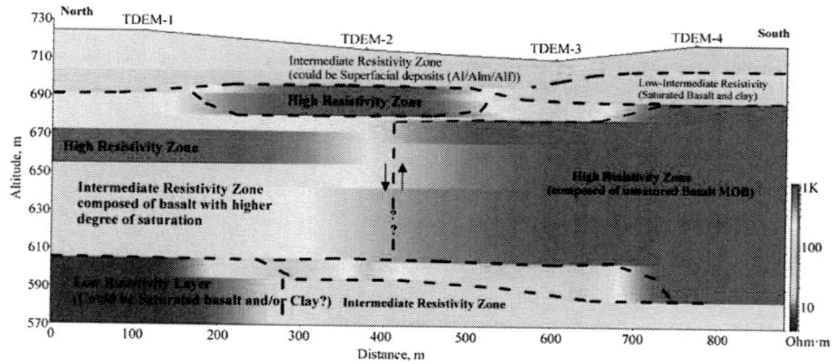

Figure 5. Interpreted hydro-geophysical resistivity model derived from TDEM's soundings, available geological information and static water level map for the study area (Figure 4(b)).

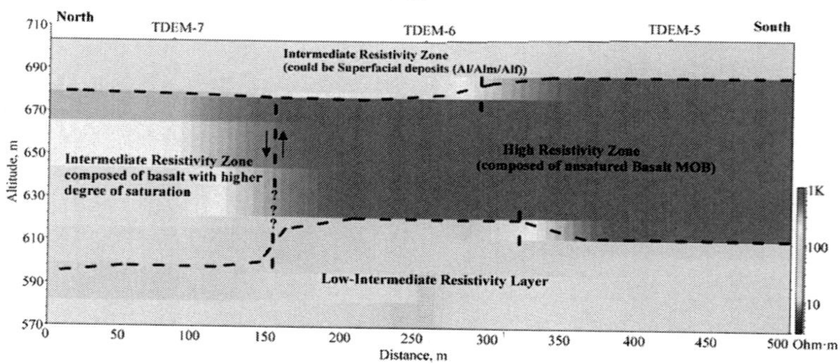

Figure 6. Interpreted hydro-geophysical resistivity model derived from TDEM's soundings, available geological information and static water level map for the study area (Figure 4(b)).

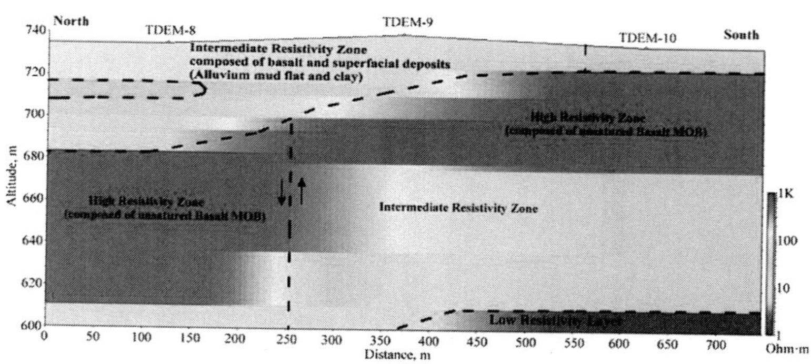

Figure 7. Interpreted hydro-geophysical resistivity model derived from TDEM's soundings, available geological information and static water level map for the study area (Figure 4(b)).

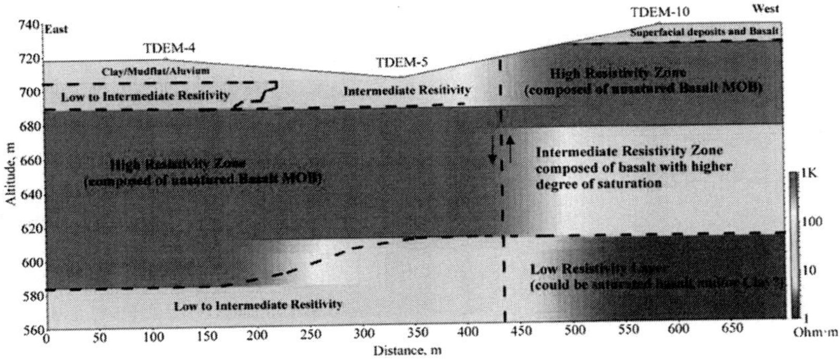

Figure 8. Interpreted hydro-geophysical resistivity model derived from TDEM's soundings, available geological information and static water level map for the study area (Figure 4(b)).

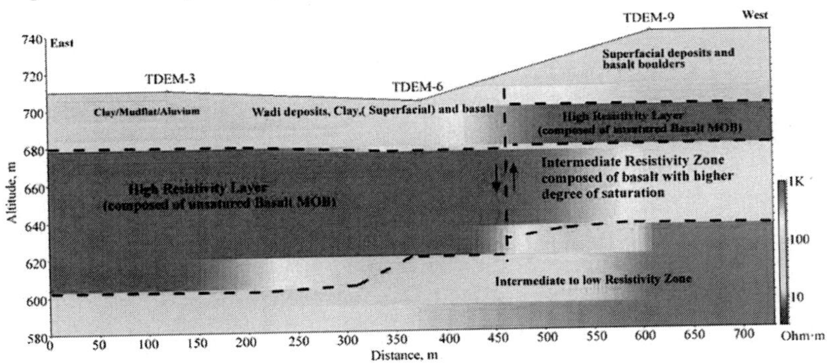

Figure 9. Interpreted hydro-geophysical resistivity model derived from TDEM's soundings, available geological information and static water level map for the study area (Figure 4(b)).

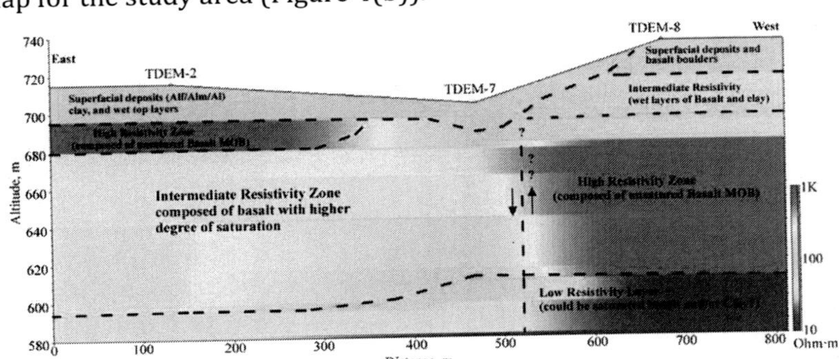

Figure 10. Interpreted hydro-geophysical resistivity model derived from TDEM's soundings, available geological information and static water level map for the study area (Figure 4(b)).

5. RESULTS, DISCUSSION AND CONCLUSIONS

TDEM's sounding points have been individually interpreted as a 1-D layering model using the available surface and subsurface geological information and the static water level map of basaltic aquifer (Figure 4(a) and Figure 4(b)). Table 2 lists the multi-layering model interpretation for each TDEM sounding point. Figures 5-7 show N-S 2-D resistivity cross-sections and Figures 8-10 show an E-W 2-D resistivity cross-section over the investigated area. In order to facilitate the geological interpretations, the inverted models were typically represented in thematic maps and cross-sections [40] . The resistivity distributions at selected depths (20 m, 75 m and 120 m) are outlined in Figures 11-13 respectively. Figure 14 shows a 3-D resistivity visualization model of TDEM's results. The TDEM's results suggest three principal subsurface layers. The top surface layer has intermediate resistivity values which could be attributed to super facial deposits such as alluvial mud flat (Alm), alluvial fans (Alf), wadi deposits and clay intercalated with basalt boulders. The thickness of this layer is ranging from a few meters to around 50 m at TDEM-8 and TDEM-4 (Figure 7 and Figure 5). The second subsurface layer has high resistivity values and is composed of unsaturated massive basalt layer. It could belong to MOB of Asfar Group. It is characterized by a variable thickness as shown in different 2-D cross-sections. For instance, it has a thickness of about 100 - 110 m at TDEM-4 (Figure 5 and Figure 8), 70 m at TDEM-5 (Figure 6 and Figure 8), and 80 m at TDEM-8 (Figure 10). Moreover, significant lateral variations in the resistivity of this layer were apparent. It has a resistivity of 90 - 100 Ω·m at TDEM-1, TDEM-2 (Figure 5), TDEM-7 (Figure 6), TDEM-9 and TDEM-10 (Figure 7) and it was identified as an intermediate resistivity zone composed of basalt with higher degree of saturation. Additionally, it was found that the same layer has a resistivity of 1000 Ω·m at TDEM-3 and TDEM-4 (Figure 5), TDEM-5 and TDEM-6 (Figure 6) and TDEM-8 (Figure 7). The large variation in resistivity could be ascribed to the degree water saturation as a result of groundwater recharge from the nearby harvested water dam, or due to lithological variations (clay content) or structural control. The third subsurface layer was with low resistivity values and observed in the majority of TDEM's soundings and the bottom of the hydro-geophysical cross-sections (Figures 5-10). It was also observed in the 3-D model visualization (Figure 14). This layer was found at depths ranging from 120 to 150 m below ground surface in the investigated area, and has resistivity values ranging from <10 Ω·m to 40 Ω·m. This layer could be interpreted as a saturated basalt layer and/or containing high ratio of clay. In addition, several lenses-like zones have been identified with different resistivities at different depths, which reflect the nature of basalt flow successions and subsurface structures.

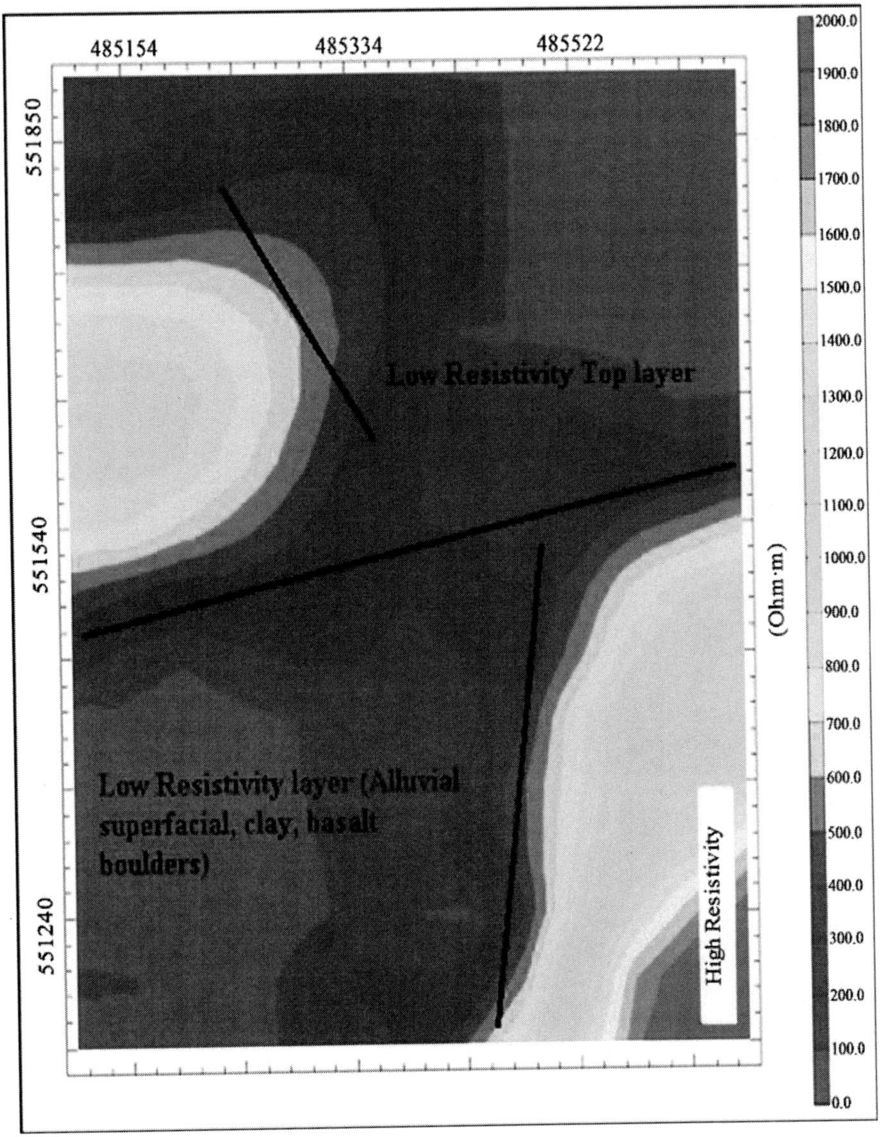

Figure 11. Resistivity map at a depth of 20 m below ground surface. Coordinates System: JTM.

Table 2. Multilayer interpretation of TDEM's soundings.

Sounding	Layer	Resistivity (Ohm·m)	Thickness (m)	Suggested Geological Interpretation
TDEM-1	1	145.8	20.2	
	2	80.6	12.3	Intermediate resistivity top soil (could be superfacial deposits, Alm/Alf/Al and clay)
	3	165.4	20	
	4	2000	17.5	High resistivity layer composed of unsaturated basalt of MOB
	5	82.4	10	
	6	4.06	20	Low resistivity layer (Saturated basalt and clay) Unknown thickness
	7	4.06		
TDEM-2	1	97.99	19.6	Intermediate resistivity top soil (could be superfacial deposits, Alm/Alf/Al and clay)
	2	2000	15.24	High resistivity layer composed of unsaturated basalt of MOB
	3	229.13	38.17	Intermediate resistivity layer composed of Basalt with higher degree of saturation
	4	309.56	10	
	5	37.85	30	Low-intermediate resistivity layer
	6	37.85		Low-intermediate resistivity layer (unknown thickness)
TDEM-3	1	57.8	17.7	
	2	42	4	Intermediate resistivity top soil (could be superfacial deposits, Alm/Alf/Al and clay)
	3	68.2	5.53	
	4	94.2	4.57	
	5	1672	12.5	High resistivity layer composed of unsaturated basalt of MOB
	6	760.8	63.1	
	7	33.6	10	Intermediate to low resistivity top soil (could be saturate basalts and clay)
	8	9.87		Low resistivity with unknown thickness
TDEM-4	1	64.4	13.8	
	2	29.6	3.78	Intermediate to low resistivity top soil (could be saturate basalts and clay)
	3	32.3	8.14	
	4	38.9	4.49	
	5	1430	11.2	High resistivity layer composed of unsaturated basalt of MOB
	6	2000	23.8	
	7	19.6		
TDEM-5	1	70	18.1	Intermediate resistivity
	2	2000	76	High resistivity layer composed of unsaturated basalt of MOB
	3	38	7.58	
	4	26	20	Low-intermediate resistivity layer (thickness > 30 m)
	5	23.3		

TDEM-6	1	133.3	27.18	Intermediate resistivity top soil (could be superfacial deposits, Alm/Alf/Al and clay)
	2	2000	56	High resistivity layer composed of unsaturated basalt of MOB
	3	32.56	26.3	Low-intermediate resistivity layer (thickness > 30 m)
	4	10.93	10	
	5	3.1	20	Low resistivity layer (thickness > 20 m)
	6	3.1		
TDEM-7	1	106.5	24.4	Intermediate resistivity top soil (could be superfacial deposits, Alm/Alf/Al and clay)
	2	311.7	13.7	
	3	207	22	
	4	123.8	27	
	5	95.5	17.8	Intermediate resistivity layer composed of Basalt with higher degree of saturation
	6	43.3	10	
	7	14.4	10	Low-intermediate resistivity (resistivity > 15 m)
	8	14.4		
TDEM-8	1	117	18.6	Intermediate resistivity zone composed of basalt and superfacial deposits (alluvium mudflat and clay)
	2	42	16.7	
	3	64	6.54	
	4	188.6	10.6	High resistivity layer composed of unsaturated basalt of MOB
	5	2000	50	
	6	10.1		Low resistivity layer with unknown thickness (saturated layer and clay)
TDEM-9	1	87.3	24	Intermediate resistivity zone composed of basalt and superfacial deposits (alluvium mudflat and clay)
	2	78.5	5.84	
	3	267.2	10.4	High resistivity layer composed of unsaturated basalt of MOB
	4	2000	21.6	
	5	141	10	Intermediate resistivity
	6	7.57	20	Low resistivity layer (saturated layer) Unknown thickness
	7	7.57		
TDEM10	1	34.39	11.2	Intermediate resistivity zone composed of basalt and superfacial deposits (alluvium mudflat and clay)
	2	2000	49	High resistivity layer composed of unsaturated basalt of MOB
	3	33.34	10	Low resistivity layer (saturated layer) Unknown thickness
	4	1.31	30	
	5	1.31		

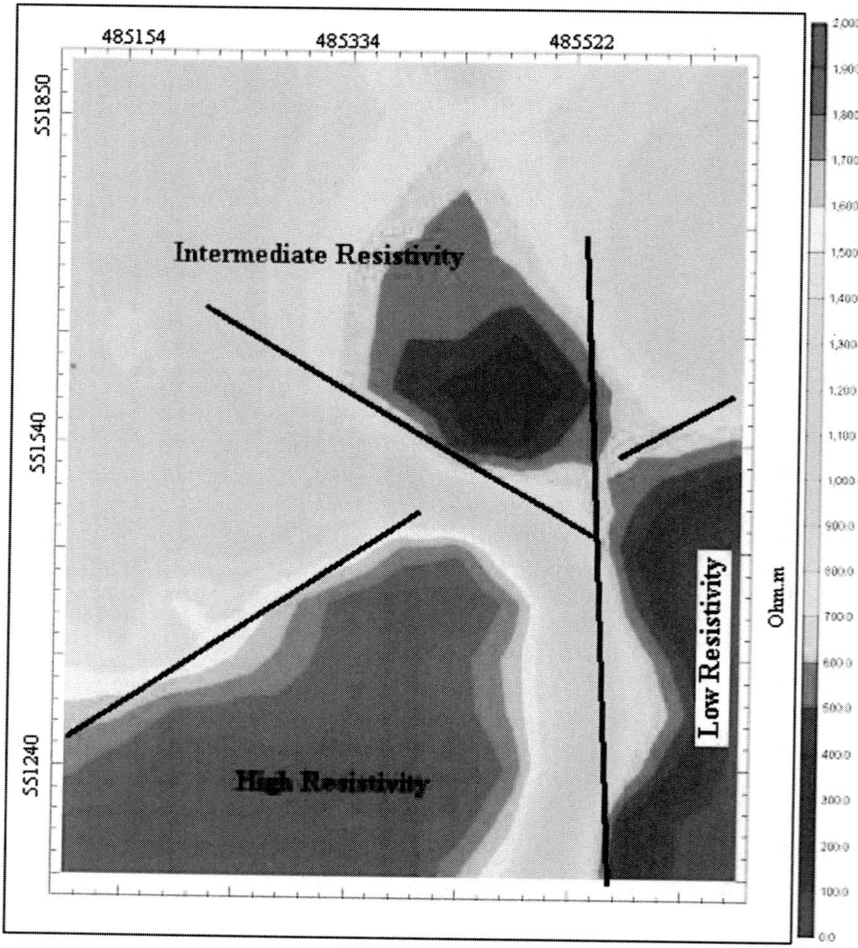

Figure 12. Resistivity map at a depth of 75 m below ground surface. Coordinates System: JTM.

There is probably a significant NNW-SSE fault identified at different cross-sections, between TDEM-7 and TDEM-8 (Figure 10), between TDEM-6 and TDEM-9 (Figure 9), and between TDEM-5 and TDEM-10 (Figure 8). It can be seen also in the thematic maps (Figures 11-13). This fault is extending along the course of the downstream valley of Alasra dam site. Moreover, it could be the extension of the well-known Fuluq fault when compared with the surface geological map (Figure 2). This fault is likely to play an important role in groundwater recharge through percolation of surface water into the groundwater aquifer through conduits and fractures. A principal NW-SE fault is probably found between the TDEM-9 and TDEM-10 soundings (Figure 7), between TDEM-6 and TDEM-7 (Figure 6), and below TDEM-2 (Figure 5). It can also be seen in the the matic maps at different depths (Figure 12 and Figure 13). A NE-SW fault extending over the investigated area has been identified and is clearly shown in thematic maps (Figures 11-13).

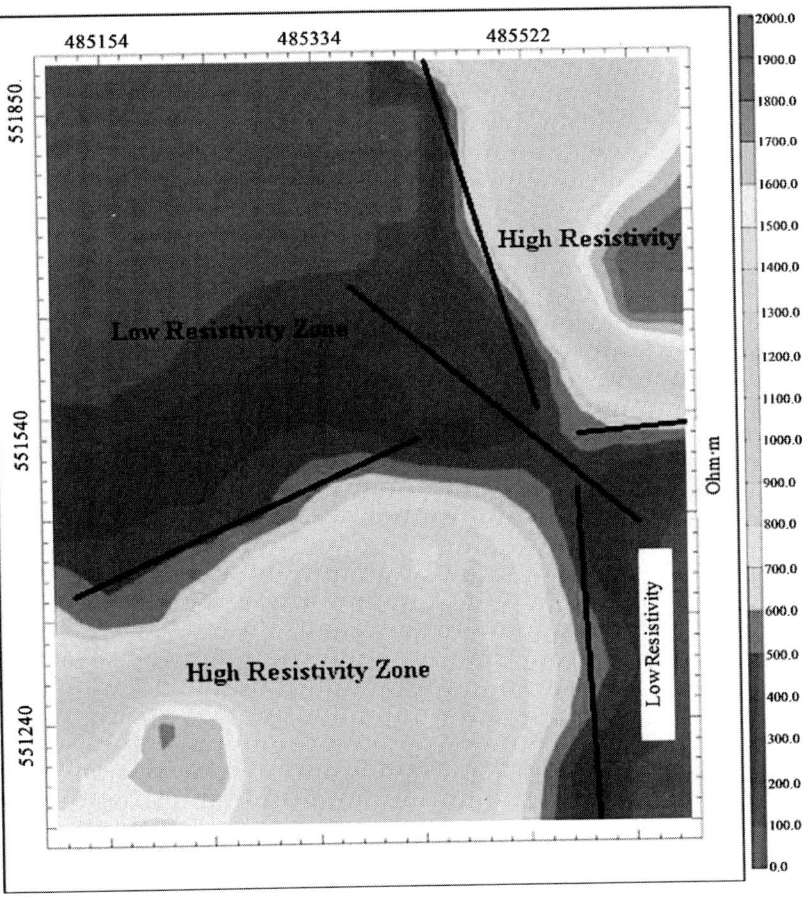

Figure 13. Resistivity map at a depth of 120 m below ground surface. Coordinates System: Jordan Transverse Mercator (JTM).

Artificial Groundwater Recharge Potentiality

There are several artificial groundwater techniques that have been developed and applied in various parts of the world. The description of these methods is mentioned in [41] - [43] . Based on the resulted hydro-geophysical models, structural pattern and interpreted multi-layering models at downstream of Alasra dam site, the techniques that could be utilized within the study area are:

1) Direct Surface Recharge: It is the simplest and most widely applied technique. In this method, water moves from the land surface to the aquifer by means of percolation through the soil. It has relatively low construction costs and is easy to operate and maintain. However, this technique cannot be applied in the study area for the following reasons:

a) It requires longer time for the recharge water to reach the aquifer. This might lead to evaporation problem, especially in arid environment as that of the study area, where the evaporation rate is high.

b) The groundwater aquifer in the study area is deep (above 150 m) which means that the surface water will need more time to reach the groundwater.

c) There are indications of a clay layer above the water table in the study area which prevents the groundwater recharge using such technique.

2) Direct Subsurface Recharge: This technique is used for recharging deeper aquifers which is the case within the study area. It is based on digging injection wells when aquifers are deep and separated from the land surface by low permeability materials. This technique is expensive but might be a solution to replenish groundwater in the study area. The problem with the clay layer above the groundwater table can be overcome by digging injection wells to penetrate this layer and establish a path between surface and the groundwater.

3) Combining Surface-Subsurface Methods: Both techniques mentioned above could be combined to have a better groundwater recharge. The surface water dam in the area could help in applying a direct surface recharge through the wadi bed. Also, having injection wells close to the dam will accelerate and maximize the potentials for replenishing the groundwater in the area.

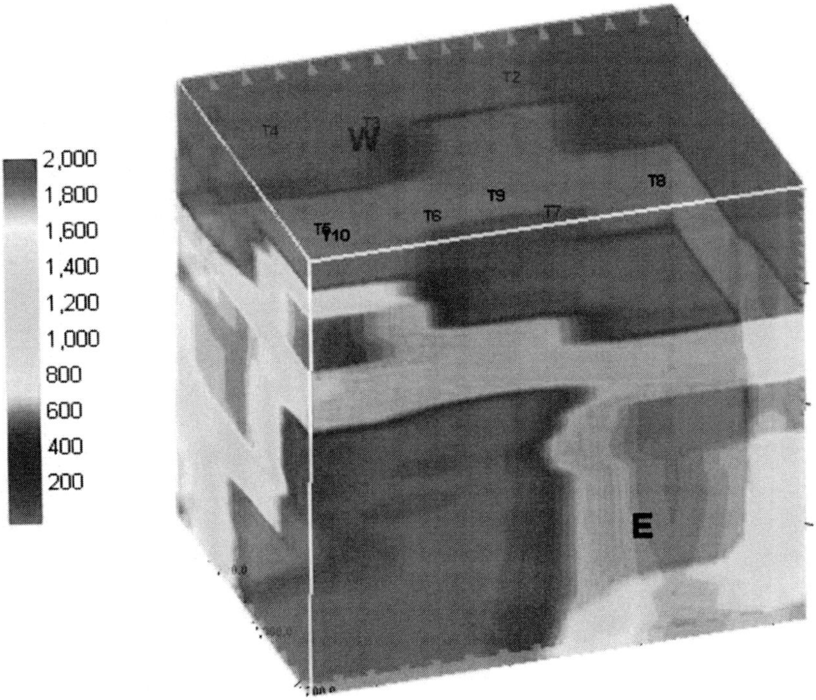

Figure 14. A 3-D resistivity model of the investigated area (downstream of Alasra Dam site). Coordinates System: JTM.

ACKNOWLEDGEMENTS

This research was carried out by Hani Al-Amoush while on sabbatical leave from Al al-Bayt University for the academic year (2015/2016). The authors would like to thank Al al-Bayt university presidency, the dean of Institute of Earth and Environmental Sciences. The authors also acknowledge the help and supports of the dean and staff of The Scientific Research Deanship at Al al-Bayt University. Critical and constructive review of the manuscript by the reviewers is greatly treasured.

CITE THIS PAPER

Hani Al-Amoush, Abdel Rahman Al-Shabeeb, Saad Al-Ayyash, Rida Al-Adamat, Majed Ibrahim, A'kif Al-Fugara, Jaafar Abu Rajab (2016)

Geophysical and Hydrological Investigations of the Northern Wadis Area of Azraq Basin for Groundwater Artificial Recharge Purposes. *International Journal of Geosciences*,**07**,744-760.

REFERENCES

1. Salameh, E. and Bannayan, H. (1993) Water Resources of Jordan. Present Status and Future Potentials, Fridrich Ebert Stiftung, Amman, Jordan.
2. Salameh, E. (1996) Water Quality Degradation in Jordan. Friedrich Elbert Stifing (FES) and Royal Society for Conservation of Nature (RSCN), Amman.
3. Al Ayyash, S., Al-Adamat, R., Al-Amoush, H.R., Al-Meshan, O., Rawjefih, Z., Shdeifat, A., Al-Harahsheh, A. and Al-Farajat, M. (2012) Runoff Estimation for Suggested Water Harvesting Sites in the Northern Jordanian Badia. Journal. Water Resources and Protection, 4, 127-132.
4. Al-Amoush, H.R. (2012) Hydro-Geophysical Investigations for the Purposes of Groundwater Artificial Recharge in Wadi Al-Butum Area, Jordan. Journal of Water Resource and Protection, 4, 507-515.
5. Batayneh, A., Al Zoubi, A. and Abueladas, A. (2001) Geophysical Investigations for the Location of a Proposed Dam in Al Bishriyya (Al Aritayn) Area, Northeast Badia of Jordan. Environmental Geology, 40, 918-922.
6. Al-Adamat, R.A., AlAyyash, S., Al-Amoush, H.R., Al-Meshan, O., Rawajfih, Z., Shdeifat, A., Al-Harahsheh, A. and Al-Farajat, A. (2012) The Combination of Indigenous Knowledge and Geo-Informatics for Water Harvesting Siting in the Jordanian Badia. Journal of Geographic Information System, 4, 366-376.
7. Bedrosian, M., Burgress, M. and Nishikawa, T. (2013) Faulting and Groundwater in Desert Environment: Constraining Hydrogeology Using Time-Domain Electromagnetic Data. Near Surface Geophysics, 11, 545-555.
8. Al-Amoush, H.R. (2006) Hydro-Geophysical Investigations for the Purposes of Groundwater Artificial Recharge in the Jordan Valley Area. Ph.D. Dissertation, University of Jordan, Amman.
9. Al-Amoush, H. (2010) Integration of Vertical Electrical Sounding and Aeromagnetic Data Using GIS Techniques to Assess the Potential of Unsaturated Zone and Natural Basalt Caves for Groundwater Artificial Recharge in NE-Jordan. Jordan Journal for Civil Engineering, 4, 389-408.
10. Hammouri, N., Al-Amoush, H., Al-Raggad, M. and Harahsheh, S. (2013) Groundwater Recharge Zones Mapping Using GIS: A Case Study in Southern Part of Jordan Valley, Jordan. Arabian Journal of Geosciences.
11. Al-Amoush, H.R., Al-Adamat, A., AlAyyash, S., Al-Meshan, O., Rawjefih, Z., Shdeifat, A., Al-Harahsheh, A. and Al-Farajat, M. (2012) Preliminary Geophysical Investigation for Suggested Water Harvesting Sites in the Northern Jordanian Badia. Research Journal of Environmental and Earth Sciences, 4, 560-569.
12. Al-Shabeeb, A. (2015) A Modified Analytical Hierarchy Process Methods to Select Sites for Groundwater Recharge in Jordan. Unpublished PhD Thesis, Leicester University, Leicester, 204 p.
13. Salameh, E. (2001) The Potential of Groundwater Artificial Recharge in the Jordan Valley Area/Jordan: Selected Contributions to Applied Geology in the Jordan Rift Valley. Freiberger Forschungshefte C494, Freiberg, 1, 63-81.
14. Fitterman, D. and Stewart, M. (1986) Transient Electromagnetic Sounding for Groundwater. Geophysics, 51, 995- 1005.
15. Bauer-Gottwein, P., Gondwe, B., Christiansen, L., Herckenrath, D., Kgotlhang, L. and Zimmermann, S. (2009) Hydrogeophysical Exploration of Three-Dimensional

Salinity Anomalies with the Time-Domain Electromagnetic Method (TDEM). Journal of Hydrology, 380, 318-329.

16. Duque, C., Calvache, M., Pedrera, A., Martin-Rosales, W. and Lopez-Chicano, M. (2008) Combined Time Domain Electromagnetic Soundings and Gravimetry to Determine Marine Intrusion in a Detrital Coastal Aquifer (Southern Spain). Journal of Hydrology, 349, 536-547.
17. Ferre, T., Bentley, L., Binley, A., Linde, N., Kemna, A., Singha, K., Holliger, K., Huisman, J. and Minsley, B. (2009) Critical Steps for the Continuing Advancement of Hydro Geophysics. EOS, Transactions, American Geophysical Union, 90, 200-201.
18. Goldman, M., Gilad, D., Ronen, A. and Melloul, A. (1991) Mapping of Seawater Intrusion into the Coastal Aquifer of Israel by the Time Domain Electromagnetic Method. Geoexploration, 28, 153-174. http://dx.doi.org/10. 1016 /0016-7142(91)90046-F
19. Herckenrath, D., Odlum, N., Nenna, V., Knight, R., Auken, E. and Bauer-Gottwein, P. (2013) Calibrating a Salt Water Intrusion Model with Time-Domain Electromagnetic Data. Groundwater, 51, 385-397.
20. Al-Amoush, H., Al-Tarazi, E., Abu Rajab, J., Al-Dwyeek, Y., Al-Atrash, M. and Shudiefat, A. (2015) Geophysical Investigation Using Time Domain Electromagnetic Method (TDEM) at Wadi Deir Al-Kahaf Area/Jordan for Groundwater Artificial Recharge Purposes. Journal of Water Resource and Protection, 7, 143-151.
21. Minsley, B.J., Ajo-Franklin, J., Mukhopadhyay, A. and Morgan, F.D. (2011) Hydro Geophysical Methods for Analyzing Aquifer Storage and Recovery Systems. Groundwater, 49, 250-269.
22. Lartet, L. (1896) La geologie de la Palastine. Thesis, University of Paris, Messon, 292 p.
23. Van Den Boom, G. and Sawan, O. (1966) Report on Geological and Petrological Studies of the Plateau Basalts in NE-Jordan. German Geological Mission, Amman, 42 p.
24. Bender, F. (1974) Geology of Jordan. Borntraeger, Berlin, 196 p.
25. Baraberi, F., Capaldi, G., Gasperini, P., Marinelli, G., Santacroce, R., Scandone, R., Treuil, M. and Varet, J. (1979) Recent Basaltic Volcanism of Jordan and Its Implications on the Geodynamic Evolution of the Afro-Arabian Rift System. Accademia Nazionale Dei Lincei, Atti Del Corvegni Lincei, 74, 667-683.
26. Ibrahim, K. (1987) Geochemistry and Petrology of Some of the Basaltic Outcrops in Central Jordan. MSc Thesis, University of Jordan, Amman, 164 p.
27. Moffat, D. (1988) A Volcano Tectonic Analysis of the Cenozoic Continental Basalts of Northern Jordan; Implications for Hydrocarbon Prospecting in the Block B Area. ERI Jordan EJ88-1, 73 p.
28. Al-Malabeh, A. (1989) The Volcanic Succession of Jebel Aritayn Volcano, NE-Jordan: A Field Petrographic and Geochemical Study. MSc Thesis, Yarmouk University, Irbid, 182 p.
29. Fakhoury, I. (1991) Mineralogy, Phase-Petrology and Geochemistry of Mafic and Ultramafic Xenoliths in the Quaternary Basalts, NE Jordan. MSc Thesis, Yarmouk University, Jordan, 155 p.
30. Khalil, I. (1991) Geochemische und petrographische untersuchungen an Tertaren bis Quarteren kontinintalen intraplattenbassalten Nordost-Jordanien. PhD Thesis, Technische Universit?t Clausthal, Clausthal-Zellerfeld, Germany, 119 p.
31. Al-Malabeh, A. (1993) The Volcanology, Mineralogy, and Geochemistry of Selected Pyroclasric Cones from NE- Jordan and Their Evaluation for Possible Industrial Applications. PhD Thesis, Universitat Erlangen, Nurnberg, 300 p.
32. Al-Malabeh, A. (1994) Geochemistry of Two Volcanic Cones from the Intra-Continental Plateau Basalt of Harra El Jabban, NE-Jordan. Geochemical Journal, 28, 517-540.
33. Ibrahim, K. (1993) The Geological Framework for Harrat Ash-Shaam Basaltic Super-Group and Its Volcano Tectonic Evolution. Natural Resources Authority, Geological Mapping Division, Bulletin 25, 33 p.

34. Ibrahim, K. (1996) The Regional Geology of Al Azraq Area Map Sheet No. 3553I. Natural Resources Authority, Geological Mapping Division, Bulletin 36, 67 p.
35. Khalil, I. (1997) The Geology of AlBishriyya (Al Arityn) Area, Map Sheet No. 3354 II (with Special Reference to the Geology of the Economic Zeolite Deposits). Geological Mapping Division, Bulletin 39.
36. Ibrahim, K., Rabba, I. and Tarawneh, K. (2001) Geological and Mineral Occurrences Map of The Northern Badia Region, Jordan, Scale 1:250.000. Geological Mapping Division.
37. AlAyyash, S., Nnadi, F. and Shatnawi, R. (2008) A Surface Water Management Tool for Arid Lands Using GIS—A Jordan Badia Case Study. In: Dutton, R.W. and Shahbaz, M.I., Eds., Jordan's Arid Badia: Deepening Our Understanding, Smith-Gordon and Company, UK.
38. Reynolds, J. (1998) An Introduction to Applied and Environmental Geophysics. John Wiley and Sons, Ltd., Chichester, 796 p.
39. Kirsch, R., Ed. (2006) Groundwater Geophysics: A Tool for Hydrogeology. Springer-Verlag, Berlin Heidelberg, 493 p.
40. Jogensen, F., Sandersen, P., Auken, E., Lykke-Andersen, H. and Sorensen, K. (2005) Contributions to the Geological Mapping of Mors, Denmark—A Study Based on a Large Scale TEM Survey. Bulletin of the Geological Society of Denmark, 52, 53-75.
41. Todd, D. (1980) Groundwater Hydrology. 2nd Edition. John Wiley & Sons, New York, 535 p.
42. Huisman, L. and Olsthoorn, T. (1983) Artificial Groundwater Recharge. Pitman Publishing Inc., Marshfield, 320 p.
43. Asano, T. (1985) Artificial Recharge of Groundwater. Butterworth Publishers, Boston, 767 p.

CHAPTER 4

Four Color Theorem and Applied Geophysics

Lev V. Eppelbaum

Department of Geophysical, Atmospheric and Planetary Sciences, Tel Aviv University, Tel Aviv, Israel

ABSTRACT

A selection of a number of geophysical methods to solve different geological, geodynamical, environmental, archaeological and other problems usually has no theoretical substantiation. The solution to this "four color" mathematical problem is able to assume that two independent geophysical methods are sufficient theoretically to characterize the geological-geophysical peculiarities of the area under study.

Keywords: Logical-Heuristic Model; Geophysical Method Integration; Geophysical Map Coloring

1. INTRODUCTION

It is well-known that the majority of the inverse problem solutions in geophysics are ill-posed (e.g., [1] [2]). It means, according to Hadamard [3] , that the solution does not exist, or is not unique, or is not a continuous function of observed geophysical data (when small perturbation in the observations will cause arbitrary mistake in the solution). This fact, in particular, calls to wide application of informational and probabilistic methodologies in applied geophysics. Main results of geophysical data measurements, processing and interpretation are usually reflected in various maps of different kinds, patterns and scales. Thus, the informational and probabilistic methods application in geophysical maps could have a greatest importance.

2. FORMALIZATION OF GEOLOGICAL-GEOPHYSICAL INVESTIGATIONS

Geophysical maps are one of the most important tools for the condensed and effective geophysical-geological (seismological, ecological, atmospheric etc.) data presentation (Figure 1). The compiled maps are employed at different stages (levels) of common and concrete investigations. It is necessary to underline that the same data may be used for different aims; at the same time different data may be applied for solving the same problem.

Geophysical observations are notoriously complicated by numerous factors (complex surrounding media, uneven topography, oblique polarization, instrumental errors, etc.) [4] [5] . To eliminate many of these disturbances modern interpretational methodologies have been developed (e.g., [4] [6] [7]). However, at times the complexity of the geological environment (extreme variability in lateral and vertical physical properties), the presence of several anomalous targets in close proximity and additional disturbances makes it impossible or unfeasible to apply these methodologies. In such cases information-probabilistic methods are effective tools to recognize and classify targets, estimate the potential information value of geophysical methods and decide upon a workable solution. The objective of geophysical surveys application is to obtain qualitative and quantitative information about the geometric and physical characteristics of buried objects; e.g., to develop physical-geological models (PGM) of target objects. PGMs of varying degrees of complexity (the simplest PGMs are simply target identification and complex PGMs can be 3D models of the objects under study) can be used for substantiation of different types of industrial (drilling, excavation, economic minerals exploitation, building, etc.) and scientific (geophysical monitoring, geological-geophysical mapping, construction of new PGMs, etc.) activity, and generation of future strategies for geological-geophysical investigations in the areas under study.

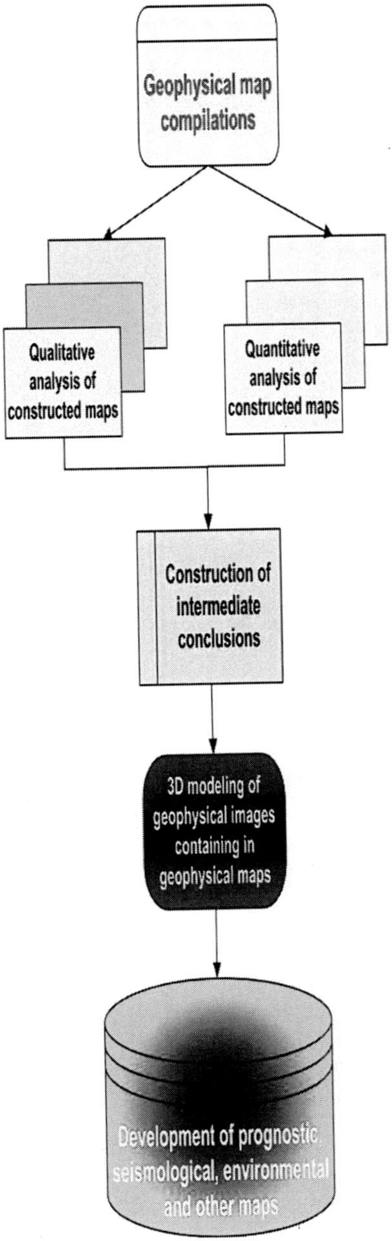

Figure 1. Simplified scheme of geophysical data processing.

Estimating the information value of geophysical and other means can be formalized on the basis of the following criteria [8] :

(1) Informativeness of the application (informational criterion G);

(2) Cost of implementing the method (cost criterion C);

(3) Time required to carry out the method (time criterion criteria C and T are easy to calculate directly, but criterion G is a non-trivial research problem. A simplified algorithm can be written as:

$$\Omega = \Gamma \cup C \cup T, \tag{1}$$

where $\cup$ is the symbol of unification.

All the available geological-geological information can be represented as the classic three-level model (Figure 2): 1) syntactic-quantity of information, 2) semantic-content of information, and 3) pragmatic-value of information. The logical-heuristic model for describing environmental information thus takes the following form:

$$\Gamma = I \cup R \cup V, \tag{2}$$

where I is the quantitative estimation of information, R is the estimation of informational reliability corresponding to the semantic criterion, and V is the estimation of informational value in terms of feasibility according to the pragmatic criterion.

Algorithm (2) is based on the fundamental terms of information theory and is combined with the structural (hierarchical) approach. This approach defines each indicator as a structure reflecting a set of typical situations and is then used to calculate the value of each estimator using the informational measure. Parameters V and R should be estimated geologically and by informational way, but there are beyond the scope of this paper. Here parameters V and R will be neglected, and it is assumed that $\Gamma = I$.

Thus, the main aim of geophysical prospection is to obtain the maximum effectiveness by minimum expenditures.

3. EVALUATING THE EFFICIENCY OF GEOPHYSICAL METHODS WITH INFORMATIONAL-STATISTICAL PROCEDURES

Choosing the right method (or number of methods) can be based on a quantitative estimate (Figure 2). For this purpose, reliable informational and statistical criteria are needed. The first issue is the quantity of information that can be obtained by a single method or a set of methods. The second is to define a criterion to express the decision-making risk as a function of the geophysical data. Nevertheless, informational criteria are preferable, because geophysical prospecting is a permanent process of acquisition and analysis of information [8] .

The classic information theoretic works by Shannon [9] and Brillouin [10] prompted Khalfin [11] to apply these criteria to geophysics. It has been shown [4] [5] that informational and statistical approaches represent two aspects of a shared approach. For instance, the solution to an identification problem using the criterion of minimal average risk or that of maximum information (minimal residual uncertainty) under certain conditions results in the same expressions [12] [13] .

The applied geophysical fields usually have maximal and minimal intensity within studied areas. The difference between the maximal and minimal intensities may be subdivided into some intervals (gradations). Gradations of indicators can be also used to obtain information about the types of the desired objects. Physical fields, geochemical analyses, some geological features, etc. can serve as indicators.

3.1. Some General Considerations

Here are considered some key elements of the theory of information conformably to geophysical studies. Illustrating how these methods can yield valuable results should highlight commonalities in geophysical theory and hence optimize PGM development and successful solving geological-geophysical problems.

As shown in countless publications on geophysical prospecting (e.g., [5] -[7] [14] [15]), there are typically no more than two or three geophysical methods employed to solve problems of detection, contouring and the development of 3-D models of the objects under study. Let us consider the results of the application of two methods (e.g., gravity and magnetic surveys). They are designated in Table 1 as follows: 1 = negative field, 2 = positive field, 3 = roughly zero field, 4 = alternating field, and 5 = high-gradient field.

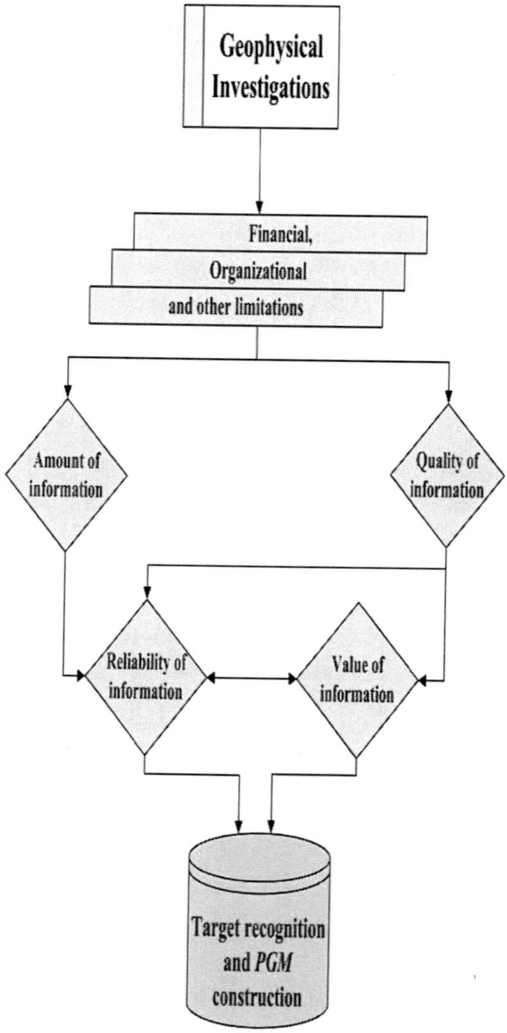

Figure 2. A scheme of desired geophysical target recognition and PGM construction with elements of information theory.

Four combinations of two parameters can represent four classes of desired targets, each ranging from 1 to 5. The number of possible combinations of two parameters divided into five categories is 25. The geological features (each class) can be characterized by one of these 25 combinations. It is also necessary to take into account concrete geological situation. For example, in Table 1 value 11 can be obtained for aqueous marl (instead of gabbroid), and this may lead to an erroneous conclusion in the process of geological-geophysical interpretation. Therefore, a certain abundance of the set of geophysical methods sometimes is advisable.

Each class of geological features can be characterized by one of these 25 combinations. The number of combinations can be increased at the expense of secondary parameters arising due to certain transformations of the fields (e.g., downward and upward continuation, various derivatives). Thus, measurements of two geophysical fields can provide sufficient data to resolve mapping problems.

Let's assume that there is only one anomalous target (AT) in the area. This area is divided into N equal cells. For simplicity each cell is assumed to have the same probability of containing the AT. Thus, the probability of finding the AT is equal to $P = 1/N$ in each cell. Hence, the entropy of experiment b (discovery of AT) is $\log N$. The entropy is determined using following expression:

$$H(\beta) = -\sum_{i=1}^{N} P(B_i) \cdot \log P(B_i), \tag{3}$$

where $P(B_i)$ is the probability of a B_i outcome (B is the range of outcomes of experiment b).

In this situation $H(\beta)$ value $(\log N)$ is maximum possible uncertainty. Experiment a (geophysical ob

Table 1. An example of integrated interpretation.

Typical field combination										Class of targets
Magnetic					Gravity					
1	2	3	4	5	1	2	3	4	5	
	◆					◆				Marl
◆					◆					Gabbroid
			◆						◆	Salt
	◆					◆				Fracture zone

servation) gives additional information. The A value is a range of outcomes for experiment a.

The difference between uncertainties in the b results before and after experiment a serves to estimate the information in a as related to b:

$$I(\alpha,\beta)=H(\beta)-H(\beta|\alpha), \tag{4}$$

where $H(\beta|\alpha)$ is the conditional entropy for experiment b (provided that experiment a has been conducted).

The conditional entropy is the average value of a random variable taking a $H(\beta_i|\alpha_i)$ value with a probability of $P(A_i)$:

$$H(\beta_i|\alpha_i)=\sum_{i=1}^{N}P(A_i)H(B_i|A_i). \tag{5}$$

3.2. Estimating the Efficiency of Individual Geophysical Methods

When selecting methods for integration, it makes sense to evaluate the amount of information provided by each geophysical method.

Starting from a well-investigated site (with an equal distance between observation points) typical of the area under study, it is assumed that 1/50 of it contains an AT. It is known that in the AT part the magnetic field is always positive, whereas in the empty part of the area it may be either positive or negative with equal probability. In other words, it is known a-priori that 2% and 98% of the area are target-containing and empty, respectively, and in 49% and 51% of this area the magnetic fields, respectively, are negative and positive. The results of experiment b can be designated as follows: B—AT occupying part of the area, $\overline{B}$ —empty part of the area. Thus, $P(\overline{B})=0.98.\ P(B)=0.02,$ According to expression (3) $H(\beta)\cong 0.14$.

The result of experiment a is expressed as follows: A is a positive field, $\overline{A}$ is a negative field. The relative partial entropy (after recording the positive magnetic field at the measurement point) can be calculated in the following way: $P(\bar{B}|A)=1-P(B|A)=49/51$. $P(B|A)=2/51$, $P(\bar{A})=0.49$; $P(A)=0.51$; Thus, $H(\beta|\alpha)\cong 0.24$. According to the recorded negative magnetic field, the area is certainly empty: $H(B|\bar{A})=0.$ Consequently, the partial entropy for experiment b under the conditions of a, as stated in Equation (5), is $H(\beta|\alpha)\cong 0.12$. Thus, the uncertainty of the determination of AT decreases after magnetic field measurement from 0.14 to 0.12.

3.3. Advantages of Geophysical Method Integration

What information can be obtained from geophysical field measurements? Assume that a magnetic field is observed in a range D, and the measurement precision is given by t. The next step is to fit an integer number of small intervals x into the intervals D and t. It is known after measurement that the value of the field in x units (with a precision up to x) fits into the interval t. Using Equation (4) and considering that the entropy in this expression before and after measurement is expressed by logarithms D/ξ and τ/ξ, respectively, one easily obtains (on the basis of [16]):

$$I(\alpha,\beta)_{\xi}=\log\left(\frac{D}{\tau}\right). \tag{6}$$

With decreasing x, these entropies increase unlimitedly, but the information remains unchanged. With unrestricted increase in measurement accuracy the information also increases unlimitedly, but slowly: an n-fold increase in accuracy leads to only $\log N$ of information units.

An essential property of composite experiments is as follows. If certain tests y, j and q are independent, experiments y and q can have zero information about j. However, an integrated experiment y and q can completely determine the outcome of experiment j. Consequently, whereas separate geophysical methods give no information on the presence (or absence) of a target, this can be obtained by integrating these methods. It follows from Equation (6) that the development of integrated investigations is more effective than increasing the accuracy of separate methods.

4. MINIMIZING THE NUMBER OF COMBINED METHODS BY THE "FOUR COLORS THEOREM"

How many geophysical methods should be applied for searching economic deposits, localization of archaeological remains or revealing some dangerous geological phenomena (e.g., karst terrane)? Extending a set of methods could be considered to be at variance with its economic efficiency, and complicated from both an organizational and a technical point of view. In addition, there is a basic limitation imposed on the number of methods. As noted Duda and Hart [17] , a growing number of target indicators require larger amounts of standard information. However, sufficient standards are only available in well-explored areas, where quantitative prediction is obviously less urgent. Therefore, a survey set should involve the minimum number of methods.

The intuitive use of a small number of integration elements in practice can be theoretically substantiated applying the well-known mathematical and cartographic "four colors" solution [18] -[21] for integrating geophysical methods by solving different geological, environmental and other problems.

Geophysical investigation is usually a multistage procedure and for simplicity it is assumed that the goal of each prospecting stage is the selection of an area for more detailed operations at the next stage. The result of prospecting is primarily a substantiated evaluation of the areas under investigation and their classification into two groups: those worthy and unworthy of further study. The objective of prospecting is to obtain the maximum information at a given cost.

Let us now examine the "four colors theorem" from this standpoint. Using elementary notions of graph theory the problem can be formulated as follows: prove that all vertices of an arbitrary planar graph can be colored with four colors in such a way that no two vertices joined by a common edge are the same color. It was proved as early as the middle of the 19th century that four colors suffice to color different counties on the map of England. However, a solution to this theorem was only found more than 100 years later [18] [19] . The authors subdivided all possible maps into almost 2000 types and developed a computer program for their investigation. For each type the problem—whether a map which cannot be colored with four colors can be found among the variety of maps—was solved. After lengthy investigations, an answer of "no" was obtained for all types, and this fact confirms the above solution.

A new (general) proof of the theorem was put forward recently [20] . Figure 3 shows seventeen of 633 reducible configurations which are displayed using the indicated convention. The whole set can be found in [20] .

Any area under study can be divided into separate sub-areas according to certain indicators. The following system of prospect classification has been adopted in the USA [22] : high (H), medium (M), low (L) and unknown (U). The objective is to single out promising areas (if any) from the entire set by an integrated geophysical survey. The colors refer to different combinations of geophysical methods. A positive conclusion for a certain prospecting method is labeled (+), and a negative (–).

$$H_i(x,y,z,t)=\begin{cases} - & |H_i|<\varphi \\ + & |H_i|>\varphi \end{cases}, \tag{7}$$

where j is some assumed value indicating the split between negative and positive values of geophysical field Hi, x,y,z are the space coordinates and t is the time. On the right side of Equation (7) Hi is assigned an absolute value since often anomalous targets may be reflected by negative geophysical anomalies.

Clearly a combination of at least two independent geophysical methods is necessary for the first three gradations (H, M and L); gradation U implies no application of the method set (on a scale or not at all) in the area under investigation (Table 2).

The geophysical methods employed are a priori assumed to be of equal significance. The threshold field values (the split between plus and minus) and specific types of geophysical investigations are determined according to the prospecting results for similar objects investigated previously and other geological and geophysical considerations. The split refers to the threshold for field values representing specified physical characteristics. These physical characteristics may, for example, include amplitudes of observed fields, field gradients or indicators of field variability. The necessary condition is that we assume the geophysical anomalies are produced by the same targets.

Let us demonstrate an applicability of this approach on a simple example. In an area of porphyry copper deposit in Kazakhstan (Figure 4), marginal values of electrical resistivity of 50 Ohm×m, and magnetic field of 0 nanoTesla have been selected (deposits of this type are characterized by low resistivity (<50 Ohm×m) and negative magnetic field) [23] . Figure 4(a) shows isolines of resistivity, 3b—isolines of magnetic field, and 3c—coloring in four colors according to Table 2. The contour of the deposit is clearly revealed on the basis of the data of two methods.

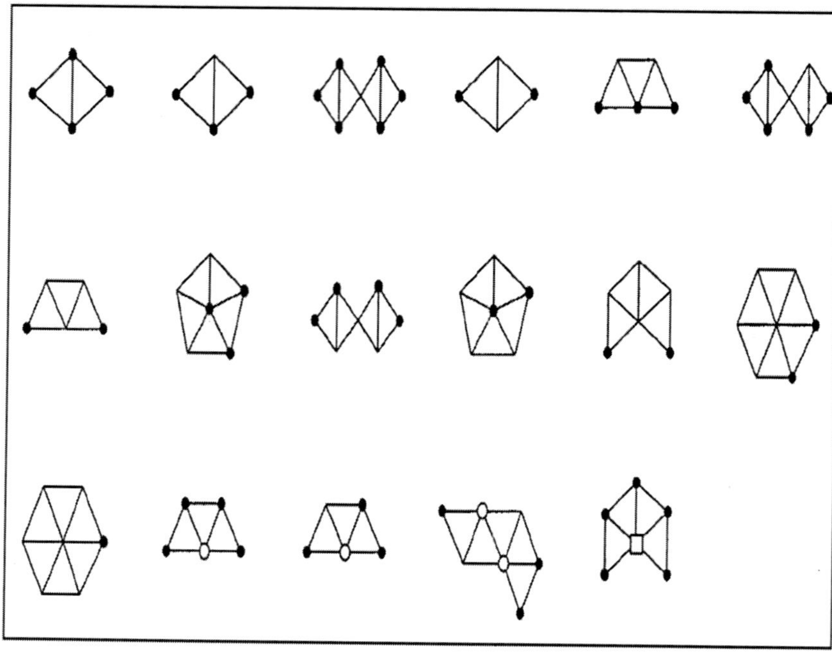

Figure 3. An example of possible configurations (after [18]).

Table 2. Subdivision of an area according to geophysical survey results.

Level of knowledge of the area	Geophysical method		Combination number (color)
	First	Second	
High (*H*)	+	+	1
Medium (*M*)	+	–	2
Low (*L*)	–	–	3
Unknown (*U*)	No necessary data		4

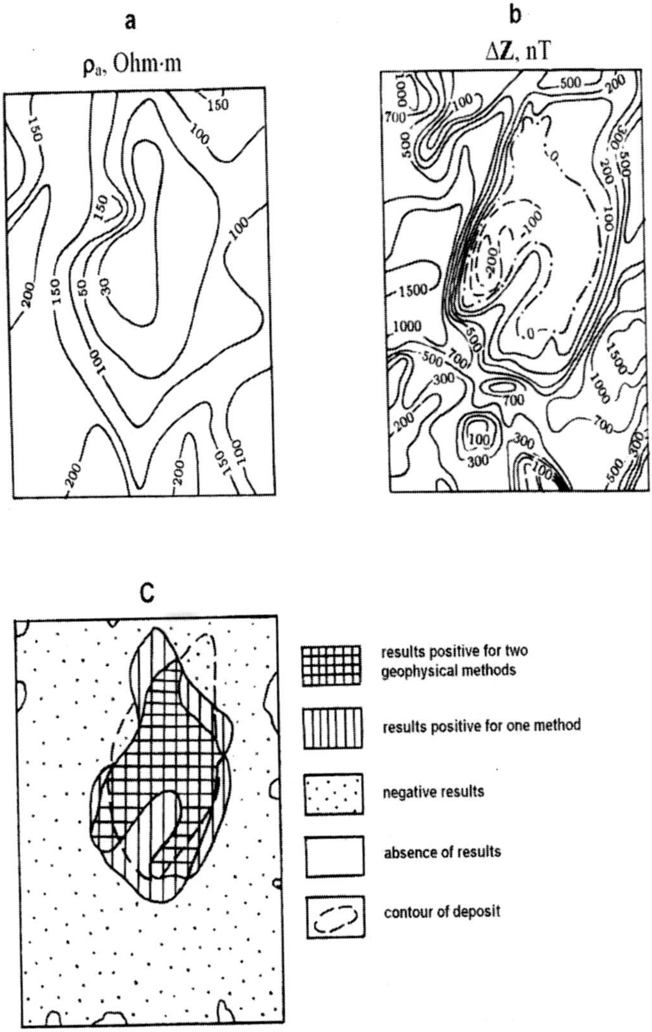

Figure 4. Electric (a) and magnetic (b) prospecting results at Benkaly porphyry copper deposit and their processing according to four colors problem solution (c) (a), (b) and contour of deposit from [23] .

5. CONCLUSIONS

It can be concluded from the foregoing that an optimum geophysical set consists of two independent geophysical methods. A map of geophysical results colored with four colors by the above technique serves as a basis for more detailed investigation. In this connection, a certain redundancy of the set is needed. It should, however, be kept in mind that the employed geophysical set is usually oriented to a particular problem and substantiated by a corresponding physical-geological model of the medium. Any change in the problem (e.g. an increase in the necessary depth of investigation) or in geological and geophysical pattern of the area may bring a change in the set of methods. In this context, this division of the theory of graphs can be attributed to information theory.

Lev V. Eppelbaum (24,33)">24] . Therefore, it is generally a good practice to use three geophysical methods which are effective under given conditions. For instance, a radiometric channel combined with aerial magnetic and electromagnetic surveys will give a negligible increase in cost, but increasing in informativeness may be a significant one.

REFERENCES

1. Tikhonov, A.N. and Arsenin, V.Y. (1977) Solutions of Ill-Posed Problems. Winston, New York.
2. Zhdanov, M.S. (2002) Geophysical Inverse Theory and Regularization Problems. In: Ser.: Methods in Geochemistry and Geophysics, 36, Elsevier, Amsterdam.
3. Hadamard, J. (1902) Sur les Problèmes aux Dérivées Partielles et Leur Signification Physique. Princeton University Bulletin, 13, 49-52.
4. Khesin, B.E., Alexeyev, V.V. and Eppelbaum, L.V. (1996) Interpretation of Geophysical Fields in Complicated Environments. In: Bookseries: Modern Approaches in Geophysics, Kluwer Academic Publishers, Dordrecht.
5. Eppelbaum, L.V., Khesin, B.E. and Itkis, S.E. (2001) Prompt Magnetic Investigations of Archaeological Remains in Areas of Infrastructure Development: Israeli Experience. Archaeological Prospection, 8, 163-185.
6. Parasnis, D.S. (1997) Principles of Applied Geophysics. Chapman & Hall, London.
7. Telford, W.M., Geldart, L.P. and Sheriff, R.E. (1990) Applied Geophysics. 2nd Edition, Cambridge University Press, Cambridge.
8. Eppelbaum, L.V., Eppelbaum, V.M. and Ben-Avraham, Z. (2003) Formalization and Estimation of Integrated Geological Investigations: Informational Approach. Geoinformatics, 14, 233-240.
9. Shannon, C.E. (1948) A Mathematical Theory of Communication. Bell System Technical Journal, 27, 379-432, 623-656.
10. Brillouin, L.N. (1962) Science and Information Theory. 2nd Edition, Academic Press, New York.
11. Khalfin, L.A. (1958) Information Theory of Geophysical Interpretation. Doklady Akademii Nauk SSSR, 122, 1007- 1010.
12. Svetov, B.S. (1992) Information Theory Basis of Geophysics. In: Bookseries: Advanced Geophysics. Russian Experience, Issue No. 1, Electromagnetic Research Centre, Moscow.
13. Nikitin, A.A. (1993) Statistical Processing of Geophysical Data. In: Bookseries: Advanced Geophysics. Russian Experience, Issue No. 2, Electromagnetic Research Centre, Moscow.
14. Gadallah, M.R. and Fisher, R. (2009) Exploration Geophysics. Springer, Berlin.

15. Eppelbaum, L.V. and Khesin, B.E. (2012) Geophysical Studies in the Caucasus. Springer, Berlin.
16. Ventsel, E.S. (1969) The Probability Theory. 3rd Edition, Nauka, Moscow. (in Russian)
17. Duda, R.O. and Hart, P.E. (1973) Pattern Classification and Scene Analysis. John Wiley & Sons, Hoboken.
18. Appel, K. and Haken, W. (1977) Every Planar Map Is Four Colorable: Part 1, Discharging. Illinois Journal of Mathematics, 21, 429-490.
19. Appel, K., Haken, W. and Koch, S. (1977) Every Planar Mar Is Four Colorable: Part 2, Reducibility. Illinois Journal of Mathematics, 21, 491-567.
20. Robertson, N., Sanders, D.P., Seymour, P.D. and Thomas, R. (1997) The Four Color Theorem. Journal of Combinatorial Theory, Series B, 70, 2-44.
21. Thomas, R. (1998) An Update on the Four-Color Theorem. Notices of the AMS, 45, 848-859.
22. Hansen, W.R. (1978) Suggestions to Authors of the Reports of the United States Geological Survey. 6th Edition, US Government Publishing Office, Washington DC.
23. Nikitin, A.A. (1986) Foundation of Integrated Interpretation of Geophysical Data. In: Brodovoi, V.V. and Nikitin, A.A., Eds., Integration of Geophysical Methods, Nedra, Moscow, 42-57. (in Russian)
24. Eppelbaum, L.V. (1989) The Development of Methods for Processing and Interpretation of Natural Geophysical Fields in Prospecting for Pyrite Ores in Mountainous Conditions. PhD Thesis, Institute of Geophysics (Georgia Academy of Science), Tbilisi. (in Russian)

CHAPTER 5

Clustering of Arc Volcanoes Caused by Temperature Perturbations In The Back-Arc Mantle

Changyeol Lee[1] & Ikuko Wada[2]

[1] *Faculty of Earth Systems and Environmental Sciences, Chonnam National University, 33 Yongbong-no Bukgu, Gwangju 61186, Republic of Korea.*

[2] *Department of Earth Sciences, University of Minnesota, 310 Pillsbury Dr. SE, Minneapolis, Minnesota 55455, USA*

ABSTRACT

Clustering of arc volcanoes in subduction zones indicates along-arc variation in the physical condition of the underlying mantle where majority of arc magmas are generated. The sub-arc mantle is brought in from the back-arc largely by slab-driven mantle wedge flow. Dynamic processes in the back-arc, such as small-scale mantle convection, are likely to cause lateral variations in the back-arc mantle temperature. Here we use a simple three-dimensional numerical model to quantify the effects of back-arc temperature perturbations on the mantle wedge flow pattern and sub-arc mantle temperature. Our model calculations show that relatively small temperature perturbations in the back-arc result in vigorous inflow of hotter mantle and subdued inflow of colder mantle beneath the arc due to the temperature dependence of the mantle viscosity. This causes a three-dimensional mantle flow pattern that amplifies the along-arc variations in the sub-arc mantle temperature, providing a simple mechanism for volcano clustering.

INTRODUCTION

In some parts of subduction margins, such as Northeast (NE) Japan[1] and Cascadia[2], arc volcanoes form in clusters (our definition of volcanic clusters excludes those of relatively small monogenetic volcanoes) at variable spacing that ranges from a few tens to hundreds of kilometers. Arc volcanoes are surface expressions of magmatic processes that are induced by plate subduction. In the mantle wedge that overlies the subducting plate, advection of heat by slab-driven mantle wedge flow and addition of aqueous fluids from the subducting slab provide the thermal and chemical conditions required for flux melting[3,4]. Therefore, volcanic clustering indicates spatial variation in the physical condition within the mantle wedge. However, the mechanism that causes such spatial variation remains unclear. Understanding the mechanism of volcano clustering is critical to the studies of subduction geodynamics and volcanic hazard.

In NE Japan, the dense seismic networks and abundant earthquakes allow high-resolution imaging of seismic velocity structures, which indicate low-velocity zones (LVZs) in the mantle wedge that extend from the back-arc to the sub-arc region where volcanoes occur in clusters[1]. Because of their strong spatial correlation with volcanic clusters, the LVZs have generally been interpreted as hot regions that feed melts to the overlying clustered volcanoes and have commonly been referred to as hot fingers[1]. Another interpretation of the LVZs is that they represent regions with free fluids and have also been referred to as wet fingers[5]. This interpretation is based on the lack of strong contrast in seismic attenuation between LVZs and the surrounding mantle because temperature is known to have a relatively large effect on seismic attenuation, and high attenuation is expected to occur in hot regions[6]. However, whether the LVZs are hot and/or wet is yet to be verified.

In the mantle wedge, solid-state mantle flow is driven largely by viscous coupling between the subducting slab and the overlying mantle[7,8]. This slab-driven mantle wedge flow brings in hot mantle from the back-arc into the sub-arc region. Geophysical observations and numerical modelling indicate that the shallow portion of the mantle wedge is decoupled from the slab, and the trench-ward extent of mantle wedge flow depends on the slab-mantle decoupling depth[9]. Along-arc fluctuation in the slab-mantle decoupling depth is a possible mechanism to cause localized mantle inflow; the overlying mantle is stagnant and cold in regions with deepened decoupling depths whereas the overlying mantle is flowing and hot in regions with shallower decoupling depths[10]. However, this contrasts with global thermal modelling studies that indicate a relatively uniform decoupling depth of 70–80 km (refs 11, 12). Further, detailed thermo-petrologic modelling results for NE Japan with a uniform decoupling depth show a strong correlation between the predicted petrologic structure and the distribution of intraslab earthquakes, supporting the idea that the presence of hydrous minerals or their dehydration is critical for the generation of intermediate-depth intraslab earthquakes[13]. Along-arc fluctuation in the decoupling depth would, however, impact the thermal structure of the subducting slab, requiring an alternative explanation for the intraslab earthquake distribution.

Small-scale mantle convection beneath the arc driven by thermal buoyancy in a low-viscosity mantle wedge is proposed as a possible mechanism for along-arc variation in the sub-arc mantle temperature based on numerical simulations[14,15,16]. The convection results in alternating regions of mantle upwelling and downwelling, and volcanoes are more likely to form above the regions of hot mantle upwelling, forming volcanic clusters. Another proposed mechanism is the occurrence of isolated small mantle return flow in the mantle wedge[17]. This type of mantle flow has been simulated within a thin low-viscosity layer of 4–8 km thickness imposed at 50–70 km depths immediately above the subducting slab in a numerical subduction model, whereby the along-arc variation in the extent of the return flow is initiated by numerical noise in the thermal field[17]. These mechanisms require relatively low viscosity in the mantle wedge or in a thin mantle layer. Mantle viscosity depends strongly on temperature and water content[18,19]. The possible extensive or (vertically) localized occurrence of mantle weakening is attributed to high water content. However, such condition combined with generally high temperature of the convecting mantle is likely to promote widespread melting.

The mantle beneath the arc is brought in from the back-arc by slab-driven mantle wedge flow, and thus the sub-arc mantle temperature depends on the thermal structure of the back-arc. In contrast to the sub-arc region, the relatively large space and vertical temperature variation in the back-arc mantle are likely to promote small-scale buoyancy-driven flow, and this flow has been proposed as the key mechanism that maintains the relatively hot state of the back-arcs that have not undergone significant extension in recent geological history[20]. The upwelling and downwelling of the mantle associated with such buoyancy-driven flow result in lateral temperature perturbations in the back-arc[14,16,21]. This also results in variations in mantle viscosity, which can affect the dynamics of the subjacent sub-arc mantle. In fact, such effects can easily be simulated in three-dimensional numerical models (see Methods section) using a similar approach to previous studies[14,16,21]. Further, presence of mantle plumes[22,23] in the back-arc can cause variations in the back-arc mantle temperature, which affect the temperature and viscosity of the mantle that is flowing into the sub-arc region. However, the effects of back-arc temperature perturbations on sub-arc mantle flow and temperature were not investigated previously.

The primary objective of this study is to quantify the effect of back-arc temperature perturbations on the slab-driven mantle wedge flow beneath the arc using a three-dimensional numerical subduction model. In the model, we impose high-temperature anomalies of a specified size and magnitude at a specified depth on the back-arc-side vertical boundary to simulate the effect of back-arc temperature perturbations. The mantle that originates from the high-temperature anomaly is less viscous, resulting in along-arc variations in mantle viscosity and leading to a complex three-dimensional mantle wedge flow pattern that amplifies along-arc variations in sub-arc mantle temperature.

RESULTS

Single high-temperature anomaly

Simulating the small-scale mantle convection (Supplementary Figure 1) in the back-arc in the model and quantifying their effects on sub-arc mantle flow and temperature would provide consistency within the model (see Methods). However, simulation of small-scale mantle convection depends on a number of factors, such as the assumed mantle viscosity, vertical temperature and density gradients (through the use of Rayleigh number), and back-arc geometry, and the resulting back-arc temperature perturbations are inherently highly variable in simulations, making it difficult to provide systematic analyses of the effects of back-arc temperature perturbations on sub-arc mantle flow and temperature. For the purpose of this study, therefore, we choose to impose a high-temperature anomaly on the back-arc-side vertical boundary (Figure 1). The anomaly is represented by an elliptic region defined by its height and width and a prescribed magnitude of increase in temperature towards its center (Supplementary Figure 2), following the approach developed by Lee and Lim22 (See Methods). We use a model with one high-temperature anomaly as our reference model (Figure 1a); its height and width are 40 km centered at a 120-km depth, and its peak temperature is 200 °C higher than the background mantle potential temperature. In the model, we use a temperature-dependent mantle wedge rheology. The model is symmetric at the right-hand-side side boundary at z=0 km, and thus only the left half is shown.

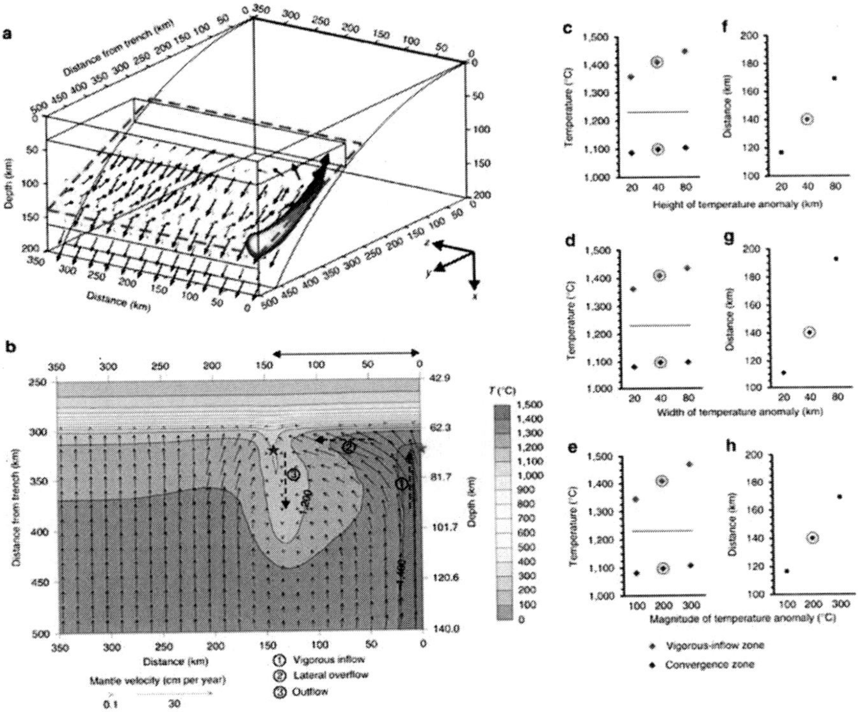

Figure 1. Three-dimensional subduction model with a back-arc high-temperature anomaly. (**a**) Three-dimensional model geometry with one high-temperature anomaly (the reference model). Black arrows indicate the calculated mantle flow field. Colour indicates temperature in the vigorous inflow zone from 1,400 °C (dark red) to 1,500 °C (bright yellow). (**b**) Temperature distribution (colour) and mantle flow field (black vectors) on the dipping plane across the mantle wedge (indicated by the red dashed plane in **a**) calculated from the reference model. The temperature contours are at every 100 °C. The general three-dimensional flow pattern of mantle wedge can be described by combination of vigorous-inflow (1), lateral overflow (2) and outflow in the convergence zone (3). Red and blue stars at a distance of 320 km from the trench indicate the probe locations for mantle temperature in the vigorous-inflow and convergence zones, respectively. (**c–e**) Variations in the sub-arc mantle temperature measured at the tip of the vigorous-inflow (red star in **b**) and convergence zones (blue star in **b**) with varying height, width and magnitude, respectively, of the high-temperature anomaly on the back-arc-side vertical boundary. Red circles indicate values from the reference model. Black line indicates the mantle temperature from the model without the high-temperature anomaly. (**f–h**) Distance between the vigorous-inflow (red star in **b**) and convergence zones (blue star in **b**) with varying height, width, and magnitude, respectively, of the high-temperature anomaly on the back-arc-side vertical boundary. Red circles indicate values from the reference model.

Here, we display the mantle flow and temperature distributions on a dipping plane across the core part of the mantle wedge to illustrate the effects of the high-temperature anomaly in the back-arc (Figure 1b). Mantle inflow extending from the high-temperature anomaly is more vigorous than the ambient inflow due to its low viscosity. Once the vigorous inflow reaches the wedge corner, some of the mantle is entrained down-dip by the down-going slab as in typical mantle wedge return flow predicted by two-dimensional subduction models, but the rest overflows laterally along the margin, inducing three-dimensional mantle flow. The lateral mantle overflow discourages the ambient inflow, and even in some cases, it leads to shallow outflow of the mantle towards the back-arc (Figure 1b). This lateral overflow plays a critical role in controlling the mass and heat transfer of sub-arc mantle. Hereafter, we refer to the zone of vigorous inflow that extends from the high-temperature anomaly as the vigorous-inflow zone, and the zone where the lateral overflow and the ambient inflow collide as the convergence zone.

Temperature in the vigorous-inflow zone is elevated relative to that in the ambient sub-arc mantle because the flow originates from the imposed high-temperature anomaly and also due to enhanced vigor of mantle inflow. By contrast, temperature in the convergence zone is reduced as the inflow of hot mantle from the back-arc is subdued by the effect of lateral overflow from the vigorous-inflow zone. This development of the cool convergence zone amplifies the along-arc variations in the sub-arc mantle temperature (Figure 1c–e). Because of the amplification, a back-arc temperature anomaly that is modest in size and/or magnitude results in relatively large along-arc variations in the sub-arc mantle temperature; for example, a temperature anomaly of 100 °C can result in along-arc temperature variations by ~250 °C (Figure 1e). The along-arc variations in the sub-arc mantle temperature increase with the size and magnitude of the high-temperature anomaly (Figure 1c–e).

A single high-temperature anomaly imposed on the back-arc-side vertical boundary allows the development of the convergence zone at a characteristic distance from the vigorous-inflow zone (Figure 1f–h). For the reference model, this distance is 140 km. The characteristic distance increases with the size and magnitude of the high-temperature anomaly. In the presence of multiple high-temperature anomalies with small spacing, interference among the vigorous-inflow and convergence zones occurs through the interactions of opposing lateral overflows. Based on a series of simulations varying the distance between two high-temperature anomalies in the back-arc mantle (see Methods section), the vigorous-inflow and convergence zones evolve without interference when the distance between two high-temperature anomalies is about three times the characteristic distance (Figure 2). Smaller spacing leads to smaller along-arc variations in the sub-arc mantle temperature due to mutual interference (Figure 2d). For example, when the spacing is about the half of the characteristic distance (that is, 70 km), the resultant along-arc temperature variation is about the same as the magnitude of the imposed temperature anomaly (that is, 200 °C), which is smaller than that without interference (~310 °C).

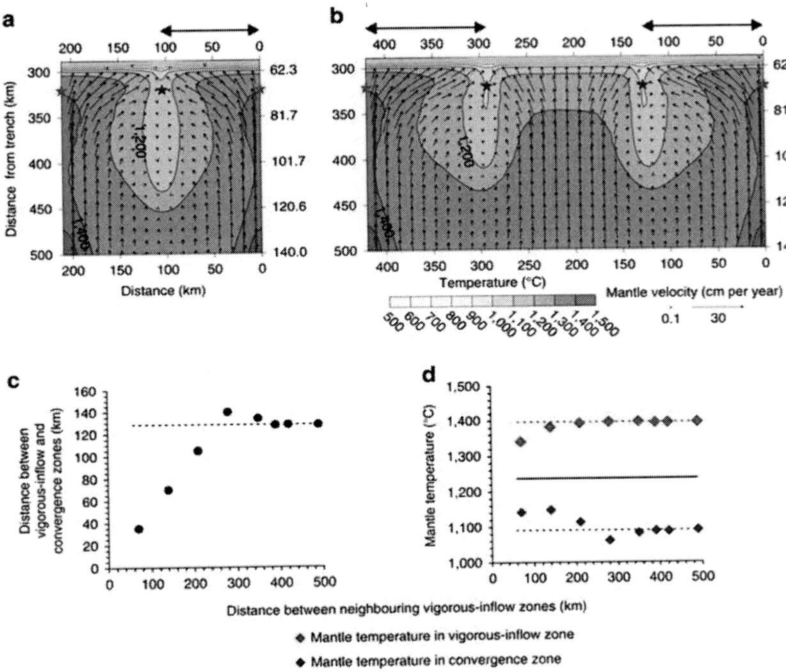

Figure 2. Temperature and mantle flow fields calculated from models with two back-arc high-temperature anomalies. (**a** and **b**) Temperature distribution (colour) and mantle flow field (black vectors) on the dipping plane (indicated by the red dashed plane in Figure 1a) calculated from models with two high-temperature anomalies that are separated by a distance of 210 and 420 km, respectively. The temperature contours are at every 100 °C. Red and blue stars indicate the probe locations for mantle temperature in the vigorous-inflow and convergence zones, respectively, shown in **d**. Double-arrow-headed lines at the top indicate the distance between the vigorous-inflow and convergence zones. (**c**) Change in the distance between the vigorous-inflow zone (red stars in **a** and **b**) and the convergence zone (blue stars in **a** and **b**) with the distance between the two vigorous-inflow zones (i.e., the two red stars). Black dashed line indicates the characteristic distance (128.3 km) from the reference model. (**d**) The sub-arc mantle temperature at the tip of the vigorous-inflow zones (red diamonds) and convergence zones (blue diamonds). Black line indicates the mantle temperature from the model without the high-temperature anomaly. Red and blue dashed lines indicate the mantle temperatures of the vigorous-inflow and convergence zones, respectively, from the reference model.

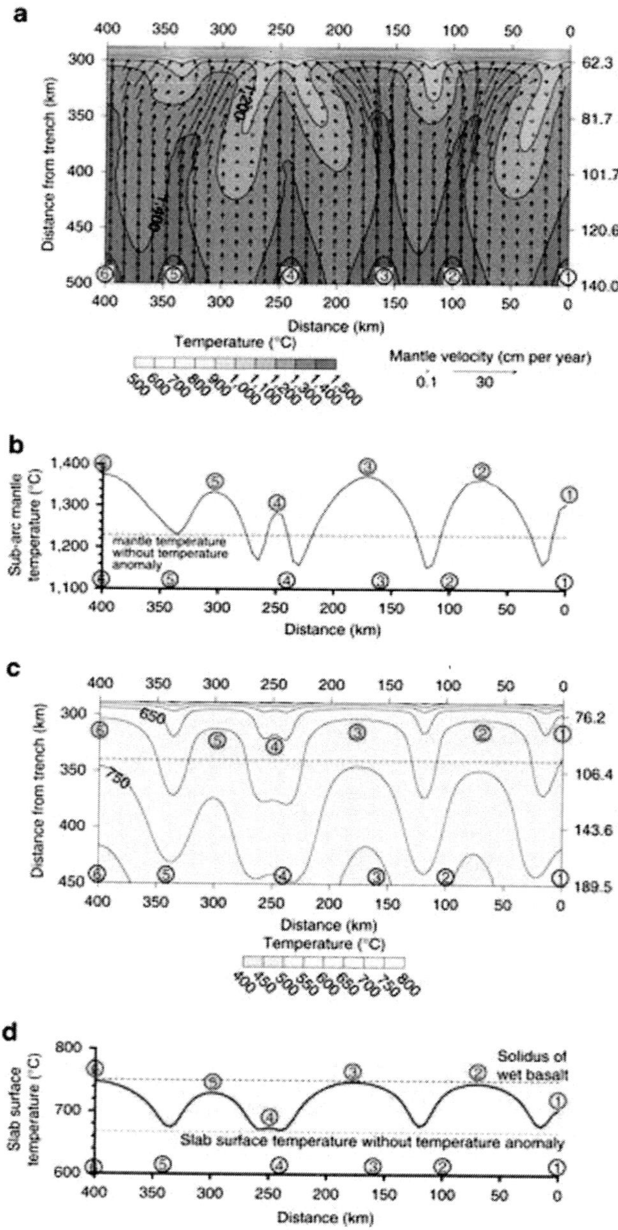

Figure 3. Temperature and mantle flow field calculated from a model with six back-arc high-temperature anomalies. (**a**) Temperature distribution (colour) and mantle flow field (black vectors) on the dipping plane

(indicated by the red dashed plane in Figure 1a) calculated from a model with six high-temperature anomalies at 0, 100, 160, 240, 340 and 400 km distance from the right side-wall (numbered from 1 to 6 enclosed by the blue circles, respectively). The temperature contours are at every 100 °C. (**b**) Temperature (blue line) along the dipping plane at a depth of 70 km (white dashed line in **a**). (**c**) Temperature distribution (colour) on the slab surface. The temperature contours are at every 50 °C. (**d**) Slab-surface temperature (red line) measured at a depth of 100 km (red dashed line in **c**). The black dashed line indicates the solidus of wet basalt34·35. The green dashed line indicates slab-surface temperature from the model without the back-arc high-temperature anomaly. Numbers enclosed by the blue and red circles in **b–d** indicate the locations of the back-arc high temperature anomalies and peak sub-arc mantle temperatures, respectively.

In NE Japan, the distance between the centers of volcanic clusters ranges from ~50 and ~120 km with an average of ~80 km (ref. 1). High-temperature anomalies in the back-arc at this spacing are likely to interfere with one another although the degree of interference depends on the size and magnitude of the anomalies. We impose in our model several high-temperature anomalies with variable spacing ranging from 60 to 100 km to reflect the variability in the volcanic cluster spacing in NE Japan (Figure 3). The details of the mantle wedge flow pattern depend on a number of other factors, such as the local geometry of the subducting slab and the distribution of fluids and melts, and this model is not intended to reproduce the exact volcano spacing in NE Japan but to provide a better understanding of the effect of possible back-arc temperature anomalies on the overall mantle flow pattern and thermal structure in NE Japan.

The relatively close spacing of the back-arc high-temperature anomalies results in strong interference between the vigorous-inflow and convergence zones (Figure 3a,b). Further, the non-uniform spacing gives rise to diverse variability in the mantle flow pattern and thermal condition beneath the arc even though all the high-temperature anomalies are of the same size and magnitude. Some of the convergence zones do not extend fully into the sub-arc region or deflected laterally due to the strong interference between neighbouring vigorous-inflow zones, resulting in an along-arc variation in the trench-ward extent and orientation of hot vigorous-inflow zones. The variation in the location of the tip of vigorous-inflow zones is likely to contribute to small along-arc variability of the location of the volcanic clustering within a given subduction zone. The resultant along-arc variations in the sub-arc mantle temperature can give rise to an along-arc temperature variation of ~100 °C at the sub-arc slab surface (Figure 3c,d). Our modelling results illustrate the importance of the interference between inflowing mantle of different temperatures and viscosities in controlling the three-dimensional mantle wedge flow pattern and thermal structure of subduction zones.

DISCUSSION

The three-dimensional mantle wedge flow pattern beneath the arc predicted by our model is characterized by typical corner flow that consists of shallow inflow and down-dip outflow, along-arc lateral overflow, and shallow outflow toward the back-arc. The occurrence of typical corner flow beneath the expected region of volcanism is consistent with the mass transfer from the sub-arc region towards the back-arc inferred from across-arc geochemical variations of arc lavas in NE Japan24. The occurrence of lateral overflow and outflow discourages the inflow of hot ambient mantle into the sub-arc region and is likely to contribute to the absence of arc volcanism between regions with the typical return flow.

In our model, the effects of free fluids on mantle wedge flow are not included. Although temperatures in the convergence zones are likely high enough to cause flux melting in the mantle wedge, higher temperatures in the vigorous-inflow zones promote higher degree of partial melting for a given water content4. The higher degree of partial melting leads to lower viscosity, which allows more vigorous mantle inflow, and the mantle temperature in the hot inflow zone becomes even higher. Increased vigor of the inflow subdues ambient inflow and results in a cooler condition in the convergence zone. This feedback between temperature, melting, viscosity, and mantle flow is likely to generate temperature contrast between the vigorous-inflow and convergence zones that is larger than what is predicted by our numerical model. This conceptual model provides a simple mechanism for generating relatively large along-arc variations in temperature and melt distributions beneath the arc, satisfying both the hot and wet conditions of the LVZs that are inferred from geophysical observations in NE Japan1,5.

Previous numerical modelling studies invoked small-scale convection and cold-plumes in the sub-arc mantle driven by thermal and/or compositional buoyancy to explain along-arc variations in the physical conditions of the sub-arc mantle14,16,21,25,26,27. Our modelling results indicate that slab-driven mantle flow alone can cause such variations without invoking buoyancy-driven mantle flow beneath the arc when temperature anomalies are present in the back-arc. This is not to say that there is no small-scale convection or a cold plume, but even in their presence, slab-driven corner flow can dominate the mantle flow field beneath the arc as reported by numerical28 and laboratory analog23 modelling studies.

Sub-arc mantle temperature strongly influences the thermo-petrologic structure of the upper portion of the subducting slab29. At a given depth, the slab beneath the vigorous-inflow zones is hotter than that beneath the convergence zones, experiencing greater degree of dehydration and releasing more fluids into the overlying vigorous-inflow zones. This may lead to lower mantle viscosity and greater degree of flux melting in the mantle wedge, both of which are likely to enhance the vigor of inflow. Further, higher degree of flux melting in the vigorous-inflow zones results in higher permeability than that in the convergence zones30, promoting the preferential development of the upward migration pathways. Lower viscosity due to higher temperature and higher water content also facilitates faster fluid migration and melt segregation31. These conditions may allow the initial localization of fluids within the vigorous-inflow zone followed by coalescence of fluid transport networks30,32 that further focuses fluids beneath the clustered volcanoes.

Petrological and geochemical studies of arc lavas in NE Japan indicate little evidence of slab melting33. Our model predicts that the slab surface beneath the vigorous-inflow zone (Figure 3d) is generally cooler than the solidus of wet basalt34,35. Back-arc temperature anomalies that are greater in size and/or magnitude would result in higher slab-surface temperatures and slab melting. Thus, the size and magnitude of the temperature anomalies are unlikely to be significantly greater than those prescribed in our model.

At the trench-ward end of the LVZs beneath the arc, the S-wave velocity in the mantle becomes relatively uniform along the arc36. Slab-driven mantle wedge flow in the presence of back-arc temperature perturbations, small-scale convection or cold plumes beneath the arc cannot explain the uniform sub-arc S-wave velocity, and additional mechanism may be at work, possibly reducing S-wave velocities in the convergence zones and obscuring the along-arc temperature variation, such as the presence of hydrous phases in the cold convergence zones or the effect of seismic anisotropy.

Shear-wave splitting measurements indicate that the sub-arc and forearc mantle in several subduction zones, including NE Japan, exhibit trench-parallel seismically fast polarization directions37,38. The lattice preferred orientation (LPO) of the olivine fabrics that are dominant in the upper mantle (that is, A- and E-types) develops parallel to the direction of maximum shear39. Assuming that the shear direction parallels the mantle flow direction, the trench-parallel mantle flow from the tip of the vigorous-inflow zones towards the convergence zones causes trench-parallel fast polarization direction. The shape preferred orientation (SPO) of partial melts also parallels the shear direction but results in fast polarization direction normal to the shear direction after melt segregation40. In vigorous-inflow zones where mantle flows normal to the trench, partial melting and melt segregation can therefore result in trench-parallel fast polarization direction. The combination of the LPO of the laterally flowing mantle and the SPO of partial melts in the vigorous-inflow zones may therefore explain the observed trench-parallel fast polarization direction.

The temperature distribution in the back-arc upper mantle is likely non-uniform as inferred from seismic velocity structures. Although the spatial variations in the back-arc mantle temperature at the length scale of our interest is difficult to constrain at present, our modelling results indicate relatively small high-temperature anomalies of 100–200 °C and a few tens of kilometers in diameter in the back-arc mantle can lead to three-dimensional mantle wedge flow that amplifies the along-arc variations of the sub-arc mantle and slab temperature. In the presence of nonlinear feedback between temperature, viscosity, melting, and mantle flow, the size and magnitude of anomalies in the back-arc that are required to cause such temperature variations are likely even smaller, making it easier for volcano clustering to occur.

METHODS

Three-dimensional kinematic-dynamic subduction model

Our subduction model22 is developed using the commercial finite element package COMSOL Multiphysics. The governing equations consist of equations of conservation of mass, momentum, and energy,

$$0 = \nabla \cdot \mathbf{v}, \quad (1)$$

$$0 = \nabla \cdot \sigma' - \nabla P + \rho \mathbf{g}, \quad (2)$$

$$\rho_c c_p \frac{DT}{Dt} = \nabla \cdot (k \nabla T), \quad (3)$$

respectively, where $\mathbf{v}$ is velocity (m s^{-1}), P is pressure (Pa), σ' is deviatoric stress tensor (Pa), ρ is density defined as $\rho=\rho_c(1-\alpha T)$ (kg m^{-3}), ρ_c is reference density (kg m^{-3}), α is thermal expansivity (K^{-1}), T is temperature (K), $\mathbf{g}$ is gravitational acceleration vector (m s^{-2}), c_p is specific heat at constant pressure (J kg^{-1} K^{-1}), t is time (s), and k is thermal conductivity (W m^{-1} K^{-1}). Radiogenic heat production is neglected because its contribution to the thermomechanical behaviour in the subduction zones is relatively small41. Viscous dissipation is also neglected; it has been reported that viscous dissipation systematically increases the slab-surface temperature up to ~50 °C compared with the incompressible model experiments42. Equations (1)–(3), , are non-dimensionalized using reference values indicated in Supplementary Table 1 for density, temperature contrast, specific heat, thermal conductivity, thermal expansivity, depth of the fluid layer, and viscosity at a depth of 200 km, following the approach of King *et al.*43

To illustrate the role of small-scale convection in the back-arc in generating the back-arc temperature perturbations, we first incorporate the effect of buoyancy in a model using the governing equations described above and the Rayleigh number of 2.33 × 10^6derived from the model parameter reference values in Supplementary Table 1. The model is 800-km long, 350-km wide and 200-km deep (Supplementary Figure 1a) and consists of two sub-domains: forearc-arc and back-arc domains. The geometry of the subducting slab is based on the Wadati-Benioff zone estimated by Syracuse and Abers44 for NE Japan, but its exact geometry is not critical for this study. The overlying crust is 35-km thick and non-deforming. Subduction velocity of 8.3 cm per year is applied to the slab surface normal to the trench of the model. In NE Japan subduction zone, the Japan trench is largely straight except near its northern and southern ends, and the convergence direction is nearly perpendicular to the trench, a situation similar to our model setup. Previous studies suggested that the corner of the mantle wedge is decoupled from the subducting slab and is stagnant31,44. To approximate the decoupling effect, we simply impose non-deforming corner to a depth of 70 km (ref. 45).

The trench-side vertical boundary and the bottom boundary of the model are stress-free, and the back-arc-side vertical boundary and side boundaries are insulated and a free-slip. The subducting Pacific plate at the Japan trench is ~130 Myr old46. The geotherm applied to the trench-side vertical boundary and the initial temperature distribution of the entire model domain (including the geotherm applied to the back-arc-side vertical boundary) are calculated by using the half-space cooling model47 for a 130-Myr-old and 50-Myr-old plates, respectively, with the mantle potential temperature of 1,350 °C and the mantle adiabat of 0.35 °C km^{-1}. The temperature at the top surface of the model is fixed at 0 °C. The basal temperature of the overlying crust in the back-arc domain is fixed at 1,000 °C to account for the relatively thin lithosphere and high surface heat flow (~88 mW m^{-2}) in the back-arc20,48. We impose a mantle temperature of 1,420 °C at the bottom of the back-arc domain from 550 to 800 km.

For the mantle wedge and the back-arc mantle, we use a diffusion creep rheology,

$$\eta = \frac{\mu}{2A}\left(\frac{b}{d}\right)^{-m}\exp\left(\frac{E+PV}{RT}\right) \quad (4)$$

with rheological parameters reported for diffusion creep of dry olivine18(Supplementary Table 2). To approximate the effect of fluid addition from the dehydrating slab into the mantle wedge, the mantle viscosity calculated based on equation (4) is reduced by 1/20 at the corner of the mantle wedge, and the viscosity reduction diminishes linearly with increasing distance away from the corner of the mantle wedge towards the back-arc over ~120 km distance[21,49]. Based on the comparison with the model without the viscosity reduction, we found that the viscosity reduction has a modest effect on the general pattern of mantle wedge flow, and the vigorous inflow, lateral overflow, and outflow to be discussed in the following sections are robust features that result from the presence of back-arc high-temperature anomalies.

The model domain consisted of 969,583 elements including tetrahedral, pyramid, prism, hexahedron, triangular and quadrilateral elements; most of the element are tetrahedrons of which sizes range from 5 to 10 km. Mesh refinements are applied around the boundaries between the mantle wedge and its surrounding sub-domains. The maximum size of the elements in the model is 10.00 km. The mesh around the boundaries between different components of the model is refined, and the smallest size of the element is 2.8 km at the tip of the mantle wedge. Time stepping of the generalized-α is applied to solve the time-evolving governing equations. The Stokes and energy equations are coupled by using the segregated time-dependent solver. For a parallel computation, the Multifrontal Massively Parallel sparse direct Solver (MUMPS) is used. These models with the effect of buoyancy were run for 120 Myr to minimize the effects of initial conditions.

In the model, the subducting slab induces corner flow in the overlying mantle wedge. Because the model includes the effect of thermal buoyancy, the temperature contrast between the top and the bottom of the back-arc mantle drives small-scale convection that generates lateral variations in the shallow mantle temperature (Supplementary Figure 1d). In the region just behind the arc (at 500 km distance from the trench), the shallow part of the back-arc mantle is dominated by the trench-ward inflow of the back-arc mantle with weakened small-scale convection, and the deeper part is dominated by the down-dip outflow portion of the slab-induced corner flow (Supplementary Figure 1c). The shallow convection-dominated part of the back-arc mantle is drawn into the forearc-arc region by the corner flow and causes three-dimensional flow beneath the forearc-arc region (Supplementary Figure 1b). This occurs because the mantle that flows in from the upwelling region in the shallow back-arc mantle is hotter and less viscous than the neighbouring down-welling regions, experiencing more vigorous flow towards the arc. Once the mantle reaches the mantle wedge corner, the vigorous flow spreads laterally to interfere with the neighbouring inflowing mantle, leading to cooler regions of sub-subdued mantle inflow beneath the arc.

This study focuses on the effects of back-arc mantle temperature perturbations on the slab-driven mantle wedge flow pattern. To perform systematic analyses of the effects, we choose to prescribe the back-arc temperature perturbation by imposing a prescribed size and magnitude of high-temperature anomalies at a prescribed depth on the back-arc-side vertical boundary. Since there is no need to simulate small-scale mantle convection in the back-arc, the back-arc domain is removed from the model (Supplementary Figure 2a), and the back-arc-side vertical boundary is placed at 500 km distance from the trench. Further, in models with a weak mantle wedge that allows small-scale convection, mass and heat transfer in the sub-arc mantle become dominated by the convection, and it is difficult to quantify the effect of back-arc mantle temperature perturbations on the 3D slab-driven mantle wedge flow pattern beneath the arc. Thus, to focus on the slab-driven mantle wedge flow, we exclude the buoyancy term $(\rho'\vec{g})$ in equation 2. The width of the models is varied depending on the number of high-temperature anomalies imposed on the back-arc-side vertical boundary (discussed below). The geotherm on the back-arc-side vertical boundary is calculated by using the half-space cooling model for a 50-Myr ago plate and is applied from the surface down to a 160-km depth, where the mantle inflow-outflow transition occurs. At depths >160 km, no temperature boundary condition is prescribed. The boundary conditions for the rest of the boundaries of the model are the same as those of the forearc-arc domain in the model with the effect of buoyancy.

On the back-arc-side vertical boundary, we implemented single, two or six high-temperature anomalies defined by the modified distribution function (Supplementary Figure 2b,c)

$$T = T_{\text{back-arc}} + T_{\text{anomaly}} \exp\left(- \frac{\left(x - x_{\text{anomaly}}\right)^2}{\left(\sigma_{\text{height}}/2\right)^2} \right) \exp\left(- \frac{\left(z - z_{\text{anomaly}}\right)^2}{\left(\sigma_{\text{width}}/2\right)^2} \right) \tag{5}$$

where $T_{\text{back-arc}}$ is the temperature boundary condition on the back-arc-side vertical boundary (°C); T_{anomaly} is the magnitude of the high-temperature anomaly (°C); x_{anomaly} and z_{anomaly} are the depth from the top surface (km) and distance from the right side-wall boundary (km) of the high-temperature anomaly, respectively; σ_{height} and σ_{width} correspond to the height and width (km) of the high-temperature anomaly, respectively. At the half of the height and width from the center (red dashed circle in Supplementary Figure 2c), temperature decreases to 36.79% of the magnitude of the high-temperature anomaly. Slab-driven corner flow causes the hot mantle to flow into the mantle wedge through the high-temperature anomalies on the back-arc-side vertical boundary, causing lower mantle viscosity. As discussed in the main text, the LVZs in the mantle wedge are likely hot and/or wet. Here, we are only explicitly incorporating the temperature effect on mantle viscosity to model hot LVZs. However, wet conditions also reduce mantle viscosity50, and the presence of wet LVZs are likely to cause similar effects on mantle wedge flow patterns through mantle viscosity reduction.

The model domain consisted of 255,780 elements including prism, hexahedral, triangular and quadrilateral elements with mesh refinements applied around the boundaries between the mantle wedge and its surrounding sub-domains (Supplementary Figure 2a). The maximum size of the elements in the mantle wedge is 6.12 km, and that in the rest of the domain is 10 km. The smallest element is 0.4 km at the tip of the mantle wedge. Model calculations with the prescribed back-arc thermal anomalies converged to near steady-state solutions at ~30 Myr. The models presented in this study were run for 60 Myr to assure a statistical steady state, minimizing the effects of initial conditions.

Temperature anomalies in the back-arc are likely to change in their positions and magnitudes with time. For example, hot upwelling regions caused by small-scale convection may migrate laterally along the arc14. While such migration of back-arc temperature anomalies is likely to be critical to periodic repositioning of volcanic clusters, for the purpose of the present study to better understand the first-order effect of back-arc temperature anomalies, we assume stationary back-arc high-temperature anomalies and present the modelling results obtained for a statistical steady-state.

Single high-temperature anomaly

In the reference model, magnitude ($T_{anomaly}$), depth ($x_{anomaly}$) and distance ($z_{anomaly}$) of the high-temperature anomaly are assumed to be 200 °C, 120 and 0 km, respectively. Both height (σ_{height}) and width (σ_{width}) of the high-temperature anomaly are assumed to be 40 km (Supplementary Figure 2c).

We tested the effect of the height, width, magnitude and depth of the high-temperature anomaly by changing one parameter at a time relative to the reference model (Supplementary Figure 3). The changes in the sub-arc mantle temperature and the distance between the vigorous-inflow and convergence zones with the height, width, and magnitude of the anomaly are summarized in Figure 1c–h, and those with the depth of the anomaly are shown in Supplementary Figure 4. In these figures, the maximum and minimum sub-arc mantle temperatures at a depth of 70 km and a distance of 320 km from the trench occur in the vigorous-inflow (right black star in Figure 1b) and convergence zones (left black star in Figure 1b), respectively. As discussed in the main text, an increase in height, width and magnitude of the high-temperature anomaly results in greater influx of the hot mantle into the mantle wedge, leading to higher sub-arc mantle temperatures and a larger distance between the vigorous-inflow and convergence zones. The velocity at which the mantle enters from the back-arc into the mantle wedge varies with depth. When the depth of the high-temperature anomaly coincides with peak mantle inflow velocities at the back-arc-side vertical boundary, the sub-arc mantle temperatures and the distance between the vigorous-inflow and convergence zones become maximized due to the high temperature and great vigor of the inflow. We found that the peak mantle inflow on the back-arc-side vertical boundary occurs at about 120 km depth (Supplementary Figure 4b). Deviation of the high-temperature anomaly from this depth results in reductions in the sub-arc mantle temperatures and the distance between the vigorous-inflow and convergence zones, but the effect is relatively small.

The resultant complex temperature field in the mantle wedge leads to along-arc variations in the slab-surface temperature. The maximum and minimum slab-surface temperatures under the vigorous-inflow and convergence zones, respectively, at a depth of 100 km and a distance of 340 km from the trench are shown in Supplementary Figure 5. Changes in the slab-surface temperature with the aforementioned parameters follow the same trends observed for sub-arc mantle temperatures.

Two high-temperature anomalies

Here, we implemented two back-arc high-temperature anomalies centered at the right and left side-wall boundaries and evaluated their interference (Figure 2a,b). We changed the spacing between the two high-temperature anomalies by changing the width of the model domain. Those with spacing of 210 and 420 km are shown in Figure 2a,b, respectively, and the results of those with other spacing are summarized in Figure 2c,d. In these models with two high-temperature anomalies, the depths of both temperature anomalies are set to 140 km. The other parameters that define the high-temperature anomalies except for the depth are the same as those used for the reference model.

For spacing of < ~250 km, the lateral overflows from the two vigorous-inflow zones merge together to form a single convergence zone between the two vigorous-inflow zones, and therefore, the distance between the vigorous-inflow and convergence zones is half of the spacing (Figure 2c). The interference between the two lateral overflows results in the development of less prominent vigorous-inflow and convergence zones, and the temperature contrast between the two zones is smaller than in the reference model (Figure 2d). For spacing of >~250 km, the lateral overflows from the two vigorous-inflow zones do not interfere with each other, and they each form a convergence zone at a characteristic distance (128.3 km) from the respective vigorous-inflow zone (Figure 2c). When the spacing is ~280 km, the largest temperature contrast (338.4 °C) occurs between the vigorous-inflow and convergence zones, and as we increase the spacing to 390 km, the temperature contrasts converges to 310.2 °C (Figure 2d), the same temperature contrast observed in the reference model. These results indicate that the two vigorous-inflow zones develop independently when the spacing is larger than 390 km, around three times the characteristic distance of the vigorous-inflow and convergence zones.

As shown in the model using a single high-temperature anomaly, slab-surface temperature reflects the along-arc variations in the sub-arc mantle temperature (Supplementary Figure 6). Because of the subdued development of the vigorous-inflow and convergence zones at spacing of <280 km, the temperature contrast between the slab surfaces under the vigorous-inflow and merged convergence zones is smaller than that in the reference model. The temperature contrast between the slab surfaces under the vigorous-inflow and merged convergence zones is maximized when the distance between the two high-temperature anomalies is 280 km (Supplementary Figure 6c). When the distance between the two high-temperature anomalies is the three times the characteristic distance between the vigorous-inflow and convergence zones, the vigorous-inflow and convergence zones independently develop (Supplementary Figure 6b).

Six high-temperature anomalies

We impose six high-temperature anomalies on the back-arc-side vertical boundary at 0, 100, 160, 240, 340 and 400 km from the right side-wall boundary of the model, with the spacing of 100, 60, 80, 100 and 60 km (Figure 3a). The depth of temperature anomalies is set to 140 km. The other parameters that define the high-temperature anomalies are the same as those used for the reference model.

The strong interference between the vigorous-inflow and convergence zones occurs in this model (Figure 3a). For example, the vigorous-inflow zones that extend from Anomalies 2 and 3 are deflected away from each other (Figure 3b). In general, greater degree of deflection occurs when spacing between high-temperature anomalies is narrower. The vigorous-inflow zone that extends from Anomaly 5 is deflected significantly to the right due to the vigorous-inflow zone that develops from Anomaly 6 along the side boundary at z=400 km with no-flow boundary condition. The interaction among vigorous-inflow zones and development of convergence zones results in complex mantle flow and temperature distributions. The vigorous-inflow zones that extend from Anomalies 2, 3 and 6 result in high sub-arc mantle temperatures, which in turn cause high slab-surface temperatures, compared with those beneath the other vigorous-inflow zones.

DATA AVAILABILITY

The model input required to reproduce the results are available within the article. The modelling results referenced in this study are available on request from C.L.

ADDITIONAL INFORMATION

How to cite this article: Lee, C. & Wada, I. Clustering of arc volcanoes caused by temperature perturbations in the back-arc mantle. *Nat. Commun.* **8,** 15753 doi: 10.1038/ncomms15753 (2017).

Publisher's note: Springer Nature remains neutral with regard to jurisdictional claims in published maps and institutional affiliations.

ACKNOWLEDGEMENTS

C.L. acknowledges the financial support from the National Research Foundation of Korea (35B-2011-1-C00043 and 2016K2A9A1A01946968). I.W. acknowledges the financial support from the University of Minnesota in the form of start-up funds.

REFERENCES

1. Tamura, Y., Tatsumi, Y., Zhao, D., Kido, Y. & Shukuno, H. Hot fingers in the mantle wedge: new insights into magma genesis in subduction zones. Earth Planet. Sci. Lett. 197, 105–116 (2002).
2. Gao, H. & Shen, Y. Upper mantle structure of the Cascades from full-wave ambient noise tomography: evidence for 3D mantle upwelling in the back-arc. Earth Planet. Sci. Lett. 390, 222–233 (2014).
3. Gaetani, G. A., Grove, T. L. & Bryan, W. B. The influence of water on the petrogenesis of subductionrelated igneous rocks. Nature 365, 332–334 (1993).
4. Grove, T. L., Chatterjee, N., Parman, S. W. & Médard, E. The influence of H2O on mantle wedge melting. Earth Planet. Sci. Lett. 249, 74–89 (2006).
5. Nakajima, J. et al. Seismic attenuation beneath northeastern Japan: constraints on mantle dynamics and arc magmatism. J. Geophys. Res. 118, 5838–5855 (2013).

6. Faul, U. H. & Jackson, I. The seismological signature of temperature and grain size variations in the upper mantle. Earth Planet. Sci. Lett. 234, 119–134 (2005).
7. Honda, S. Structures and processes in subduction zones thermal structure beneath Tohoku, northeast japan. Tectonophysics 112, 69–102 (1985).
8. Furukawa, Y. Depth of the decoupling plate interface and thermal structure under arcs. J. Geophys. Res. 98, 20005–20013 (1993).
9. Wada, I., Wang, K., He, J. & Hyndman, R. D. Weakening of the subduction interface and its effects on surface heat flow, slab dehydration, and mantle wedge serpentinization. J. Geophys. Res. 113, B04402 (2008).
10. Wada, I., He, J., Hasegawa, A. & Nakajima, J. Mantle wedge flow pattern and thermal structure in Northeast Japan: effects of oblique subduction and 3-D slab geometry. Earth Planet. Sci. Lett. 426, 76–88 (2015).
11. Wada, I. & Wang, K. Common depth of slab-mantle decoupling: reconciling diversity and uniformity of subduction zones. Geochem. Geophys. Geosyst. 10, Q10009 (2009).
12. Syracuse, E. M., van Keken, P. E. & Abers, G. A. The global range of subduction zone thermal models. Phys. Earth Planet. Interiors 183, 73–90 (2010).
13. van Keken, P. E., Kita, S. & Nakajima, J. Thermal structure and intermediate-depth seismicity in the Tohoku-Hokkaido subduction zones. Solid Earth 3, 355–364 (2012).
14. Honda, S. & Yoshida, T. Application of the model of small-scale convection under the island arc to the NE Honshu subduction zone. Geochem. Geophys. Geosyst. 6, Q01002 (2005).
15. Honda, S. & Yoshida, T. Effects of oblique subduction on the 3-D pattern of small-scale convection within the mantle wedge. Geophys. Res. Lett. 32, L13307 (2005).
16. Wirth, E. A. & Korenaga, J. Small-scale convection in the subduction zone mantle wedge. Earth Planet. Sci. Lett. 357–358, 111–118 (2012).
17. Morishige, M. A new regime of slab-mantle coupling at the plate interface and its possible implications for the distribution of volcanoes. Earth Planet. Sci. Lett. 427, 262–271 (2015).
18. Karato, S.-I. & Wu, P. Rheology of the upper mantle: a synthesis. Science 260, 771–778 (1993).
19. Hirth, G. & Kohlstedt, D. Rheology of the upper mantle and the mantle wedge: a view from the experimentalists. Inside the Subduction Factory 138, 83–105 (2003).
20. Currie, C. A. & Hyndman, R. D. The thermal structure of subduction zone back arcs. J. Geophys. Res. 111, B08404 (2006).
21. Honda, S. & Saito, M. Small-scale convection under the back-arc occurring in the low viscosity wedge. Earth Planet. Sci. Lett. 216, 703–715 (2003).
22. Lee, C. & Lim, C. Three-dimensional time-evolving plume-slab interaction for the generation of the Abukuma adakite, Northeast Japan. Gondwana Res. 38, 99–112 (2016).
23. Kincaid, C., Druken, K. A., Griffiths, R. W. & Stegman, D. R. Bifurcation of the Yellowstone plume driven by subduction-induced mantle flow. Nat. Geosci. 6, 395–399 (2013).
24. Tamura, Y. Some geochemical constraints on hot fingers in the mantle wedge: evidence from NE Japan. Geol. Soc. 219, 221–237 (2003).
25. Zhu, G. et al. Three-dimensional dynamics of hydrous thermal-chemical plumes in oceanic subduction zones. Geochem. Geophys. Geosyst. 10, Q11006 (2009).
26. Gerya, T. V. & Yuen, D. A. Rayleigh-Taylor instabilities from hydration and melting propel "cold plumes" at subduction zones. Earth Planet. Sci. Lett. 212, 47–62 (2003).
27. Honda, S., Gerya, T. & Zhu, G. A simple three-dimensional model of thermo–chemical convection in the mantle wedge. Earth Planet. Sci. Lett. 290, 311–318 (2010).
28. Morishige, M. & Honda, S. Three-dimensional structure of P-wave anisotropy in the presence of small-scale convection in the mantle wedge. Geochem. Geophys. Geosyst. 12, Q12010 (2011).

29. van Keken, P. E., Hacker, B. R., Syracuse, E. M. & Abers, G. A. Subduction factory: 4. Depth-dependent flux of H2O from subducting slabs worldwide. J. Geophys. Res. 116, B01401 (2011).
30. Spiegelman, M., Kelemen, P. B. & Aharonov, E. Causes and consequences of flow organization during melt transport: the reaction infiltration instability in compactible media. J. Geophys. Res. 106, 2061–2077 (2001).
31. McKenzie, D. The generation and compaction of partially molten rock. J. Petrol. 25, 713–765 (1984).
32. Kelemen, P. B., Braun, M. & Hirth, G. Spatial distribution of melt conduits in the mantle beneath oceanic spreading ridges: Observations from the Ingalls and Oman ophiolites. Geochem. Geophys. Geosyst. 1, 1005 (2000).
33. Kimura, J.-I. & Yoshida, T. Contributions of slab fluid, mantle wedge and crust to the origin of quaternary lavas in the NE Japan arc. J. Petrol. 47, 2185–2232 (2006).
34. Schmidt, M. W. & Poli, S. Experimentally based water budgets for dehydrating slabs and consequences for arc magma generation. Earth Planet. Sci. Lett. 163, 361–379 (1998).
35. Kessel, R., Ulmer, P., Pettke, T., Schmidt, M. W. & Thompson, A. B. The water-basalt system at 4 to 6 GPa: phase relations and second critical endpoint in a K-free eclogite at 700 to 1400 °C. Earth Planet. Sci. Lett. 237, 873–892 (2005).
36. Hasegawa, A. & Nakajima, J. Geophysical constraints on slab subduction and arc magmatism. In The State of the Planet: Frontiers and Challenges in Geophysics (eds Sparks, R. S. J. & Hawkesworth, C. J.) 81–93 (American Geophysical Union, 2004).
37. Nakajima, J., Shimizu, J., Hori, S. & Hasegawa, A. Shear-wave splitting beneath the southwestern Kurile arc and northeastern Japan arc: a new insight into mantle return flow. Geophys. Res. Lett. 33, L05305 (2006).
38. Long, M. D. & Wirth, E. A. Mantle flow in subduction systems: the mantle wedge flow field and implications for wedge processes. J. Geophys. Res. 118, 583–606 (2013).
39. Karato, S.I., Jung, H., Katayama, I. & Skemer, P. Geodynamic significance of seismic anisotropy of the upper mantle: new insights from laboratory studies. Annu. Rev. Earth Planet. Sci. 36, 59–95 (2008).
40. Holtzman, B. K. et al. Melt segregation and strain partitioning: implications for seismic anisotropy and mantle flow. Science 301, 1227–1230 (2003).
41. Hall, P. S. On the thermal evolution of the mantle wedge at subduction zones. Phys. Earth Planet. Interiors 198-199, 9–27 (2012).
42. Lee, C. & King, S. D. Effect of mantle compressibility on the thermal and flow structures of the subduction zones. Geochem. Geophys. Geosyst. 10, Q01006 (2009).
43. King, S. D. et al. A community benchmark for 2-D Cartesian compressible convection in the Earth's mantle. Geophys. J. Int. 180, 73–87 (2010).
44. Syracuse, E. M. & Abers, G. A. Global compilation of variations in slab depth beneath arc volcanoes and implications. Geochem. Geophys. Geosyst. 7, Q05017 (2006).
45. Lee, C. & Lim, C. Short-term and localized plume-slab interaction explains the genesis of Abukuma adakite in Northeastern Japan. Earth Planet. Sci. Lett. 396, 116–124 (2014).
46. Sdrolias, M. & Müller, R. D. Controls on back-arc basin formation. Geochem. Geophys. Geosyst. 7, Q04016 (2006).
47. Turcotte, D. & Schubert, G. Geodynamics. 2nd edn Cambridge Univ. Press (2002).
48. Sato, T. et al. Last stage of the Japan Sea back-arc opening deduced from the seismic velocity structure using wide-angle data. Geochem. Geophys. Geosyst. 7, Q06004 (2006).
49. Lee, C. & King, S. D. Why are high-Mg# andesites widespread in the western Aleutians? A numerical model approach. Geology 38, 583–586 (2010).
50. Jung, H. & Karato, S.-I. Water-induced fabric transitions in olivine. Science 293, 1460–1463 (2001).

CHAPTER 6

Field Detection of Microcracks to Define the Nucleation Stage of Earthquake Occurrence

Y. Fujinawa,[1] Y. Noda,[2] K. Takahashi,[3] M. Kobayashi,[4] K. Takamatsu,[4] and J. Natsumeda[4]

[1]*Risk Control Association, Kanda Hirakawa Chiyoda-ku, Tokyo 101-0027, Japan*
[2]*Tierra Tecnica Ltd., Musashimurayama, Tokyo, Japan*
[3]*Communication Research Laboratories, Koganei, Tokyo, Japan*
[4]*OKI Engineering Co., Ltd., Nerima-ku, Tokyo, Japan*

ABSTRACT

Main shocks of natural earthquakes are known to be accompanied by preshocks which evolve following the modified Ohmori's law in average over many samples. Individual preshock activity, however, is far less systematic for predictive purposes. On the other hand, the microcracks in laboratory rock experiments are always preceded to final rupture. And, previous investigations of field acoustic emissions showed that the activity increases prominently before and after the main shock. But there is no detection of any phenomena to identify the nucleation stage. Here we show that a special underground electric field measurement could detect microcracks. Pulse-like variations were classified into three groups (A, B, C) by frequency. The B-type is suggested to define the nucleation period: activity increases sharply following the modified Omori's law before the main shock and there is no activity afterward. The B-type is subgrouped into three types possibly corresponding to crack-rupture modes. The variations are supposed to be induced by crack occurrence through electrokinetic effects in the elastic-porous medium. The detection distance is suggested to be several orders larger than that of the acoustic emission due to the effective smallness of dissipation rate, and the waveform can be used to infer the rupture mode.

1. INTRODUCTION

The identification of anomalous phenomena in each stage of seismic cycle is a prerequisite for understanding physical processes to be applied to earthquake forecasting (e.g., Mogi [1], Rikitake [2], and Scholz [3]). For the short-term prediction, numerous observations by seismic, geodetic, hydrologic, and electromagnetic approaches have been tried to investigate various phenomena to find precursory phenomena and to build physical and statistical models [1, 3].

Among all, the preshock has been considered to be the most direct and plausible phenomenon for identifying the earthquake nucleation stage. Previous investigations of preshock activity of natural earthquakes or microcracks activity in laboratory experiments (e.g., [1, 3], Jones and Molnar [4], Ohnaka [5], and Maeda [6]) have provided essential evidence as accelerating activity. Those evidences have been utilized to build the physical and statistical models of nucleation process [1, 5]. These models have been further developed: for instance, superimposed undulation of foreshock activity was utilized to infer dynamical critical phenomena to estimate more exact final rupture time (e.g., Sornette and Sammis [7], Kapiris et al. [8], Eftaxias et al. [9], and many papers referred in the report).

As for natural earthquakes, however, the evidence is generally limited to average images. There is large extent of variability in particular cases of earthquake with the result of no plausible clue to be applied for prediction of individual earthquake (e.g., [3]). The most crucial problem is too small number of preshocks to fit the characteristic formula of the evolution to infer the nucleation stage. For instance only four and three preshocks were observed in case of the Hyogoken Earthquake (M7.2, January 17, 1995) and the Tohoku Earthquake (M9.0, March 11, 2011), respectively. The situation has not been improved even by the most extensive microseismic network (Hi-net) of Japan.

On the other hand, the rock fracture experiment can always observe the characteristic rapid increase of acoustic emission activity in the nucleation period [1, 3, 5, 10–14]. Especially the accelerated increase of microcracks is shown to follow the modified Ohmori Law. Detailed investigation on activity in terms of spatiotemporal distribution and focal mechanism of microcracks revealed physical process of rock fracture. The investigation extends to the seismoelectromagnetic phenomena (e.g., Hayakawa and Fujinawa [15], Hayakawa [16], and Eftaxias and Potirakis [17]).

The acoustic emission technique has been widely used in the geotechnical engineering field to examine geological materials since the late 1930s (e.g., Drnevich and Gray [18]). The technique has been used for monitoring the mine and stones by using dense seismic network with spacing of several tens of meters. Concerning natural earthquakes, however, there are only several pioneering studies to investigate field acoustic emission activities (e.g., Morgunov et al. [19], Hattori [20], and Molchanov et al. [21]). Those researches showed that the acoustic emission intensity is high about half-day both before and after the earthquake occurrence. The index of the activities are limited to the strength and could not distinguish the difference between the activity before the main shock and that after the shock without providing any useful clue to define the nucleation stage.

Electromagnetic approaches have been providing many interesting clues to shed light on preparatory process of earthquakes (e.g., Park et al. [22]). As for the nucleation process of natural earthquake, Eftaxias has been investigating in detail by referring laboratory and earthquake generation model by assuming that the VLF emissions are induced by the microcracks [9, 17].

In the present paper we present an electromagnetic field observation of microcracks by using high quality instruments providing new clues to detect particular features to identify the nucleation stage.

2. METHODS

2.1. Sensor

We used electric sensors to observe microcracks instead of ordinary acoustic emission transducer. The sensor in the field is a special antenna made of a vertical casing pipe that is installed into a borehole of 200 to 1,800 m deep. The boreholes are not for electric field measurement but for intrinsic purposes as measurements of seismic motion, ground water level, and chemical contents; ground deformation or pumping of natural gas or waters for drinking; and hot spring belonging to several organizations and private persons (Fujinawa et al. [23–25]). The antenna is one of monopole type immersed underground (Figure 1) of conductive crust similar to the towing tail cable antenna for submarines' underwater communication.

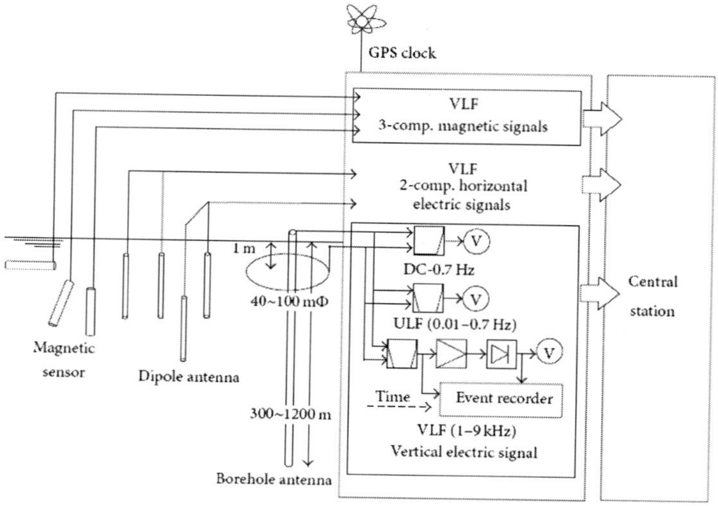

Figure 1. Schematic diagram of multiple component electromagnetic field observations [25, 28] of the full option. The electric field is measured by a pair of horizontal dipoles of 8 m in length with a span of 20~50 m and using a vertical monopole of 600–1,200 m surrounded by wire with a radius about 10–20 m buried at a depth of about 1 m. The magnetic field is measured by three induction-type magnetometers (BF-6, EMI) installed perpendicularly to each other. Triggered data in three electrical and magnetic fields are transmitted to a central station once a night. The present observation at Hasaki was conducted by the simplest model using only borehole sensor with continuous data acquisition.

The electrical circuit is composed of the preamplifier, filtering, and data logging. Additionally a simple counter surge circuit is inserted between sensors and preamplifiers to withstand electric surge induced by lighting discharge. The recording frequencies are DC band (0–0.7 Hz), ULF (0.001–0.7 Hz), and ELF/VLF (1 kHz–9 kHz). The data of VLF band is logged by an event trigger system focused to locate the VLF pulse-like signal for early stage of investigation. Later envelope of electric strength is recorded to monitor the activity of the field. Lower frequency component is recorded continuously with the sampling interval of 2 seconds. The present observation used continuous recording.

2.2. Prior Observation

Prior observation has confirmed that the system has high robustness against both meteorological and urban noise enabling us to get large signals related to the crustal activities (Fujinawa et al. [23–26], Takahashi et al.

[27]). Exceptions are the large noises induced by nearby lighting discharges and smaller "noises" by severe geomagnetic disturbances compared with the normal state. After confirming the performance of the borehole antenna the observation was conducted only by borehole antenna.

The observation network has been built and operated since 1989, and the sites situate at several kinds of geological conditions nearly uniformly in central Japan (Figure 2). Several kinds of anomalies were picked up and investigated from the points of view of finding and evaluating candidates of precursor in relation with earthquakes, volcanic eruption, and ground slides. The ULF band anomalies have been most extensively investigated because they were detected at almost cases of nearby volcanic eruption at Izu-Oshima and Miyake-Jima and seismic swarms as Hodaka in a geothermal region near volcanos (Fujinawa et al. [25]).

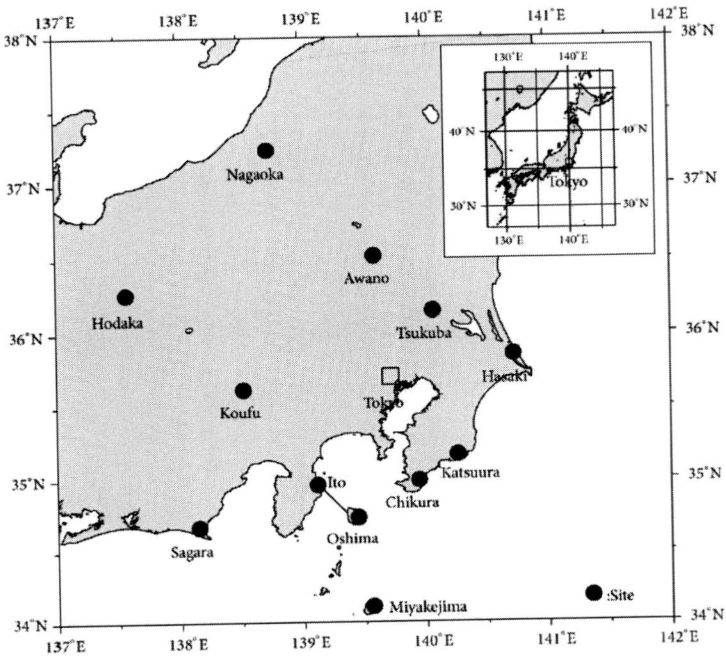

Figure 2. Locations of the electromagnetic field observation sites (●) in central Japan utilizing borehole antenna 150–1200 m long as the principal sensor. Those sites have been constructed during 1989~1996. Every site measures the vertical component of the electric field. Additionally mutually perpendicular short dipoles of 20~60 m long were used for measurement of horizontal electric component at five sites (Hasaki, Chikura, Koufu, Nagaoka and Sagara) by 2002. And the induction type magnetometers were used at two sites (Nagaoka, and Sagara) by 2002 [25, 28].

The ULF band characteristic waveform was first identified by the temporally rapid recording at the time of volcanic eruption activities in 1990 at Izu-Oshima (Figure 3, Fujinawa et al. [23]). The signals have never been detected in a normal state, but observed in almost all volcanic eruption activities and seismic swarms occurring near the observation sites: the earthquake swarm around Mt. Hodaka in 1998 (Takahashi et al. [28]) and Mt. Nikko-Shirane in 1999, and volcanic eruption at the Miyake Island in 2000. The phenomena have been suggested to be induced by an electrokinetic effect (e.g., [25, 26]; Nourbehect [29], Fitterman [30], and Ishido and Muzutani [31]. The ULF variation is called a "geyser-like ULF signals (GUV)" because of similarity of the temporal change of electric strength with water height change of the geyser. We imagined that the whole of the pulse-like variations detected by the slow speed recording are of this type. We could not analyze higher-frequency phenomena in detail because of the system's slow recording speed.

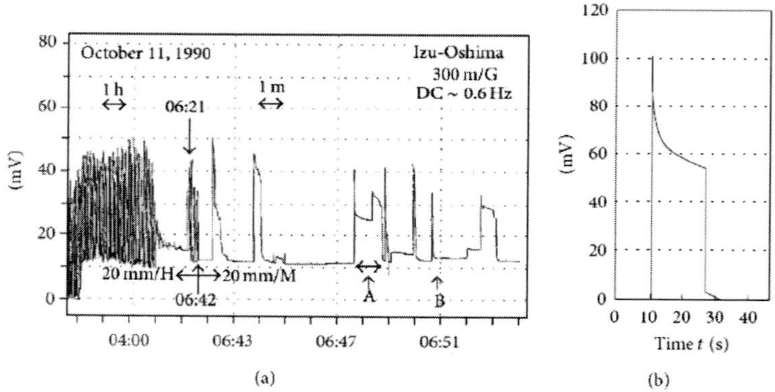

Figure 3: (a) Records of electric field anomalies on 11 October, 1990 related to the small volcanic activities in the Izu-Oshima Island by the borehole antenna [25, 28]. Slow speed recording shows numerous pulse-like signals (left part of (a) before 6:42 JST). Rapid recording by 60 times (right part of (a) after 6:42 JST) shows the particular waveform similar to the geyser height evolution. (b) A model of confined pressure change is proposed to explain the electric field changes based on the electrokinetic effect.

2.3. New Detector and Observation

The present observation was conducted using detectors having larger dynamic range. The experiment was originally to investigating the coseismic signals with propagating velocity much larger than the seismic wave (Fujinawa et al. [32] and references in the literature). We installed a

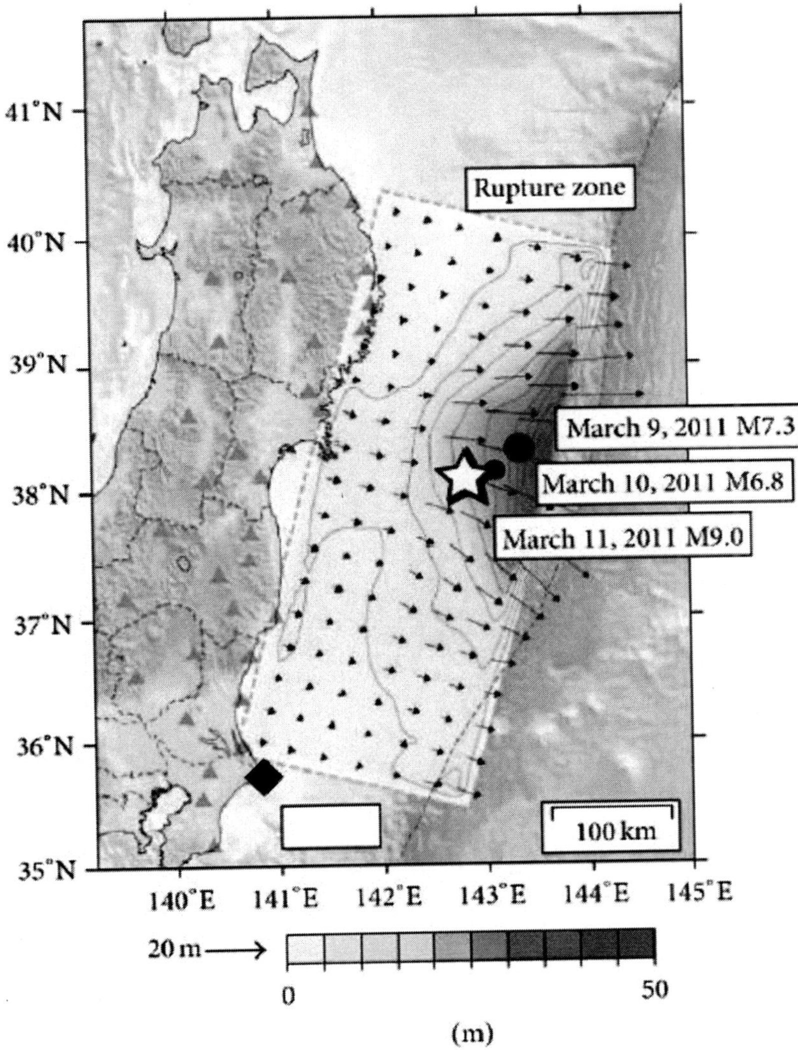

Figure 4: Geographical relationship between the observation site Hasaki (◆) and the rupture zone 500 km × 200 km (dashed line). Calculated displacement distribution (↗) of the Tohoku Earthquake (★, magnitude 9.0). Also shown are the two largest preshocks, which occurred two days

before (□, M7.3) and one day before (□, M6.8) the earthquake (revised from Suzuki et al. [33]).

new detection system on 3, March 2011 at Hasaki (Figure 2, now Kamisu), which is about 300 km south of the epicenter of the Tohoku Earthquake (Figure 4, modified from Suzuki et al. [33]). The borehole belongs to the National Research Institute for Earth Science and Disaster Prevention (NIED). The frequency bands are of DC (0–1.45 Hz) and AC (1.45 Hz–9 kHz). AC recording band is amplified by gain of 100. And the sampling rate is 4.5 kHz before March 11, 18 kHz from 24, March, 2011.

2.4. Data Analysis

To construct diurnal monitoring of DC and AC components we averaged in the interval of 1 sec and/or resampled by 150 Hz from the original data sampled 4.5 kHz (by 11th March or 18 kHz after 24th March). Figure 5(a) illustrates the result of the averaging every second for one day data on March 25. Figure 5(b) is another record for diurnal monitoring using data through both resampling by 150 Hz and averaging of 1 sec. Difference is in apparent background noise levels of 0.003 mV and 0.040 mV for the procedure without and with 150 Hz sampling, respectively. The larger noise level for 150 Hz resampling suggests the procedure without 150 Hz resampling is more suitable to monitor the activity of the events. It is noted that the averaging procedure reduce the pulse strength with the result of difference of pulse strength in the monitoring and individual events. After finding there are pulse-like events, we picked up whole events of strength larger than predetermined threshold of 2 mv from original raw data. There are no other procedures to depict meaningful signals except those induced by lightning surge in this method of measurement having high robustness.

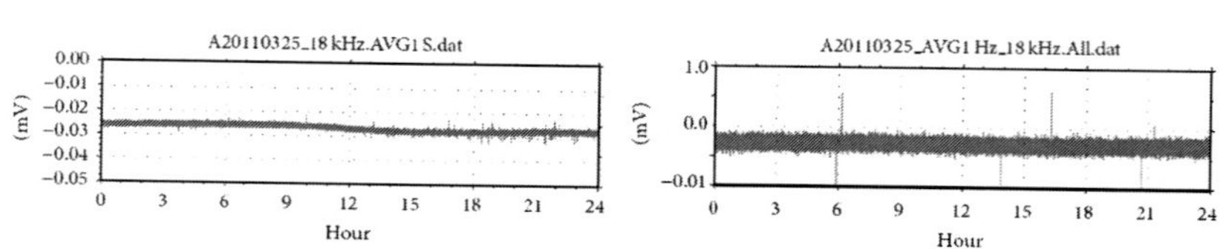

(a) Simple average (1 sec) (b) 150 Hz sampling and average (1 sec)

Figure 5: Comparison of two kinds of monitoring data. (a) Raw data on 25th March 2011 sampled by 18 kHz and averaged for 1 second. Small drift of about 0.003 mV is due to the limit of circuit performance. (b) The same raw data through resampling by 150 Hz and averaged for 1 second.

3. RESULTS

Figure 6(a) is the daily DC monitoring record on March 31, 2011 corresponding to the calm state. Except the background noises there were only diurnal variations due to the effect of earth and ocean tide (e.g., Lai et al. [34]). On the other hand the record on March 25 (Figure 6(b)) includes large pulse-like variations. Figure 6(c)is AC daily record on March 4 showing normal state for background noises of about 0.03 mV seen as a belt except short time period of activity between 17:00 and 19:00 ("Activity" in c)). Contrary to this normal record, March 9 (Figure 6(d)) had numerous pulses.

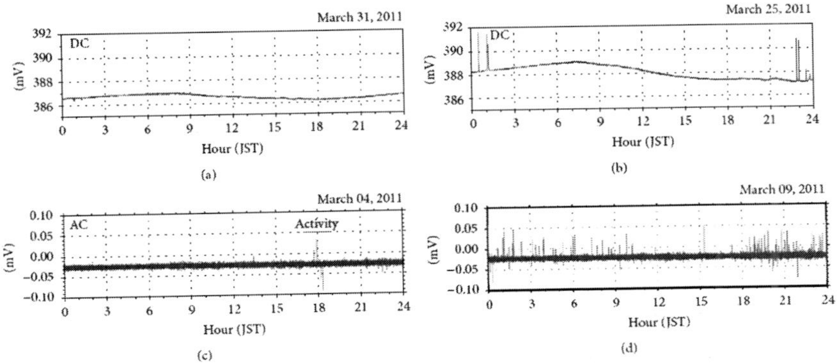

Figure 6: Monitoring records of normal state and active state. (a) DC channel record on March 31, 2011 corresponding to the normal state without any short period fluctuation of large strength except for a diurnal variation due to the earth and ocean tide. There were no conspicuous environmental noises to make it difficult to detect anomalous signals. (b) DC channel record on March 25 with pulse-like fluctuations around 1:00 and 23:00. The signals are very similar to those found in the analog record at several cases of crustal activities [25, 26]. The strength of the signal is about twice as large as the earth tide. (c) The AC channel record on March 4 corresponding to the normal state except during the time period 17:00–19:00. During the calm state there were no large variations except for the atmospherics and environmental noises comprising a horizontal belt-like part. (d) AC channel records on March 9 corresponding to the most active day. The amplitude of the pulse has a distribution, and the positive polarity dominates compared to the negative polarity because of the pulse-like signals are of B type having B-1 and B-3 types of unbalanced polarity. The peak value is diminished because of averaging procedure for dairy monitoring.

Whole pulse-like variations with strength greater than 2 mV about 40 times the background noise level of 0.025 mV were picked up from the raw data. There were a total of 225 events. The successive large pulses corresponding to nearby lightning have not been detected. Most of the events are isolated in the sampled interval of 100 ms except cases superimposed small events as described in the discussion section. The largest event is of the coseismic variation corresponding to the main shock reported previously [32]. The event is taken up later in the discussion section. Waveforms of the whole events were enlarged for the investigation of their characteristics and origin. Those events are found to be classified into three groups depending on their signal duration. The typical waveforms in each group are shown in Figure 7. Whole pulse-like signals belong to one of the three types indicating that there are no extraneous noises.

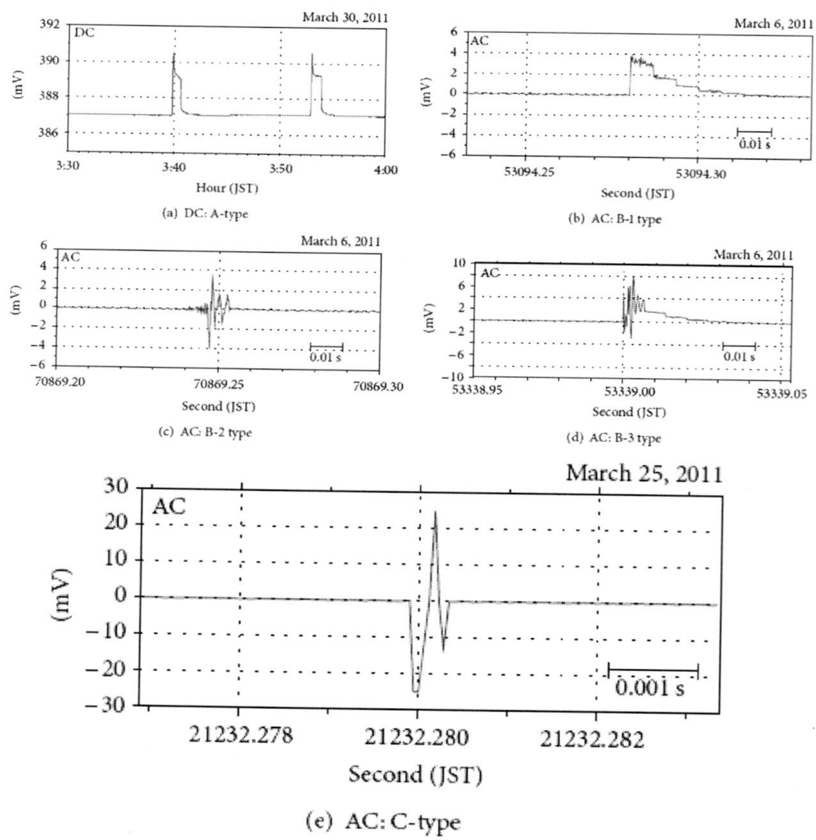

Figure 7: Waveform of the pulse-like particular signals. Every event has large strength compared with noise level with the result of none data

filtering. (a) DC band (type A). Type A has a duration of several tens of seconds to several tens of minutes and a height of about 2-3 mV. The form was first identified by the temporally rapid recording at the time of volcanic eruption activities in 1990 at Izu-Oshima [23, 25]. Afterward, we detected similar signals at almost cases of nearby volcanic eruption and seismic swarms [25]. (b) AC band (type B-1). Type B-1 waveform of the pulse-like signals in the AC band. The form is very similar to type A except that it is of smaller duration (several tens of ms) and shows several steps in the process of relaxation. There are events both with positive and negative polarities. But the majority events are positive. (c) AC band (type B-2). Type B-2 waveform of the pulse-like signals in the AC band similar with a wave-packet. There is no coda phase, in contrast to seismic waves. A similar pattern was detected in the case of wetted granite specimens with a dominant frequency of 500 kHz [12, 13]. The S-P time was about 30 ms, suggesting an epicentral distance of 270 m. The absence or very slight appearance of a P phase indicates an epicentral distance larger than a few km. (d) AC band (type B-3). Type B-3 waveform in the AC band with a waveform compounded by type B-1 and type B-2. There are different degree of compounding. (e) AC band (type C). Type C waveform in the AC band. The waveform is the same as type B-2, but frequency is about 5 kHz and amplitude is very large of 20–30 mV, some 10 times larger than those of types A and B types.

The type of an event (Figure 7(a)) was simply the so-called ULF geyser-like variation of the sort that had been recognized earlier (GUV) [23, 25, 26] by us as is shown in Figure 3. The ULF variations have been investigated also by many authors from the point of view for prediction of earthquake and volcanic eruptions by using electromagnetic measurement on the surface (e.g., [8, 9, 21–23], Nagao et al. [35], Revil and Saracco [36], and Han [37]). But there are still no consensus on the characteristics of waveform and the generation mechanism, though there are large extent of similarity in the waveforms and pulse width and uniformity of amplitudes.

Types B and C events were firstly identified in this observation of increased dynamic range in our method. They have a much smaller duration time from 10 ms to several tens of ms for type B and 0.3–0.5 ms for type C compared with that of A-type of several minutes.

Type B pulses are subdivided into three types depending on their waveform (Figures 7(b)–7(d)). Main phase of the type B-1 (Figure 7(b)) is similar to type A in its abrupt rise and maintaining polarity similar to the relaxation waveform, but is different in decaying stepwise. It is noted that almost all events of B-1 contain higher frequency components.

Similar electric variation without stepped decays had been previously reported in association with acoustic emission in rock deformation experiments [10–14]. The electric field variation associated with the acoustic emission was detected in the time period of stable sliding in the nucleation period. For the dry granite the variation has high similarity with B-1 type in the points that rapid growth followed by exponential decay with dominant positive polarity. The differences to the present case are in different time constants of 50 μ second and in compound rate of dominant phase of lower frequency with higher frequency components. In case of stick-slip movement, they found the variation similar to the wave packet type of B-2 without P phase. Moore and Glaser [38] reported similar self-potential spikes with less sharp rise of electric field change in hydraulic fracturing experiment.

The simple relaxation type electric variation was also deduced in the analytical estimation by Fenoglio et al. [39] concerning magnetic anomalies associated with the Loma Prieta Earthquake1990 (Fraser-Smith et al. [40]). In that model, confined water is assumed to eject as the Poiseulle flow during crack opening inducing electric field changes through the electrokinetic effect.

Type B-2 (Figure 7(c)) is a wave packet similar to the seismic wave and acoustic emission waveform. But the dominant frequency is 1-2 kHz several orders smaller. In the rock fracture experiment using wet granite Yoshida and Ogawa [13] detected high frequency variations of dominant frequency of 0.3 MHz having similar waveforms as B-2.

Type B-3 (Figure 7(d)) is a superposition of types B-1 and B-2 with various rates of compounding. The waveform compounded by relaxation type variation can be seen also in the results of Yoshida and Ogawa [13]. The starting time difference of the two components is short enough to suggest that they are due to the same dynamic motion. The source mechanism of the microcracks was determined using the moment tensor method by Sellers et al. [41] and Timms et al. [42]. They revealed that many cases have an isotropic component corresponding to tensile opening or compressive closure. The mixed waveform B-2 is thought to be caused by a crack of this composite type of mechanism. The catalog of all B-type variations is shown in the appendix.

Type C (Figure 7(e)) is a wave packet with a carrier frequency of about 5 kHz, about ten times larger than that of type B-2 and the shortest duration of some 0.5 ms. Amplitude reaches some 30 mV several times larger than that of A and B type. This type does not seem to be subdivided as B type under the limit of present sampling interval.

In the crack theory [3], it is known that the displacement field of a crack has three modes: tensile mode (I) and shear modes of two types: in-plane shear mode II and anti-plane shear mode III. The space and time distribution and focal mechanism of microcracks were investigated in rock experiments [38–42], Lei et al. [43]. And it was found that tensile-type cracks occur rather uniformly in space and time, whereas shear-type cracks tend to concentrate in the area of unstable movement. Based on previous and present findings, we can suggest that type A and type B-1 correspond to tensile cracks, types B-2 and C to shear modes, and type B-3 to the compounded mode.

Figure 8 shows the time evolution of the number of pulses per hour of types A (green), B (red), and C (blue). The observation of this session started 13:00, March 3, 2011 till 13:00, March 30. There are several time periods of recording interruption (horizontal grey bold line) owing to malfunction of the instrument or electric power shutoff after the main shock by 13:00, March 24. Days from 12 to 23 (vertical dotted lines) are compressed in this figure. In the following statement on time evolution of pulse-like events should be understood under the limit of this period of observation failure. Before the Tohoku Earthquake ("EQ3.11" in this figure), only B type events occurred except for only one pulse of the type A. After the earthquake, on the other hand, types A and C events occurred without type B events.

From these results we can understand the dominance of the positive polarity in diurnal monitoring record on March 9 (Figure 6(d)). On this day there are only B-type variations as seen in Figure 8. The B type consists of three types, B-1, B-2, and B-3. The B-1 and B-3 types are dominated by positive polarity as seen in Figure 7. On the other hand B-2 is not dominated by neither positive nor negative polarity. The compressed records result in the apparent domination of positive polarity.

As for the B-type variations, there are only several events per day on March 3 and 4. The number of events started to increase on the 7th, had a prominent peak on the 9th, a pronounced lull on the 10th, and recovered considerably on the morning of 11th until the earthquake occurred. The lull does not mean there are no activity of microcracks. The smaller the threshold is, the more the number of events are. The feature is in agreement with the evolution of space-time distribution of microcracks of rock fracture experiment [41–43].

The activity of the type B phenomena evolved similarly to those of the acoustic emissions just before the rupture in the rock experiment [3, 13] and to those of the foreshocks before main shocks of natural earthquakes (e.g., [4, 6, 7], Varnes [44]). The cumulative number of the foreshock period evolves following the modified Ohmori law. The characteristic evolution has been taken as one of the most direct and clear evidence to identify the nucleation stage in the preparatory process. The accelerated increase of strain causes the particular form of time dependence intrinsic for rupture in the heterogeneous crust [7, 8].

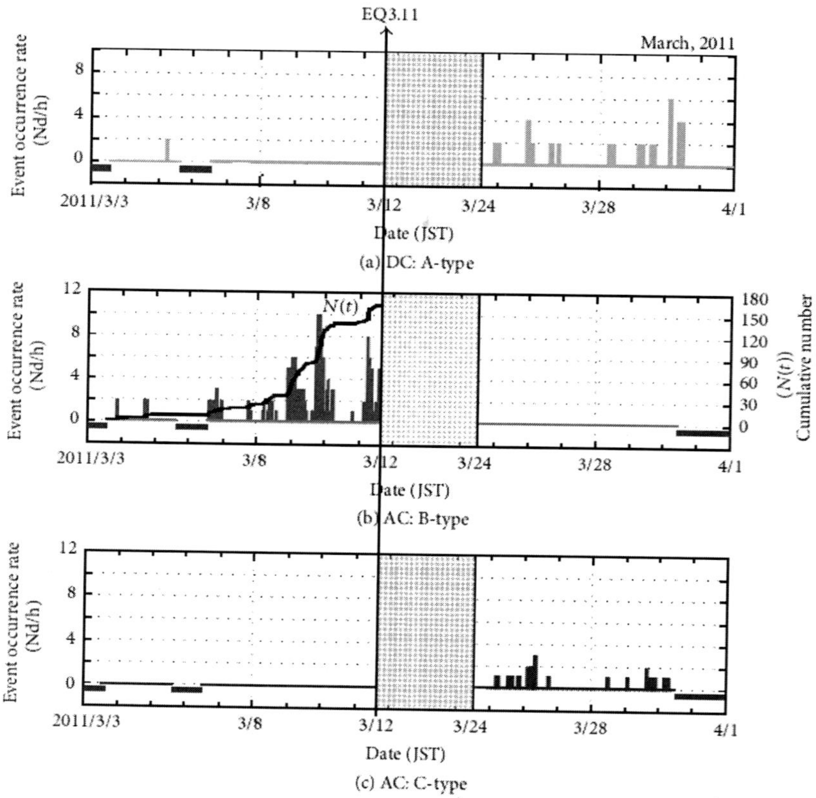

Figure 8: The Tohoku Earthquake, magnitude 9.0, occurred on March 11, 2011. There are several time periods (grey horizontal line) without data due to logger problems before the main shock and regional electric power shutoff. (a) Time evolution of number of pulse-like signal per hour of type A (green). The unit is shown on the left hand side. Almost all type A variations occurred after the main shock. (b) Time evolution of type B (red) including B-1, B-2, and B-3. The type B variation is suggested to be particularly useful for imminent prediction. The bold continuous curve attached byshows the cumulative number of B-type variations. The unit of the cumulative number is shown in the right hand side. (c) Time evolution of type C (blue). This type occurred only after the main shock when the after-effect stage of the seismic cycle.

Fitting the modified Ohmori lawto the cumulative number(a black curve in Figure 8(b)) of cracks at arbitrary time(hour) from the final rupture time results in the index number to the exponentand to . The present index number 3.0 is the largest compared with 1.7 for the foreshock [44] and 0.5 for the microcracks in the laboratory experiment [11]. The assumed rupture time agrees satisfactorily with the actual value of 11.6. The characteristic evolution suggests that the microcracks activity is related to the nucleation process of the main shock of the Tohoku Earthquake.

The three types, B-1, B-2, and B-3, occurred in nearly steady percent, 33, 46, and 21% for the whole period. However the compound type (B-3) decreased its role in the more active period (from 9 to 11, March) with the result of increase of activity of tensile type (B-1) and shear type (B-2) (Figure 9). In particular the shear type became dominant from the midnight of the 9th as approaching main rupture. The tendency is also reported in the rock experiment [41–43]. The index number to the exponent of inverse power lawfor whole B type is 3.0. Individual fitting of types B-1, B-2, and B-3 to the inverse power law in the period from March 3 till 11:00 March 8 results in , 3.0, and 2.2, respectively. The index numbers to the exponentdo not differ largely in agreement to the near steady percent of each type in comparison with previous result of 0.5 for rock experiment and 1.7 for foreshocks.

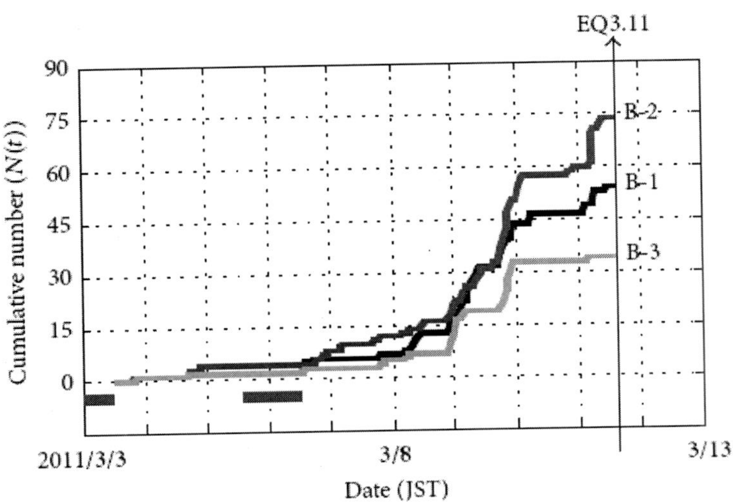

Figure 9: Evolution of the cumulative numbers , , andfor B-1, B-2, and B-3, respectively. B-2 type tend to dominate as approaching to final rupture. But there are no big changes in partial importance in the activity.

Occurrence of higher frequency events of C type and low frequency A type after the main shock can be interpreted according to the after effect stage of the seismic cycle. It is supposed that there is a strong adjustment of stress and strain in the rupture area inducing normal closure and shear slips (e.g., Yoshioka and Scholz [45]) causing high-frequency acoustic emissions [11] similar to the C-type microcracks. Around the joints, on the other hand, larger-scale confined water movements are assumed to occur, similarly to cases of seismic swarms and volcanic activities, inducing A-type electric variation of large duration of the order of minutes [25].

4. DISCUSSION

4.1. Correspondence to Foreshocks, Main Shock and Aftershock Period

4.1.1. Foreshock Period

The main shock occurred on March 11 some 50 km south of the foreshock at almost the same region as predicted and was made public several years before. There occurred two foreshocks of major class, one of magnitude 7.3 (March 9) and other of 6.8 (March 10). The microcracks might be imagined to be associated with the foreshocks that occurred two days before the Tohoku earthquake when the activity of microcracks was largest. There were no coseismic electric variations corresponding to these foreshocks themselves as detected previously in 2002 [32]. And, the detection distance of the characteristic electric variation was estimated to be at most 100 kilometers [25, 32]. Additionally the epicentral distances of these f foreshocks from the observation site, Hasaki, are about 300 km northeast (Figure 4, Suzuki et al. [33]). So that we suppose that the B-type variations are not related to the foreshocks.

4.1.2. Main Shock

It has been revealed that the great rupture consist of three subevents: main asperity and subsequent two events [33]: first in the northern part, second succeeded southward, and third propagated from southwest to north (e.g., [33]).

Anomalous electric signals corresponding to main shock are shown in the lower sheet of Figure 10. And the strong motion records of K-NET (CHB005) near Hasaki are in the upper sheet. There are coseismic variations, but the prominent peak appeared at around the peak acceleration and the second peak with opposite polarity appeared about one minutes later. But there are no significant electric variations in correlation with the origin time and P, S wave arrival time ("org", "P", and "S", in the figure, resp.) as reported in the previous reports [32]. It may be caused by complex rupture comprised of three stages [33]. We note that the time scale of the electric variation related to main shock is of a few minutes compared several tens ms of B and several ms of C-type cracks. On the other hand, the maximum electric strength corresponding to the main shock is 5 mV nearly the same as those of microcracks.

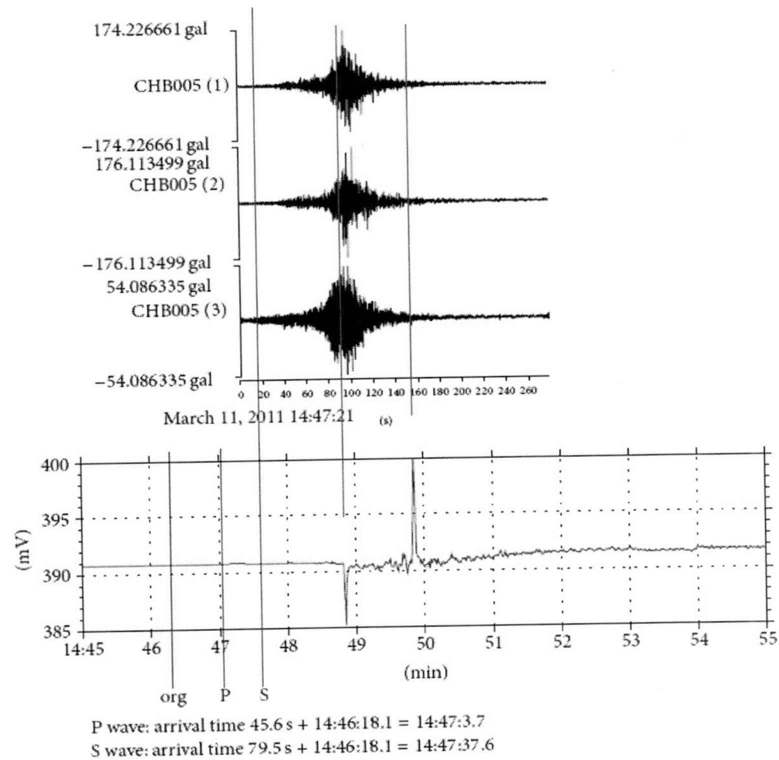

Figure 10: Coseismic electric field variation for the Tohoku earthquake (3.11) at Hasaki (lower sheet) and the strong motion seismograph (UD component) at nearest K-net observation site CHB005 (upper sheet, (1): EW, (2): NS, (3): UD). The vertical line "org", "P", and "S" are origin time of 3.11, P wave arrival time, and S wave arrival time at CHB, respectively. There is very faint electric phase corresponding P arrival time and none phase corresponding to S arrival. The first large negative peak coincides with the peak acceleration and the second prominent peak did not correspond to any phase.

The fault displacement shows a simple distribution [33] without any trace to separate rupture blocks. We can think that the whole fault including the southwestern region (Figure 4) ruptured simultaneously. The time evolution of the B-type activity suggests intimate relation to the main shock through Omori's law (Figure 8). On the basis of these evidences, we can suppose that the microcracks of B type are related to the nucleation process of the main shock which induced the fault movement amounting several centimeters near the observation site.

4.1.3. Aftershock Period

Aftershock occurred in an extensive area in and around the main rupture of 3.11. During 3 weeks after the main shock there occurred as many as 427 aftershocks with a magnitude larger than 5.0 in and around the rupture zone of 3.11. In this period there occurred A- and C-type variations of highest frequency of some 5 kHz (Figure 8). After the main shock there were no events of type B and none of coseismic variations association with aftershocks. The increased strength of AE related to the natural earthquakes after the main shock [19–21] might be caused by this kind of emission.

4.2. Radio Wave Emission

Electromagnetic emission observation in VLF bands has been widely conducted to investigate the nucleation process. It is generally reported that there are considerable correlation between occurrence time of earthquakes and those of emission: the strength of electromagnetic emission increase before and/or after earthquake (e.g., [20, 21] Fujinawa et al. [46], Parrot et al. [47]). Some reports suggest that MHz emission occur several weeks before and the VLF emission occurs only in the nucleation period [8, 9, 17] in contrary to the appearance of VLF type variation (C type) in this report. We do not have data to discuss the MHz event because of the detection frequency that is less than about 10 kHz. As for VLF emission we need supplementary data such as location information [46, 47] to know whether we are treating the same phenomena or not.

4.3. Detection Distance

The AE monitoring for rock burst is usually conducted with the spacing of several tens of meter (e.g., Hasegawa et al. [48]) suggesting low value of Q factor of about 3. They found there are two kinds of AEs, lower frequency ones of several hundred Hz and several kHz for monitoring rock burst for tunnel works. These two dominant frequency bands seem to correspond to those of B type and of C type of our observation. But we need more investigation at present to determine if there are any intrinsic reasons or not.

The detection distance of seismoelectromagnetic signals (SES) by the borehole antenna is suspected to be some three orders much larger than that of the acoustic emission method [25, 32]. Here we do not concern the radio wave in the atmosphere: they can propagate surrounding the earth. The long detection distance of the electric signal detected from the origin time [32] was assumed to be due to the high resolving power of the observation system and/or to the small attenuation rate of the seismoelectric variation itself. Another possibility is the conversion of the seismic wave to electric mode at the conductivity contrast (e.g., [31, 36], Gershenzon and Bambakidis [49]). The converted waves can propagate long distance because of power law dissipation compared with the exponential dissipation. The effective propagation speed on the order of 100 km/s or possibly electromagnetic speed in the earth is much larger than that of the diffusion spreading for the realistic conductivity of 0.1–0.001 Sm^{-1}. We can expect that the microcracks can be detected at a larger distance by the electromagnetic measurement of high quality than by the ordinary technique using acoustic emission.

4.4. Generation Mechanism

Fujinawa et al. [32] reported the remote detection of the Earth's electric field variations from the start of fault rupturing at about 100 km from the fault region using borehole antenna. Investigation of generation mechanism of the seismoelectromagnetic signals (SES) (e.g., [36, 49]) in the field indicates that the dominant mechanism is the electrokinetic effect compared with the piezomagnetic, piezoelectric, nonclassical piezoelectric, and induction effects [32]. A simple model of electric variation propagation in porous media using the basic Maxwell equations and electrickinetic interaction effects [25, 26] show that detective length reaches to several tens of kilometers.

The characteristic electric field variation induced by cracks through electrokinetic mechanism can be more systematically discussed on the basis of formulation of Pride [50] and Haartsen and Pride [51]. It has been shown by analytical approaches and numerical simulations that there are four kinds of wave, slow P wave and fast P wave (ordinary P wave), S wave (ordinary S wave), and electromagnetic wave (EM). The theory is extended to no-Newtonian fluid (Araji et al. [52]), and numerous application have been conducted on the interaction of seismic wave and electric field in the porous media (e.g., [52], Schoemaker et al. [53]). Han [37] tried to interpret the ULF geomagnetic variation induced by the slow slip by using formula of Pride [50] and fault model of Gao and Hu [54]. The A-type variation after the main shock can be interpreted as the electric field variation induced by slow movement of confined fluid though normal closure or opening in the process of rearrangement of strain or stress.

4.5. Phases

There are some events in B types which have clear primary phase preceding main phase as illustrated in Figure 11. The primary wave (P) has very small amplitude of some one-tenth to the secondary phase (S), and majority of events have no trace of primary wave. The feature is similar to the coseismic variation with the ordinary seismic wave [32]. The tendency is the same for the whole cases of detections in the field. But the feature is contrary to the simulation results (e.g., [51, 54], Ren et al. [55]). The electric strength induced by P-wave is much larger than that by S-wave. We can suppose that the tendency may change according to specimens of rocks [54, 55] or distribution of contrasts of sediment. Electromagnetic mode (EM) has been confirmed in the laboratory and simulation experiments [52, 53]. In the present observation there are occasionally small forerunners as illustrated in Figure 11. The arrival time difference E-S corresponds to the electric wave propagation velocity of some 900 times S wave velocity. But the EM phase is too vague to confirm the EM mode. The slow P wave has very small velocity of several m/s [52, 53] and has too slight strength hardly recorded in the present ability of recording in agreement with the numerical simulation results that the independently traveling EM. The wave has strength three order.smaller than those by seismic waves [54, 56].

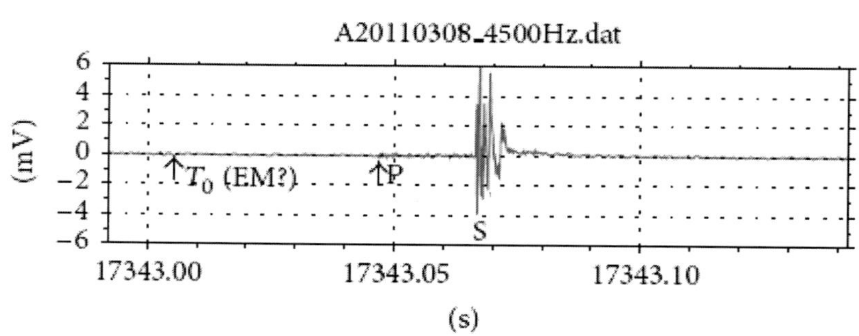

Figure 11: Sample waveform of events having the primary phase ("P" in the figure) before the main phase ("S" in the figure). The EM mode is suspected at around the origin time (EM) considering very large speed of more than one hundred times that of fast P phase.

Table 1: Catalog B-type Pulse 2011-03-032011-03-11.

Number	Second	Hour	Type	Strength (mV)	Duration (ms)	Comment
				3-March		
1	74089.15	20.58	B-2	−3.5	40	P phase?
2	76994.65	21.39	B-3	−5.5	30	P phase
3	77752.56	21.60	B-1	2	25	
				4-March		
4	63758.70	17.71	B-3	11.5	30	
5	64522.10	17.92	B-2	5	40	
6	65485.30	18.19	B-2	−11.5	10	
7	65937.20	18.32	B-2	12	10	
8	81148.30	22.54	B-2	−3.0	10	P phase
				6-March Till 14:00 (no-data)		
8	50571.1	14.05	B-1	2	30	
9	53094.25	14.75	B-1	3.8	40	
10	53336.8	14.82	B-1	16.0	40	
11	64179.30	17.83	B-1	3.3	40	Clear P phase
12	65995.55	18.33	B-2	6.5	14	
13	70869.25	19.69	B-2	3.5	20	
14	75902.15	21.08	B-2	9.5	25	
15	83206.30	23.11	B-3	−2.2	40	
16	84237.55	23.40	B-2	8.3	25	Subevents
				7-March		
16	16950.85	4.71	B-2	2	15	
17	20103.65	5.58	B-2	3.3	12	
18	64481.80	17.91	B-2	2.1	13	
19	68579.95	19.05	B-1	2.2	37	
20	69144.05	19.21	B-3	4.2	40	
21	69559.45	19.32	B-2	2.1	9	
22	79177.40	21.99	B-3	7.9	45	
				8-March		
23	17343.05	4.82	B-2	8.0	35	P, S phase
24	21232.45	5.90	B-3	12.4	45	P, S phase
25	21512.30	5.98	B-1	4.1	37	HF. At 1st step
26	26099.85	7.25	B-2	2.1	4	
27	26871.50	7.46	B-3	5.7	37	
28	31388.05	8.72	B-1	7.0	45	Subevent in coda
29	33307.20	9.25	B-1	4.1	60	P, S phase?
30	33624.30	9.34	B-1	8.5	60	
31	39163.85	10.88	B-1	5.1	43	
32	39727.80	11.04	B-1	4	40	
33	42297.30	11.75	B-2	2	7	
34	44458.00	12.35	B-2	2	15	
35	75881.30	21.08	B-2	2.2	8	
36	79821.30	22.17	B-3	5.8	30	
37	80090.35	22.25	B-3	−3.8	45	
38	80203.15	22.28	B-2	6.2	20	
39	82117.00	22.81	B-3	11.5	50	
40	82234.60	22.84	B-1	3	40	
41	82617.35	22.95	B-3	5.8	40	
42	83904.35	23.31	B-2	2.7	10	
43	83904.40	23.31	B-2	5	10	

44	85307.20	23.7	B-2	4.1	10	
45	85885.95	23.86	B-2	−11.7	25	
46	86122.75	23.92	B-3	−2.8	30	P, S phase?
47	86738.30	24.09	B-2	2.2	10	
48	86953.40	24.15	B-2	4.2	15	
49	87379.20	24.27	B-2	2.1	10	
50	87590.00	24.33	B-2	6.0	10	
51	89027.35	24.73	B-1	5	45	
				9-March		
52	2493.70	0.69	B-1	2.5	47	
53	3817.85	1.06	B-2	3.7	33	High freq.com at 1st step 1.5 kHz
54	4109.35	1.14	B-3	8.3	43	
55	5327.10	1.48	B-3	6	44	
56	5374.60	1.49	B-1	2.2	32	
57	6221.80	1.73	B-3	2.1	40	
58	6272.40	1.74	B-3	3.5	50	Complex
59	6454.95	1.79	B-3	5.4	60	
60	6684.75	1.86	B-3	8.9	30	P, S phase?
61	7091.50	1.97	B-3	18.3	40	P, S phase?
62	8501.15	2.36	B-1	3	30	
63	11404.10	3.17	B-1	3.5	35	P, S phase?
64	12040.80	3.34	B-1	3.5	40	P, S phase?
65	12185.25	3.38	B-1	8.0	20	
66	12983.65	3.61	B-1	7.5	50	P, S phase?
67	14067.05	3.91	B-2	6.3	30	P, S phase?
68	14206.80	3.95	B-1	5	40	P, S phase?
69	16891.85	4.69	B-3	11.3	30	P, S phase?
70	16892.00	4.69	B-2	4.1	10	
71	19113.05	5.31	B-2	4.5	10	P, S phase?
72	19198.05	5.33	B-1	3	35	P, S phase?
73	21171.70	5.88	B-1	2	30	
74	21657.55	6.02	B-1	3.5	40	
75	22136.65	6.15	B-1	4	35	
76	26479.25	7.36	B-1	3.5	35	
77	27832.15	7.73	B-1	2.2	40	
78	29191.10	8.11	B-1	3	30	
79	30824.20	8.56	B-2	2.2	20	
80	31511.55	8.75	B-2	2.2	10	
81	33641.20	9.34	B-2	4.3	10	
82	33649.35	9.35	B-1	2	35	
83	35323.25	9.81	B-1	4	45	
84	37181.60	10.33	B-2	−2.1	10	
85	39193.76	10.89	B-2	4	10	
86	42674.60	13.8	B-2	4	40	
87	54068.55	15.02	B-2	2	10	
88	55135.70	15.32	B-1	11.3	40	
89	66156.10	18.38	B-2	−5.1	10	
90	66775.25	18.55	B-2	5.1	30	
91	67365.35	18.43	B-1	4	40	
92	67574.65	18.77	B-2	3	27	Complex
93	67651.80	18.79	B-2	2.8	10	

94	67766.85	18.82	B-1	3.7	33	
95	67999.75	18.89	B-1	3	30	Most complex
96	68063.10	18.91	B-3	10.8	45	
97	68211.65	18.95	B-2	5	15	Most complex
98	68801.6	19.11	B-3	11.9	60	
99	69211.45	19.23	B-1	3.5	35	
100	69769.60	19.38	B-3	−4.0	35	
101	70573.30	19.6	B-1	3.1	35	
102	71162.05	19.77	B-2	2.9	28	Complex
103	71496.80	19.86	B-1	4.1	50	Subevent
104	71694.70	19.92	B-2	3.5	10	
105	71939.50	19.98	B-2	2.8	10	
106	72573.75	20.16	B-3	−4.2	25	Complex
107	72754.65	20.21	B-3	−16.0	40	
108	73382.65	20.38	B-2	4	10	
109	73409.05	20.39	B-2	2.2	8	
110	73619.15	20.45	B-1	3	40	Complex
111	73893.35	20.53	B-2	3.9	10	
112	74918.70	20.81	B-3	9.2	35	
113	75750.80	21.04	B-1	−4.0	30	Subevent
114	76353.05	21.21	B-2	9.7	10	Subevent
115	76732.30	21.31	B-2	3.9	10	Subevent
116	77958.25	21.66	B-2	4.3	15	
117	78709.60	21.86	B-3	10.5	40	
118	78908.15	21.92	B-2	−7.6	10	Typical B-2 type
119	79258.25	22.02	B-3	2	33	P? subevent?
120	80042.45	22.23	B-2	2.1	9	P? subevent?
121	82451.20	22.9	B-2	4	10	
122	82632.30	22.95	B-3	6.5	30	P? subevent?
123	83138.70	23.09	B-2	−10.1	10	
124	84541.40	23.48	B-3	11.0	30	
125	84795.30	23.55	B-3	6.5	35	
126	84841.35	23.57	B-1	4	30	Subevent
127	85341.90	23.71	B-1	4.4	34	
128	85352.70	23.71	B-2	6	10	
129	85374.55	23.72	B-1	3	45	
130	87677.00	24.35	B-1	5	38	P phase?
131	88837.15	24.68	B-2	3.1	10	
				10-March		
132	3800.15	1.06	B-2	4.1	20	P phase?
133	4262.10	1.18	B-2	2.2	22	P phase?
134	5869.60	1.63	B-2	3.7	15	Subevent
135	7020.15	1.95	B-2	2.5	15	
136	7870.35	2.19	B-2	2	10	
137	8116.80	2.25	B-2	2	10	
138	17156.75	4.77	B-3	12.0	40	
139	24115.35	6.70	B-1	5	50	
140	24348.45	6.76	B-1	5.5	37	
141	75609.20	21.00	B-2	10.5	35	P phase?
142	84056.55	23.35	B-2	2	15	

			11-March			
143	13276.85	3.69	B-1	3.5	45	
144	13400.30	3.72	B-1	2	35	
145	15506.5	4.31	B-3	4.2	45	P, S phase
146	18325.85	5.09	B-2	2.2	20	P, S phase
147	21801.45	6.06	B-2	4	20	P, S phase
148	22395.30	6.22	B-2	7.0	10	
149	22503.65	6.25	B-2	8.0	10	
150	22628.50	6.29	B-2	3	10	
151	22891.70	6.36	B-2	−7.0	15	
152	23432.95	6.51	B-2	5.5	10	P-S 40 ms
153	23485.85	6.52	B-2	−11.0	35	
154	25421.75	7.06	B-2	4	15	P, S phase
155	25658.20	7.13	B-1	12.5	35	
156	27287.35	7.58	B-1	2	35	
157	27319.50	7.59	B-2	−8.5	10	
158	27385.00	7.61	B-1	2	35	P, S phase
159	28639.15	7.96	B-1	3.2	36	
160	34295.7	9.53	B-2	2.6	20	
161	34564.55	9.60	B-2	4.1	10	P, S phase
162	36804.60	10.22	B-2	2.5	10	
163	45290.80	12.58	B-1	2	35	P, S phase

The S-P time of events A20110308_04h49m03s (Figure 11), A20110308 _23h55m 23s, A20110309_ 00h09m14s, and A20110309_0 1h51m25s (Table 1 in the appendix) are 20, 25, 17, and 14 ms corresponding to 160, 200, 133, and 109 m of epicentral distances, respectively. But almost all events have no or slight appearance of P phase. It is supposed that S-P times are larger than 50 ms corresponding to epicentral distance larger than a few km.

4.6. Complex Events

Just before the main shock occurrence there appeared some events superposed by numerous small events as illustrated in Figure 12. The relative occurrence time is considerably different with the different phase of the main events. It suggests that small event increased to occur in the neighborhood of larger events. Another phenomenon is that there appeared long chain of small events. We suspect that microcracks activity has changed at the last stage of nucleation phase corresponding to intermittent criticality causing fluctuation of cumulative curve fitted to the Ohmori's law. These complex and long chain events occurred only on 2 days before the main shock without appearance on 10th, 11th, March, that is, in the lull period of microcracks activity. The phenomena may be related with the intermittent criticality [6].

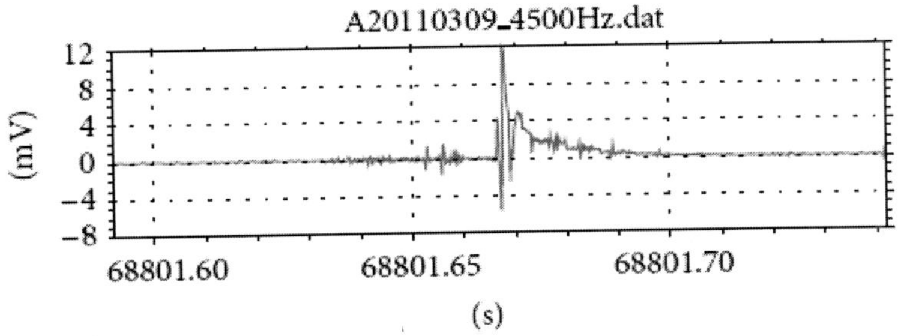

Figure 12: A sample waveform of the event of B-2 superimposed with many small events suggesting connection of groups of small scale cracks. This type appeared only on the most active day of microcrack, that is, on 9th March (Figure 8).

4.7. Intermittent Criticality

There appeared undulation of microcracks activity after the most active period around 9th March. Similar phenomena have been reported in laboratory experiments and field observation (e.g., [7–9, 23] Huang et al. [55]). The undulation has been suggested to reflect the intermittent criticality indicating entering another phase of nucleation to interpret the superposed undulation of the seismic activity as the case of the Loma Prieta Earthquake [7]. Another example concerns rapid change of magnetotelluric impedance (Honkura et al. [57]) some half hour before the Izmit earthquake (2000) in Turkey. Eftaxias and his group [8, 9] used their VLF radio wave emissions to investigate nucleation stage using their observation evidence of disappearance of the emission just before the main shock. They also tried to explain the IC based on fracture dynamics, critical phenomena, and percolation theory.

There is a disagreement in the VLF activity between some of previous results and present cases. It is reported that there is no VLF after the main shock in the report of [8, 9]. On the contrary we detect the C-type variation in frequency bands of VLF. We suppose that present events are suggested to be different from to the so-called VLF emission as those reported in [46, 47] and [7–9]. We need supplementary data to compare different observational results. For instance the location of the VLF wave rarely agrees with the epicentral region [46, 47].

Here we present the case that the activity is dominated by the B types in the final stage superimposed numerous complex microcracks, which suggests appearance of different regime of the nucleation process but without drastic phase transition.

4.8. Asperity

We suggest that the microcracks occurred in the southeastern peripheral of the great rupture of length 500 km in relation to the nucleation stage of the main rupture. We can imagine that the accelerated strain accumulation is not restricted to the so-called asperity region at around the epicenter of foreshock or main shock (Figure 4). Significant accumulation is suggested to occur at the whole area of future rupture with the result of microcrack activity though there is difference in the intensity or type of cracks. Extensive monitoring of the microcracks can reveal these points to provide a clue to answer this question.

5. CONCLUSION

We showed that the electromagnetic method can detect microcracks preceding natural earthquakes by a special underground antenna as an alternative to acoustic emission measurement. The nucleation stage of a crustal rupture can be identified by monitoring the electric field variations focusing on particular waveforms in the selected frequency ranges. These pulse-like phenomena are suggested to be plausible precursors for an imminent prediction.

Earthquakes are understood to be fracture of crustal rocks or stick slip of neighboring blocks. Preshocks occur generally in the rupture preparatory stage, and the activity has accelerated phase preceding main rupture. But that feature appears when averaged for many earthquakes. Preshock activity of individual earthquake is less regular to be used for confident predictive purpose.

Microcracks are always detected in the nucleation stage of accelerated deformation in almost all laboratory experiments. There are several efforts to investigate microcracks related to natural earthquakes by means of acoustic emission measurements. But they have not succeeded to find particular phenomena to identify the nucleation stage though the increase of acoustic emission intensity has been detected just half a day before and after main shocks.

We have been observing electromagnetic field changes for investigating the field characteristics related to earthquake activity and seismic wave in order for earthquake prediction and for application for early warning by using special underground antenna. The measurement method has been proved to be robust to the natural and man-made noises by observational investigation since 1989. Here we show that the instrument could detect microcracks appearing in the nucleation stage of the Tohoku Earthquake. Whole large pulse-like variations were grouped into three (A, B, and C) by frequency and waveforms and found to characterize different phases of the earthquake occurrence, the preparation, main shock, and after shock processes. Those variations are grouped by time constants: type A with duration of several to several ten minutes just the same as the pulse-like signals detected previously at the time of volcanic eruption and seismic swarms, B of several hundred Hz, and C of several kHz. The A type was identified in the early stage of observation at the time of volcanic eruption in 1992 and has been suggested to be induced by seismic swarms and volcanic eruption. The B type variations are subgrouped into three waveform: type B-1 of similar to the GUV except in the stepped decay in the relaxation phase after sharp rise, type B-2 of wave packet similar to acoustic emission, and type B-3 of superposition of B-1 and B-2. Different types of electric waveforms of the B type are suggested to correspond to crack rupture modes of tensile and shear and to be generated by confined water movement through the electrokinetic effects on the ground of previous investigations on laboratory experiments. The earthquake preparation period is characterized by the appearance of characteristic microcracks: beginning period of the nucleation period is characterized by rare occurrence of A types, nucleation period is by B types and particular evolution of the subgrouped types. The second kind of cracks (B type) is shown to occur only in the nucleation period: activity increases sharply before the earthquake following the modified Ohmori's law. A and C occur at the stress arrangement period after the main shock. Field monitoring of microcracks activity by EM method is shown to be practically possible to investigate the nucleation stage providing a break-through for the short term prediction method.

APPENDIX

Waveform Catalog of B-Type Variations

We picked up 224 events under the condition of larger event than three times of back ground level. All of the events are grouped in one of three groups without any undefined ones originating from urban and natural noises. It is very unique point of e-field observation by the borehole antenna. Among the three types the B type is the most interesting from the point of view of identification of the nucleation stage of natural earthquake. We present a catalog of whole B-type events consisting of 163 events.

Almost all events are compounded by B-1 and B-2. But events containing dominantly the tensile mode are grouped into B-1, and the shear modes are grouped into B-2. The grouping is done not quantitatively but qualitatively. The results of grouping exist in Table 1. The Table is constructed as follows:(1)column "Number," sequential number of events of the type B,(2)column "Second," occurrence time (JST) in second of the event,(3)column "Hour," occurrence time (JST) in hour,(4)column "Type," sub-group type of the B event,(5)column "Strength," peak height of the pulse, in milivolt,(6)column "Duration," width of pulse, in milisecond,(7)column "Comment," several features of waveforms: appearance of P phase, association of subevent, complex type, and so forth.

ACKNOWLEDGMENTS

This work was supported by JST Grant AS2121347A to OKI Engineering Co., Ltd. The authors thank T. Matsumoto and Y. Okada (NIED) for support to use the borehole and F. Freund and W. C Lai, K. Hattori, and Q. H. Huang for valuable discussions and comments. Detailed and valuable comments by Editor Professor J. P. Makris and reviewer K. Eftaxias are greatly appreciated. The authors used data of K-NET at CHB005.

REFERENCES

1. K. Mogi, Earthquake Prediction, Academic Press, Tokyo, Japan, 1985.
2. T. Rikitake, Earthquake Forecasting and Warning, Tokyo Center for Academic Publications, D. Reidel Publishing, Tokyo, Japan, 1982.
3. C. H. Scholz, The Mechanics of Earthquake and Faulting, Cambridge University Press, Cambridge, Mass, USA, 2nd edition, 2002.
4. L. M. Jones and P. Molnar, "Some characteristics of foreshocks and their possible relationship to earthquake prediction and premonitory slip on faults," Journal of Geophysical Research, vol. 84, no. 7, pp. 3596–3608, 1979. ·
5. M. Ohnaka, "A shear failure strength law of rock in the brittle-plastic transition regime," Geophysical Research Letters, vol. 22, no. 1, pp. 25–28, 1995. ·
6. K. Maeda, "Time distribution of immediate foreshocks obtained by a stacking method," Pure and Applied Geophysics, vol. 155, no. 2–4, pp. 381–394, 1999. ·
7. D. Sornette and C. Sammis:, "Complex critical exponents from renormalization group theory of Earthquakes: implications for Earthquake predictions," Journal de Physique I, vol. 5, no. 5, pp. 607–619, 1995.
8. P. Kapiris, K. Nomicos, G. Antonopoulos et al., "Distinguished seismological and electromagnetic features of the impending global failure: did the 7/9/1999 M5.9 Athens earthquake come with a warning?" Earth, Planets and Space, vol. 57, no. 3, pp. 215–230, 2005. ·
9. K. Eftaxias, S. M. Potirakisb, and T. Chelidzec, "On the puzzling feature of the silence of precursory electromagnetic emissions,"

10. X. Lei, K. Kusunose, M. V. M. S. Rao, O. Nishizawa, and T. Satoh, "Quasi-static fault growth and cracking in homogeneous brittle rock under triaxial compression using acoustic emission monitoring," Journal of Geophysical Research B, vol. 105, no. 3, pp. 6127–6139, 2000. r · ·
11. S. Yoshida, P. Manjgaladze, D. Zilpimiani, M. Ohnaka, and M. Nakatani, "Electromagnetic emissions associated with frictional sliding of rock," in Electromagnetic Phenomena Related to Earthquake Prediction, M. Hayakawa and Y. Fujinawa, Eds., pp. 307–322, Terrapub, Tokyo, Japan, 1994.
12. S. Yoshida, O. C. Clint, and P. R. Sammonds, "Electric potential changes prior to shear fracture in dry and saturated rocks," Geophysical Research Letters, vol. 25, no. 10, pp. 1577–1580, 1998. ·
13. S. Yoshida and T. Ogawa, "Electromagnetic emissions from dry and wet granite associated with acoustic emissions," Journal of Geophysical Research B, vol. 109, no. 9, Article ID B09204, 2004. r · ·
14. X. Lei, K. Masuda, O. Nishizawa et al., "Detailed analysis of acoustic emission activity during catastrophic fracture of faults in rock," Journal of Structural Geology, vol. 26, no. 2, pp. 247–258, 2004.r · ·
15. M. Hayakawa and Y. Fujinawa, Eds., Electromagnetic Phenomena Related to Earthquake Prediction, Terra Scientific Publishing Company, Tokyo, Japan, 1994.
16. M. Hayakawa, Ed., Atmospheric and Ionospheric Electromagnetic Phenomena Associated with Earthquake, Terra Scientific Publishing Company, Tokyo, Japan, 1999.
17. K. Eftaxias and S. M. Potirakis, "Current challenges for preseismic electromagnetic emissions: shedding light from micro-scale plastic flow, granular packings, phase transitions and self-affinity notion of fracture process,"
18. V. P. Drnevich and R. E. Gray, Acoustic Emissions in Geotechnical Engineering Practice, American Society for Testing and Materials (ASTM), Philadelphia, Pa, USA, 1981.
19. V. A. Morgunov, M. N. Lubashevsky, V. Z. Fubrizius, and Z. E. Fubrizius, "Geoacoustic precursor of Spitak earthquake," Volcanology and Seismology, vol. 4, pp. 104–107, 1991.
20. K. Hattori, "Results of acoustic emission observation at Matsushiro station, Japan," Final Report, International Frontier Research Group on Earthquakes (RIKEN IFREQ), Group Director Seiya Uyeda, 2003.
21. O. Molchanov, A. Schekotov, M. Solovieva et al., "Near-seismic effects in ULF fields and seismo-acoustic emission: statistics and explanation," Natural Hazards and Earth System Science, vol. 5, no. 1, pp. 1–10, 2005. ·
22. S. K. Park, M. J. S. Johnston, T. R. Madden, F. D. Morgan, and H. F. Morrison, "Electromagnetic precursors to earthquakes in the ulf band: a review of observations and mechanisms," Reviews of Geophysics, vol. 31, no. 2, pp. 117–132, 1993. ·
23. Y. Fujinawa, T. Kumagai, and K. Takahashi, "A study of anomalous underground electric field variations associated with a volcanic eruption," Geophysical Research Letters, vol. 19, no. 1, pp. 9–12, 1992. ·
24. Y. Fujinawa and K. Takahashi, "Electromagnetic radiations associated with major earthquakes," Physics of the Earth and Planetary Interiors, vol. 105, no. 3-4, pp. 249–259, 1998. ·
25. Y. Fujinawa, K. Takahashi, T. Matsumoto et al., "Electric field variations related with seismic swarms," Bulletin of the Earthquake Research Institute, vol. 76, pp. 391–415, 2001.
26. Y. Fujinawa, T. Matsumoto, and K. Takahashi, "Modeling confined pressure changes inducing anomalous electromagnetic fields related with earthquakes," Journal of Applied Geophysics, vol. 49, no. 1-2, pp. 101–110, 2002. r · ·
27. Y. Fujinawa and K. Takahashi, "Emission of electromagnetic radiation preceding the Ito seismic swarm of 1989," Nature, vol. 347, no. 6291, pp. 376–378, 1990. ·

28. K. Takahashi, Y. Fujinawa, T. Matsumoto et al., "An anomalous electric field variation with the seismic swarm(1)- underground electric field observation at Hodaka station (1995–1999)," Technical Notes 204, National Research Institute for Earth Science and Disaster Prevention, 2000,
29. B. Nourbehect, Irreversible thermodynamic effects in inhomogeneous media and their applications in certain geoelectric problems [Ph.D. thesis], Massachusetts Institute of Technology, Cambridge, Mass, USA, 1963.
30. V. Fitterman, "Electrokinetic and magnetic anomalies associated with dilatant regions in a layered Earth," Journal of Geophysical Research, vol. 83, pp. 5923–5928, 1978.
31. T. Ishido and J. Muzutani, "Experimental and theoretical basis of electrokinetic phenomena in rock-water systems and its applications to geophysics," Journal of Geophysical Research, vol. 86, no. 3, pp. 1763–1775, 1981. ·
32. Y. Fujinawa, K. Takahashi, Y. Noda, H. Iitaka, and S. Yazaki, "Remote detection of the electric field change induced at the seismic wave front the start of fault rupturing," International Journal of Geophysics, vol. 2011, Article ID 752193, 11 pages, 2011. r ·
33. W. Suzuki, S. Aoi, H. Sekiguchi, and T. Kunugi, "Rupture process of the 2011 Tohoku-Oki mega-thrust earthquake (M9.0) inverted from strong-motion data," Geophysical Research Letters, vol. 38, no. 7, pp. 1944–8007, 2011. r · ·
34. W.-C. Lai, K.-C. Hsu, C.-L. Shiehu et al., "Evaluation of the effects of ground shaking and static volumetric strain change on earthquake-related groundwater level changes in Taiwan," Earth, Planets and Space, vol. 62, no. 4, pp. 391–400, 2010. r · ·
35. T. Nagao, Y. Orihara, T. Yamaguchi et al., "Co-seismic geoelectric potential changes observed in Japan," Geophysical Research Letters, vol. 27, no. 10, pp. 1535–1538, 2000. ·
36. A. Revil and G. Saracco, "The volcano-electric effect," Journal of Geophysical Research, vol. 108, pp. 2251–2270, 2003.
37. P. Han, Investigation of ULF seismo-magnetic phenomena in Kanto, Japan during 2000–2010 [Ph.D. thesis], Chiba University, 2012.
38. J. R. Moore and S. D. Glaser, "Self-potential observations during hydraulic fracturing," Journal of Geophysical Research, vol. 112, Article ID B02204, 2007. r ·
39. M. A. Fenoglio, M. J. S. Johnston, and J. D. Byerlee, "Magnetic and electric fields associated with changes in high pore pressure in fault zones: application to the Loma Prieta ULF emissions," Journal of Geophysical Research, vol. 100, no. 7, pp. 12951–12958, 1995. ·
40. A. C. Fraser-Smith, A. Bernardi, P. R. McGill, M. E. Ladd, R. A. Helliwell, and O. G. Villard Jr., "Low-frequency magnetic field measurements near the epicenter of the Ms 7.1 Loma Prieta earthquake," Geophysical Research Letters, vol. 17, no. 9, pp. 1465–1468, 1990.
41. E. J. Sellers, M. O. Kataka, and L. M. Linzer, "Source parameters of acoustic emission events and scaling with mining-induced seismicity," Journal of Geophysical Research B, vol. 108, no. 9, pp. 1–16, 2003. ·
42. N. E. Timms, D. Healy, J. M. Reyes-Montes, D. S. Collins, D. J. Prior, and R. P. Young, "Effects of crystallographic anisotropy on fracture development and acoustic emission in quartz," Journal of Geophysical Research, vol. 115, Article ID B07202, 2010. r ·
43. X. Lei, K. Kusunose, M. V. M. S. Rao, O. Nishizawa, and T. Satoh, "Quasi-static fault growth and cracking in homogeneous brittle rock under triaxial compression using acoustic emission monitoring," Journal of Geophysical Research B, vol. 105, no. 3, pp. 6127–6139, 2000. ·
44. D. J. Varnes, "Predicting earthquakes by analyzing accelerating precursory seismic activity," Pure and Applied Geophysics, vol. 130, no. 4, pp. 661–686, 1989. r · ·
45. N. Yoshioka and C. H. Scholz, "Elastic properties of contacting surfaces under normal and shear loads: 2. Comparison of theory with experiment," Journal of Geophysical Research B, vol. 94, no. 12, pp. 17691–17700, 1989. r ·

46. Y. Fujinawa, K. Takahashi, T. Matsumoto, and N. Kawakami, "Experiments to locate sources of earthquake-related VLF electromagnetic signals," Proceedings of the Japan Academy B, vol. 73, no. 3, pp. 33–38, 1997. ·

47. M. Parrot, D. Benoist, J. J. Berthelier et al., "The magnetic field experiment IMSC and its data processing onboard DEMETER: scientific objectives, description and first results," Planetary and Space Science, vol. 54, no. 5, pp. 441–455, 2006. r · ·

48. O. Hasegawa, Y. Hosokawa, T. Mori, A. Takeichi, and H. Fujii, "Application of AE monitoring for rock burst and rock noise phenomenon under tunnel construction," Japan Civil Engineering Society, 2008.

49. N. Gershenzon and G. Bambakidis, "Modeling of seismo-electromagnetic phenomena," Russian Journal of Earth Sciences, vol. 3, no. 4, pp. 247–275, 2001.

50. S. Pride, "Governing equations for the coupled electromagnetics and acoustics of porous media," Physical Review B, vol. 50, no. 21, pp. 15678–15696, 1994. r · ·

51. M. W. Haartsen and S. R. Pride, "Electroseismic waves from point sources in layered media," Journal of Geophysical Research B, vol. 102, no. 11, pp. 24745–24769, 1997. ·

52. A. H. Araji, A. Revil, A. Jardani, B. J. Minsley, and M. Karaoulis, "Imaging with cross-hole seismoelectric tomography," Geophysical Journal International, vol. 188, no. 3, pp. 1285–1302, 2012. r · ·

53. F. C. Schoemaker, N. Grobbe, M. D. Schakel, S. A. L. de Ridder, E. C. Slob, and D. M. J. Smeulders, "Experimental validation of the electrokinetic theory and development of seismoelectric interferometry by cross-correlation," International Journal of Geophysics, vol. 2012, Article ID 514242, 23 pages, 2012.r ·

54. Y. Gao and H. Hu, "Seismoelectromagnetic waves radiated by a double couple source in a saturated porous medium," Geophysical Journal International, vol. 181, no. 2, pp. 873–896, 2010. r · ·

55. Y. Huang, H. Saleur, and D. Sornette, "Reexamination of log periodicity observed in the seismic precursors of the 1989 Loma Prieta earthquake," Journal of Geophysical Research B, vol. 105, no. 12, pp. 28111–28123, 2000. ·

56. H. X. Ren, X. F. Chen, and Q. H. Huang, "Numerical simulation of coseismic electromagnetic fields associated with seismic waves due to finite faulting in porous media," Geophysical Journal International, vol. 188, no. 3, pp. 925–944, 2012. r ·

57. Y. Honkura, M. Matsushima, Ş. Barış et al., "Rapid changes in the electrical state of the 1999 Izmit earthquake rupture zone," Nature Communications, vol. 4, article 2116, 2013. r ·

CHAPTER 7

Geophysical Contribution in the Characterization of Deep Water Tables Geometry (Sidi Bouzid, Central Tunisia)

D. Khazri and H. Gabtni

Laboratoire Géoressources, Centre de Recherches et des Technologies des Eaux (CERTE), Technopôle Borj-Cedria, BP 273, 8020 Soliman, Tunisia

ABSTRACT

Geophysical data combined with geological and hydrogeological data were analyzed to characterize the geometry of Oued El Hajel and Ouled Asker deep water tables (Sidi Bouzid). The obtained results allowed refining the geostructural schema by highlighting the individualization of the NE-SW underground convexity of Ouled Asker and the anticline of axis Es Souda-Hmaeima and Ezaouia on either sides of two hydrogeological thresholds. The geometrical analysis determined the spatial extension of Ouled Asker and Oued El Hajel subbasins. The seismic cartography of semideep and deep reservoirs (Oligo-Miocene; Eocene and upper Cretaceous) associated with the main subbasins contributed to proposing hydrogeological prospect zones for a rationalized groundwater exploitation.

1. INTRODUCTION

Agricultural development and rapid population growth contributed to an increasing demand for water resources. Consequently groundwater became a vital necessity in the Sidi Bouzid governorate in central Tunisia. This area is characterized by a semiarid to arid Mediterranean climate with irregular annual rainfall that does not exceed 350 mm/year and long periods of drought [1].

The Sidi Bouzid water table is one of the most important aquifers in Tunisia. However, the overexploitation of this aquifer requires a reassessment of deep water resources. Among these deep aquifers those of Oued El Hajel and Ouled Asker are of an ultimate interest. They are lodged into narrow NE-SW Atlasic synclines [2], situated near to Jebel Es Souda-Hmaeima on the eastern border of the Tunisian Mountains between the Atlasic block to the west and the Pelagian block to the east (Figure 1). The Jebel Es Souda-Hmaeima is a rather short Atlasic anticline which could be compared with a dome. According to Tanfous et al. [3] it is a "back folded anticline" affected by kilometric faults oriented NS, NW-SE, and EW [4].

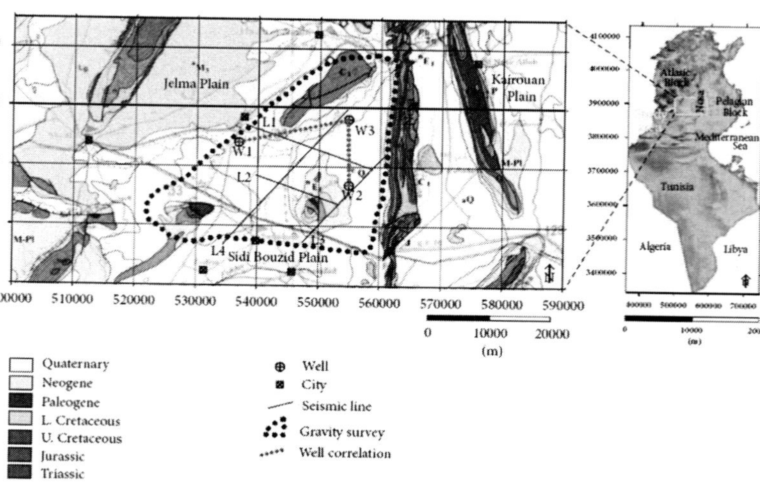

Figure 1: Location of the study area and the used geophysical data on the geological and structure map of central Tunisia [16].

In the present study, the characterization of Oued El Hajel and Ouled Asker water tables was developed and applied to deeper aquifers through the combination of geophysical, tectonic, and hydrodynamic data.

A detailed geometrical characterization of these subbasins by geophysical methods will limit each deep water table and help image the geometry of these aquifers such as the gravity survey method. In this study, gravity was selected in order to give a regional picture of the subsurface geology before making extensive surveys by the seismic reflection method.

For a better use of these deep aquifers and a rationalization of the future groundwater exploration in the arid area of central Tunisia, prognostic wells were realized based on seismic mapping and the estimation of water reservoirs depths.

2. GEOLOGICAL AND HYDROGEOLOGICAL SETTING

Geophysical and hydrogeological studies consist of an integration between modern geophysics and conventional methods (rainfall, piezometric, lithology, hydrodynamics...). To define the hydrogeological setting of the study area, a topographic map covering the Jebel Es Souda and its surroundings has been presented (Figure 2). The altitude varies between 260 m and 555 m. Jebel Es Souda-Hamaeima (540 m) is bordered to the north by Jebel Ezaouia, to the east by the N-S axis and Oued EL Hajel syncline, to the west by Ouled Asker syncline and Jebel Hamra, and to the south by the Kebar massif. The drainage system in this area is mainly characterized by the dominance of Oued El Hajel and its affluents [1].

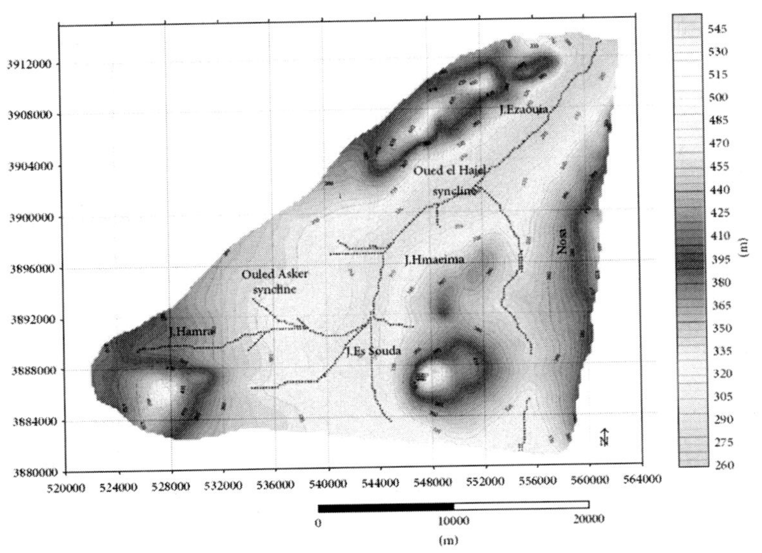

Figure 2: Topographic map of the study area.

Oued El Hajel and Ouled asker deep water tables belong to a zone characterized by a multiphase tectonic regime. The most important movements that affected this area were mainly compressive and extensional events [5] controlled by megafaults oriented NE-SW, N-S, and NW-SE. The main aquifers associated with these deep water tables can be recognized and described in the flank of Es Souda-Hmaeima anticline (Figure 1). Oued El Hajel syncline is limited to the NNW by the anticline of Ezaouia-Roua, the N-S axis to the East and Jebel Es Souda, and the Kasserine fault to the South. The aquifers corresponding to the Oued El Hajel syncline are mainly made of Mio-Pliocene and Quaternary sandstone. Its thickness varies from 1000 m to 1150 m: the first reservoir level reaches a depth of 100 m and it consists of a shallow Quaternary water table formed by an alternation of sands and clays. The second reservoir refers to a semideep water table and it is mainly made of Oligo-Miocene sand and sandstone. Its depth varies between 600 m and 800 m. The Miocene sandstones are characterized by an annual exploitation value of 0,86 Mm^3/year with a continuous fictitious debit of 27,4 L/s and a global salinity that reaches 3,90 g/L. It must be mentioned that the Upper Cretaceous carbonates were reached in a depth of 1150 m in the Garaât Ben nour hydrolic well and are considered as the deepest aquifer of the Oued El hajel water table (W3, Figure 1).

The Ouled Asker water table is a part of Ouled asker syncline which is bordered by the anticlines of Jebel Mghila to the North-West, Jebel Labaiedh to the North, and Jebel Ezaouia to the North-East and to the south by Jebel Hamra. It is a multilayer aquifer system containing Pliocene-Quaternary shallow aquifers that reach 40 m of thickness. Semideep aquifers are also present in this area and they are made of Oligo-Miocene sand and sandstone. The deepest aquifers of this water table are mainly made of Upper Cretaceous-lower to middle Eocene carbonate.

The aquifers of Oued El Hajel and Ouled Asker deep water tables are made of Beglia formation (Langhian-Serravallian), Ain Grab formation (Langhian), Fortuna formation (Oligocene), Metlaoui (lower Middle Eocene), and Abiod formations (Campanian- Maastrichtian) (Figure 3).

3. MATERIAL AND METHODS

3.1. Gravity Data

The gravity data that has been used for the current study were obtained from the Office National des Mines (ONM). The free-air and Bouguer corrections were made using the sea level as reference and the average density was fixed to 2.4 g/cm^3. In order to elaborate the complete Bouguer map of the zone, a grid was performed using the gravity data with a spacing of 1 point per Km^2.

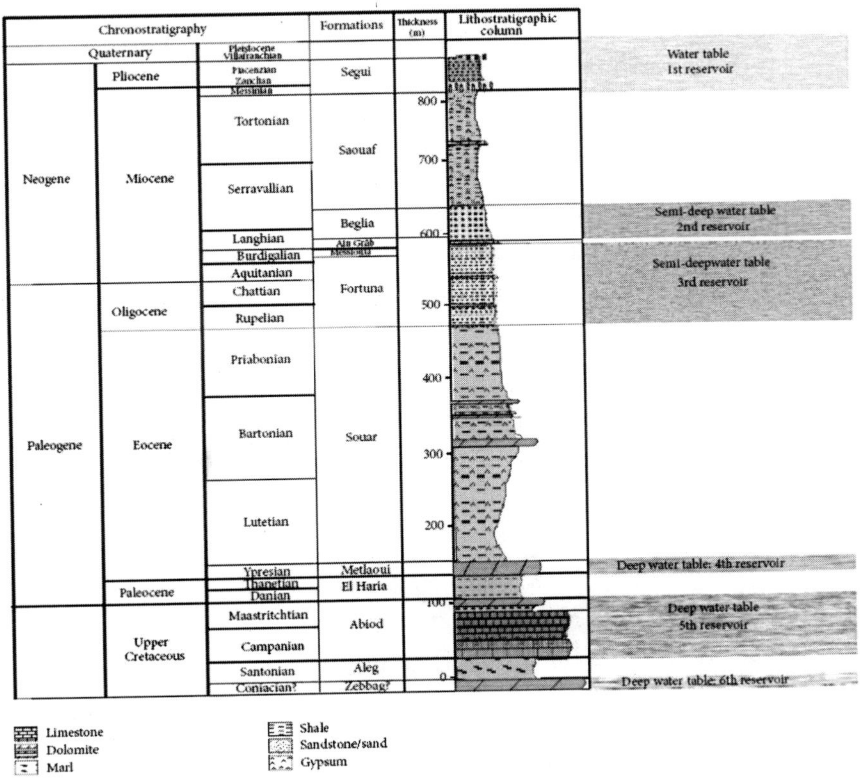

Figure 3: Lithostratigraphic column of Jebel Es Souda modified from [17].

3.2. Seismic Data

The seismic acquisition in the study area was conducted by the General Geophysical Company (GGC) for the benefit of the "Union Texas Tunisia" company in the region of Kasserine from March to August 1981. Four seismic lines L1, L2, L3, and L4 were analyzed and interpreted in this work. The latter crosses the main structures of the study area such as Oued El Hajel and Ouled Asker synclines which contain the target deep water tables (Figure 1).

4. RESULTS AND DISCUSSIONS

4.1. Gravity Analysis

(i) Bouguer and Residual Gravity Maps. Based on the gravity data, a Bouguer gravity map was performed in which the gravity anomaly varies between −46 mGal and −10 mGal (Figure 4). The study area shows a structural complexity illustrated through different gravity axes: a negative gravity axis to the extreme west and a positive gravity axis to the extreme east. This is related to the thinning of the earth's crust in Tunisia that was proved by the deep seismic refraction company Geotraverse EGT85 [6].

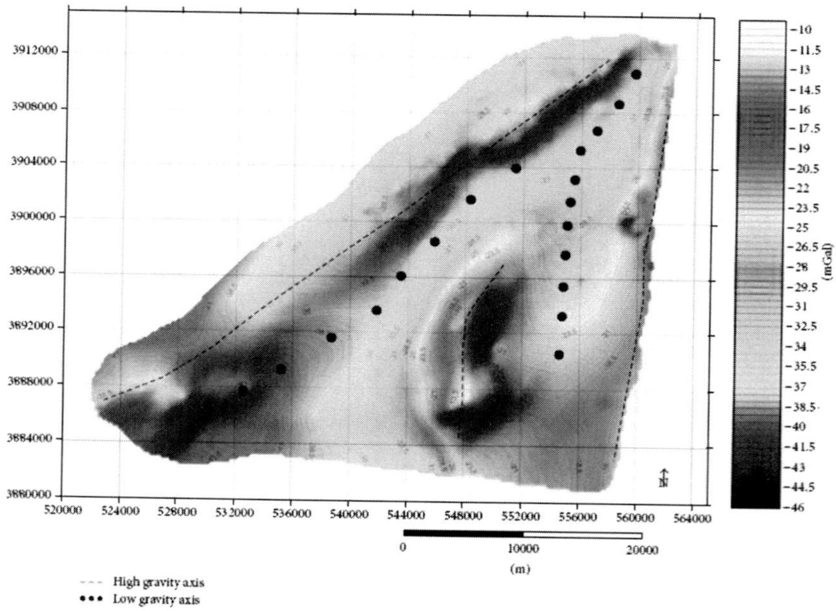

Figure 4: Bouguer gravity map of the study zone showing the positive and the negative trends of the anomaly.

These anomalies are separated by accentuated gravity gradients that indicate the presence of several discontinuities in the subsurface. These discontinuities are controlling the geometries of the different structures in this study area. The Bouguer gravity map illustrated in Figure 4 shows the subsurface structural complexity. Several structures are defined such as anticlines, synclines, major faults, and Triassic outcrops. The principal anomaly axes are illustrated as follows from the west to the east.(a)The first

NE-SW positive anomaly axis goes to −33 mGal and it corresponds to Jebel Hamra and Jebel Ezaouia.(b)The second N-S to NNE-SSW positive axis reaches −18 mGal and it corresponds to Jebel Es Souda-Hmaeima.(c)The third N-S positive axis corresponds to the N-S axis lineaments and it reaches −13 mGal.

The principal negative gravity axes are from the west to the east as follows (Figure 4).(a)A first NE-SW negative gravity axis delineates the Ouled Askar syncline and it reaches −46 mGal.(b)The second negative axis reaches −31 mGal with a N-S to NNE-SSW major direction and it corresponds to the Oued El Hajel syncline.

Several upward continuations were performed based on the Bouguer gravity grid of the studied area. At a depth of 8000 m we notice the persistence of particular anomaly sources which are considered as the deepest and most rooted. Also a gravity gradient was identified separating both the negative and the positive anomaly axes which are almost vertical, and isogal curves were well drawn. Thus, the regional gravity maps were chosen (Figure 5). The gravity gradient observed in the regional gravity map delineates two gravity domains: (1) a positive domain corresponding to the Jebel Es Souda-Hmaeima due to the deep structures affecting the ante-Triassic levels; these results confirm those determined by Tanfous et al. [3], and (2) a negative domain covering the Ouled Asker syncline with persistence of a less important negative anomaly (Oued El Hajel).

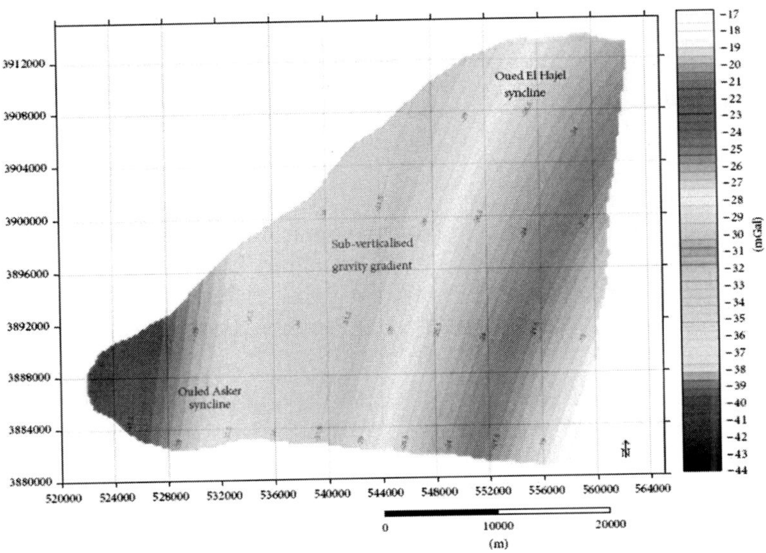

Figure 5: Regional gravity anomaly map of the study area.

Therefore the regional gravity anomaly will be subtracted from the Bouguer gravity anomaly in order to obtain the residual gravity map (Figure 6). The emplacement of the different gravity anomalies on this map coincides with the different geological lineaments determined in the surface. The anomaly values vary from −9 to 15 mGal and the major anomaly axes are NE-SW and N-S to NNE-SSW (Figure 6).

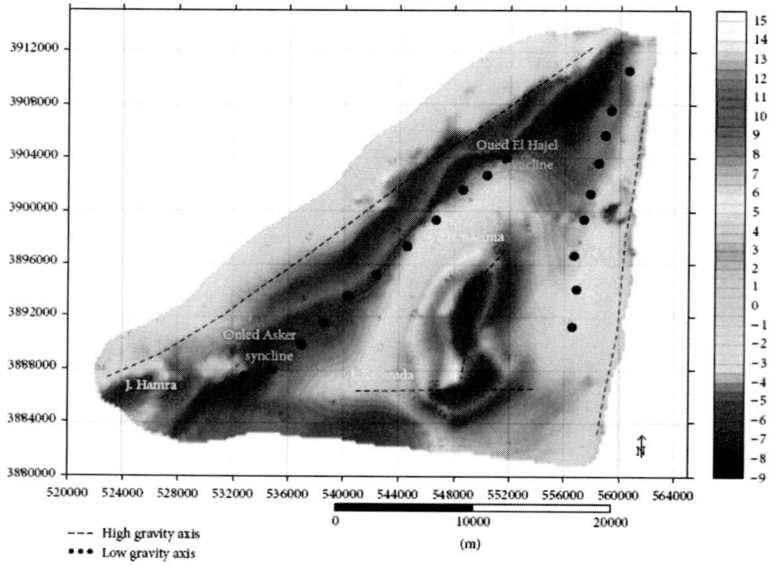

Figure 6: Residual gravity map of the study zone.

(ii) The Horizontal Gravity Gradient Magnitude Map (HGGM). The performance of the horizontal gravity gradient magnitude map will allow us to trace the emplacement of the different anomaly sources and causative bodies [7]. In fact, this map highlights the different faulting systems affecting the study area for a better understanding of the structural and tectonic movements that affected the zone.

The major lineaments and principle accidents affecting the Jebel Es Souda-Hmaeima and its neighboring structures will be highlighted using the HGGM map. These accidents affect the different hydrogeological reservoirs and water tables of Ouled Asker and Oued El Hajal according to two main directions: a first NE-SW to NNE-SSW direction associated with Jebel Hamra-Jebel Ezaouia, to Jebel Es Souda-Hmaeima, and to the underground bulge of Ouled Asker. A second N-S main direction associated with the N-S axis and the southern part of Jebel Es Souda. Other minor E-W directions were also recorded (Figure 7).

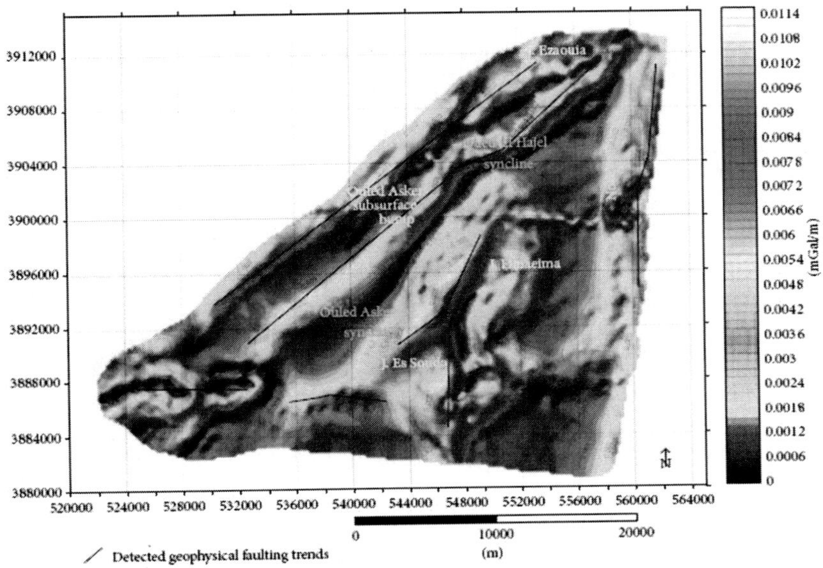

Figure 7: The structural lineaments after the horizontal gradient gravity magnitude map (HGGM).

For a better subsurface architecture visualization of this area, gravimetric profiles such as the HGGM and the residual gravity anomaly were performed and coupled along with a hydrostratigraphic section. The latter was obtained by correlating the hydrologic wells W1, W2, and W3 (Figure 1) in order to affirm the existence of faults affecting the Oued El Hajel and Ouled Asker basins.

The hydrostratigraphic section (Figure 8) illustrates the lateral variation of the different geologic units and it confirms the results earlier determined through the gravimetric profiles. In fact, it shows horst and graben structuring affected by normal and inverse faulting systems. This will be better explained and illustrated in the seismic study of this area.

(iii) Estimation of the Euler Solutions. The Euler deconvolution is a geophysical technique that consists of a qualitative and a quantitative analysis of gravity data. It highlights the main discontinuities affecting the region in the subsurface in order to detect the contact zone affecting the hydrogeological reservoirs [8]. The deepest accidents detected are of NE-SW and N-S directions with depth values reaching the 3500 m. Using this method it was also possible to confirm the presence of N-S fault corridor (Figure 9) to the west of the N-S axis [9, 10] and continues to the south to Sebkhat Enoual near Meknassy-Mezzouna [11]. Other NE-SW lineaments were detected over the Ouled Asker bulge with depth values going from 2000 m to 3500 m. These lineaments separate the subbasins of Ouled Asker and Oued El Hajel.

4.2. Seismic Analysis

The different seismic profiles were calibrated and interpreted using previous studies [12–15], outcrop data, field geology and also by an interpolation with the neighboring hydraulic wells. The results of those interpretations were used to locate the seismic horizons susceptive of being the most important aquifers in the study area such as the following:(i)the carbonate horizons of the Abiod formation (Campanian-Maastrichtian) and the Metlaoui formation (Ypresian);(ii)the sandstone horizon of the Oligocene Fortuna formation;(iii)the sandy and conglomeratic Middle Miocene (Langhian-Serravallian) horizons of the Beglia formation.

(i) The First Seismic Line (L1). It consists of a NW-SE seismic section crossing the Jilma-Ouled Asker basin and Jebel Hmaeima structure and reaches the Oued El Hajel syncline. The North western part of this seismic section shows intense fracturing of the different geologic units that delineates elevated and collapsed zones. This tectonic regime affected the flanks of the different anticlines in the area. The latter was the result of successive compressive phases (Upper Cretaceous-Eocene; Miocene until Villafranchian phase).

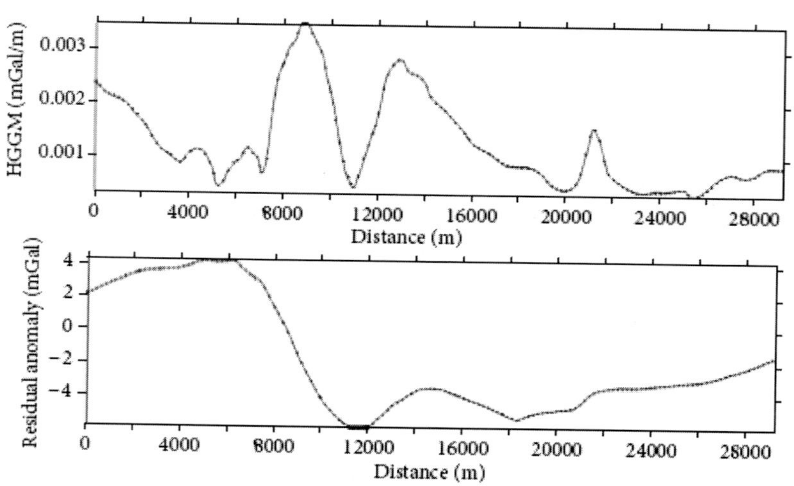

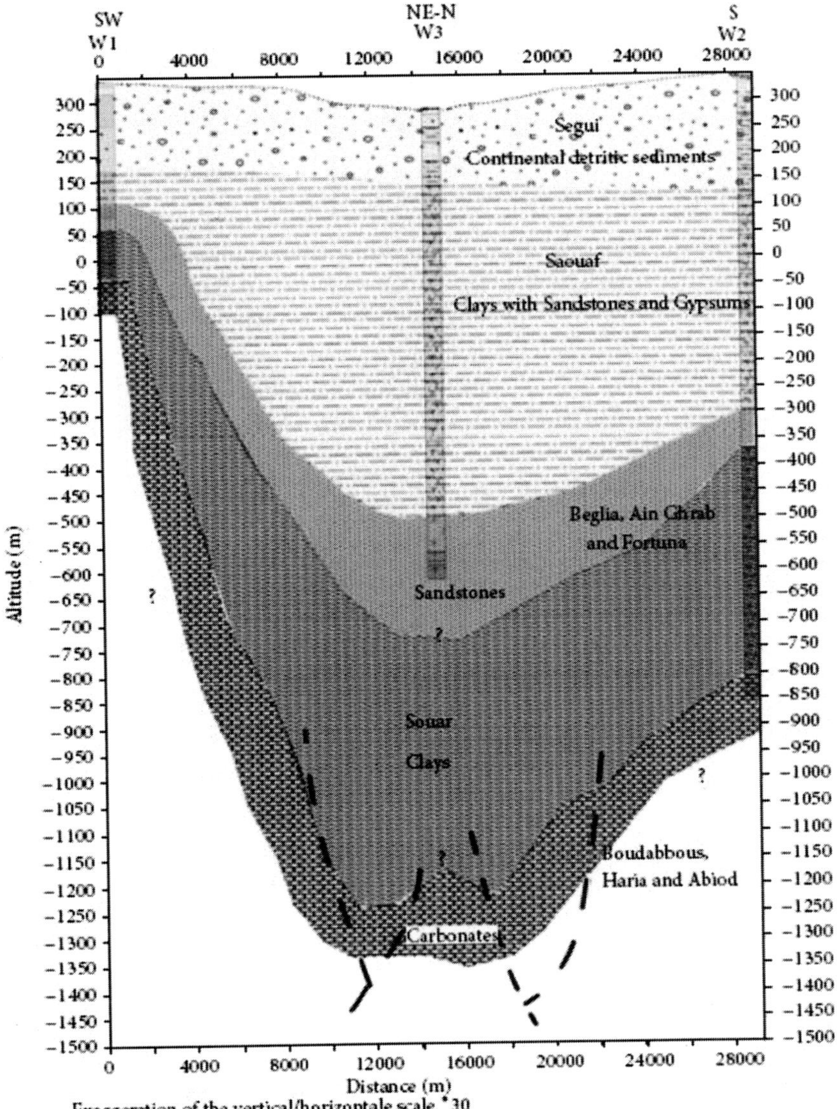

Figure 8: Subsurface structure of the study area using a hydrostratigraphic cross sections linking hydraulic well (W1, W2, and W3), horizontal gradient gravity magnitude, and residual gravity anomaly profiles.

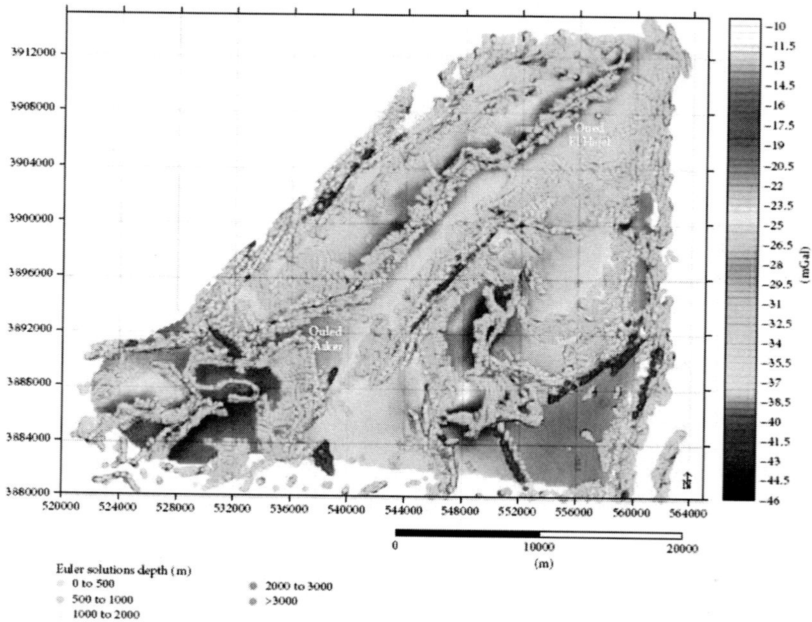

Figure 9: Euler solution to detect contact zone (structural index 0, window 1010, and error 15%).

In this seismic section, box folds are identified as well as tilted blocks due to a set of major listric faults. Significant negative flower structures were recognized in the area resulting from grafting of second order faults on the major listric faults mentioned previously. Also horst and graben system were associated with this tectonic set. In the South eastern seismic line part, a decrease in the intensity of fracturing and thickening of geological layers is noted. Also an important gutter structure covering the Oued El Hajel syncline which is a basin marked by an intensive subsidence was identified (Figure 10, L1).

(ii) The Second Seismic Line (L2). To the North-west of this profile a gutter structure is individualized corresponding to Ouled Asker syncline, which is bordered to the west by the bulge of Ouled Asker limiting the extent of this basin as compared to Oued El Hajel syncline. To the SE, the sedimentary series are beveled marking the thinning of geological layers in this area (Figure 10. L2).

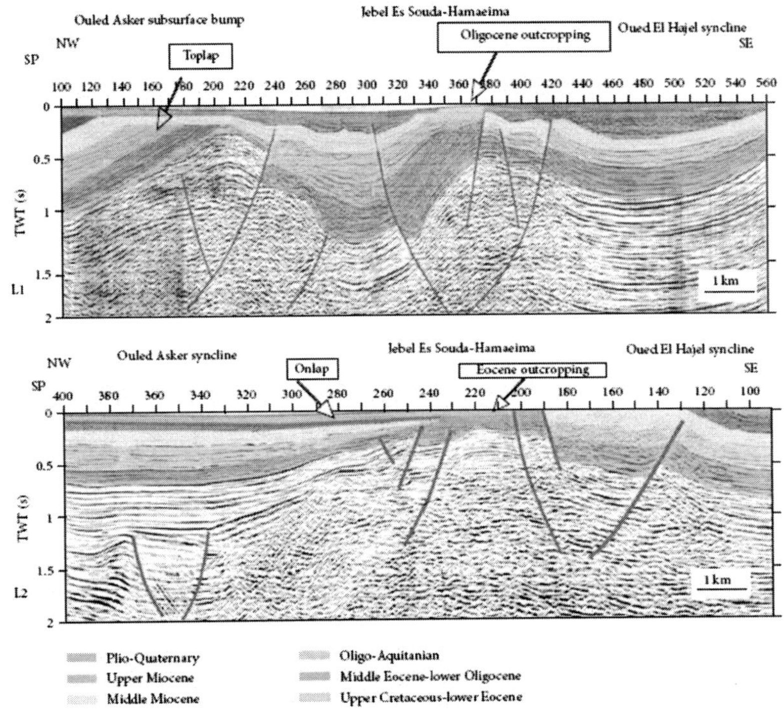

Figure 10: Interpreted seismic lines across the study area (L1 and L2).

(iii) The Third Seismic Line (Line L3). The South western part of this seismic section is marked by oblique to subparallel configuration and also by a thinning of geological layers marking the transition from a subsident domain occupied by Oued El Hajel syncline to a folded domain. Several pop-up structures (at the 370 shot point) mark this domain resulting from the action of listric faults uplifting the central compartments and sinking those borders (Figure 11, L3).

(iv) The Fourth Seismic Line (L4). The South western part of this seismic section is noticeable; it is characterized by a fan configuration related to a negative flower structure in the center of Ouled Asker syncline previously detected in the first seismic line. In this structure the sedimentary series mark a significant thickening as compared to the Es Souda anticline flank (Figure 11, L4). This flower structure is of a particular hydrogeological interest regarding its geometry and its groundwater flow direction. In addition, this structure provides a vertical groundwater flow supplying the deep aquifers of the region through the permeable layers and the deepest faults. After being confined the deep geological layers in the center of Ouled Asker basin can accumulate highly saline water (charged water), while, in the edges of the basin and to shallower depths, the water is less charged and the flow is said transient (Figure 12).

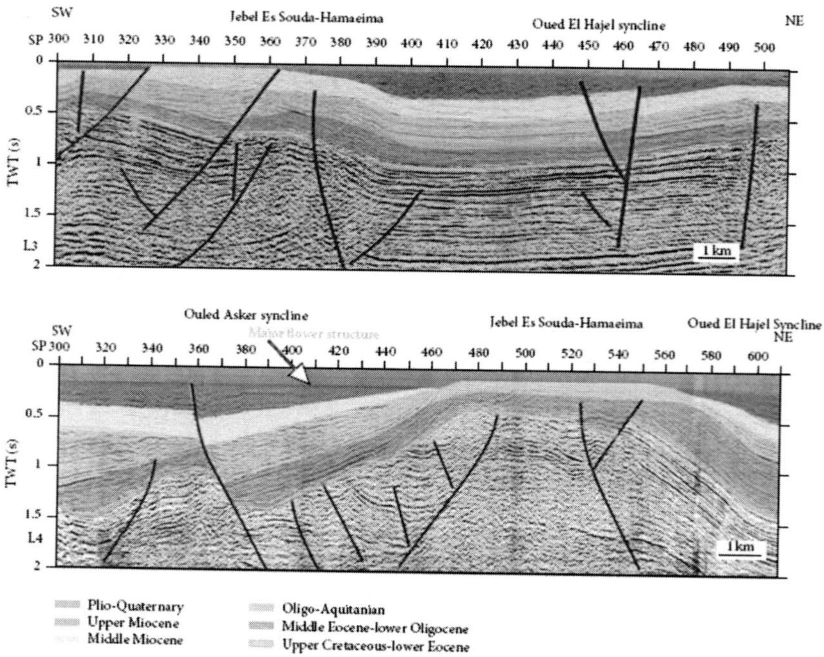

Figure 11: Interpreted seismic lines across the study area (L3 and L4).

Interval velocities calculated using the Dix formula (1955) were adopted, and profiles of average velocities were established for both lines 3 and 4 to estimate the depths of target aquifers and to provide several fictitious well (L3: W380; W480 and L4: W400; W460) in the center and edges of Oued El Hajel and Ouled Asker basins facilitating the exploitation of the main deep aquifers in this study (Figures 13 and 14).

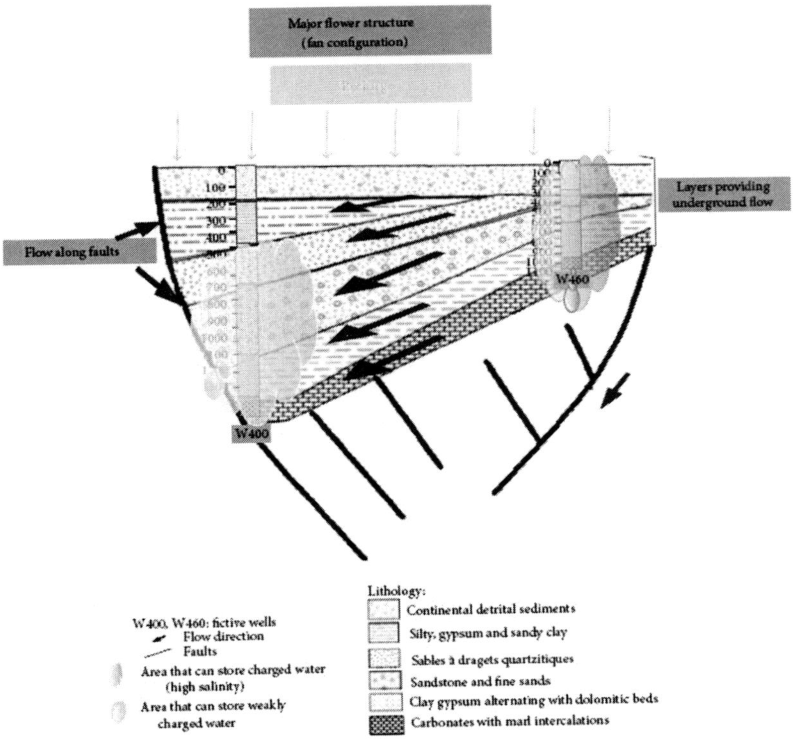

Figure 12: Hydrogeological sketch of Ouled Asker basin with W400 and W460 fictive wells.

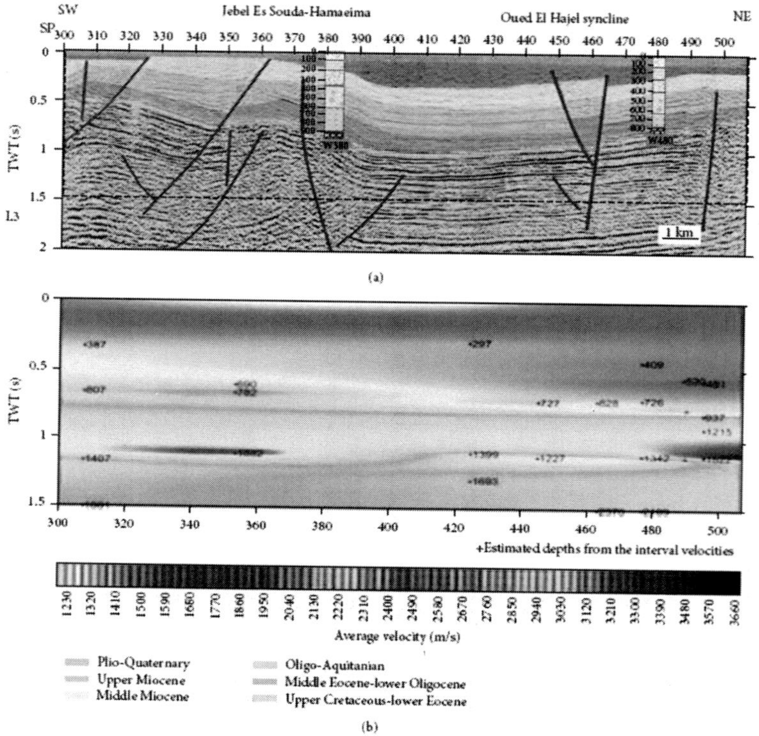

Figure 13: (a) W380 and W480 fictive wells associated with the L3 seismic line and (b) average velocity variation associated with L3 (black dashed line frame). (+) Depths estimated from the interval.

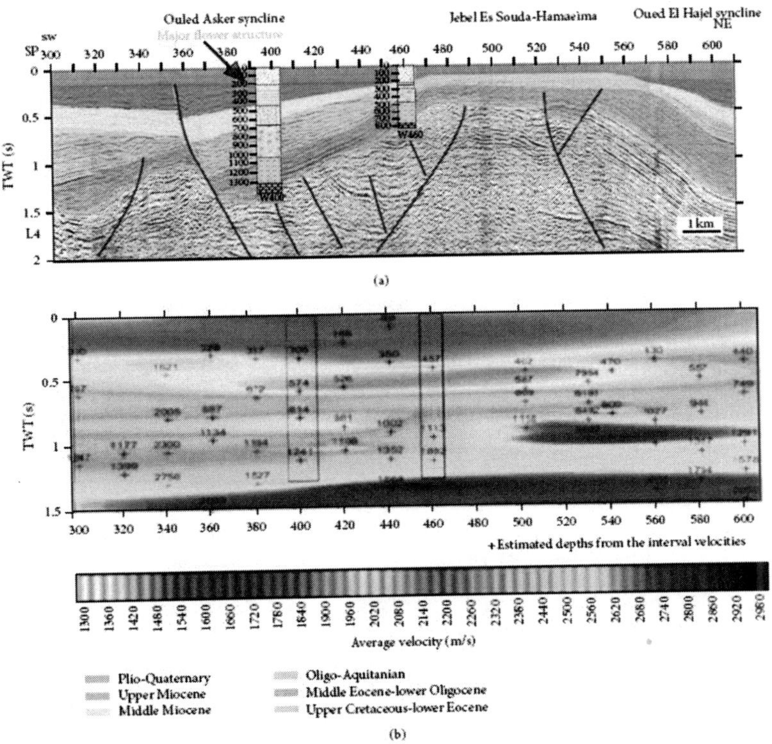

Figure 14: (a) W400 and W460 fictive wells associated with the L4 seismic line and (b) average velocity variation associated with L4 (black dashed line frame). (+) Depths estimated from the interval velocities.

Concerning the third seismic line L3 and after the distribution of depths relative to the average velocities, at a double time from 0.8 s to 1 s, the Upper Cretaceous is characterized by a velocity variation from 2700 (m/s) to 3030 (m/s) for depths up to 1215 m and at a less important double time from 0.3 s to 0.5 s, the velocities variations extend from 1600 (m/s) to 2000 (m/s) with depths ranging from 387 m to 520 m. It is also noted that the change in average velocities is proportional to the lateral variation of sedimentary sequences being structured in moderately folded to the SW higher areas, against sunken, and subsidence areas to the NE corresponding to the Oued El Hajel gutter (Figure 13). This structure has been confirmed by seismic mapping tool Isochronous and Isopaque showing the geometry of deep groundwater reservoirs and target horizons and thus to define their depth-time and thicknesses-time.

(v) 3D Isochronous Model of Semideep and Deep Aquifers in Oued El Hajel and Ouled Asker Basins. After analyzing the seismic lines, three target aquifers were detected. In order to provide concrete and realistic exploitation results for Oued El Hajel and Ouled Asker deep aquifers, a 3D mapping of these aquifers (first aquifer, second aquifer, and third aquifer) was performed to obtain a general subsurface model of the study area with folded and uplifted zone against collapsed basins forming subsiding synclines interpreted as potential areas of groundwater convergence and accumulation.(a)The first aquifer consists of a siliciclastic reservoir (Langhian-Serravallian): Ain Grab and Beglia formations (Figure 15).(b)The second aquifer is a sandstone reservoir (Oligocene-Lower Miocene (Aquitanian)): Fortuna formation (Figure 16).(c)The third aquifer consists of a carbonate reservoir (Upper Cretaceous: Campanian- Maastrichtian): "Abiod formation," and Eocene: Ypresian "Metlaoui formation" (Figure 17).

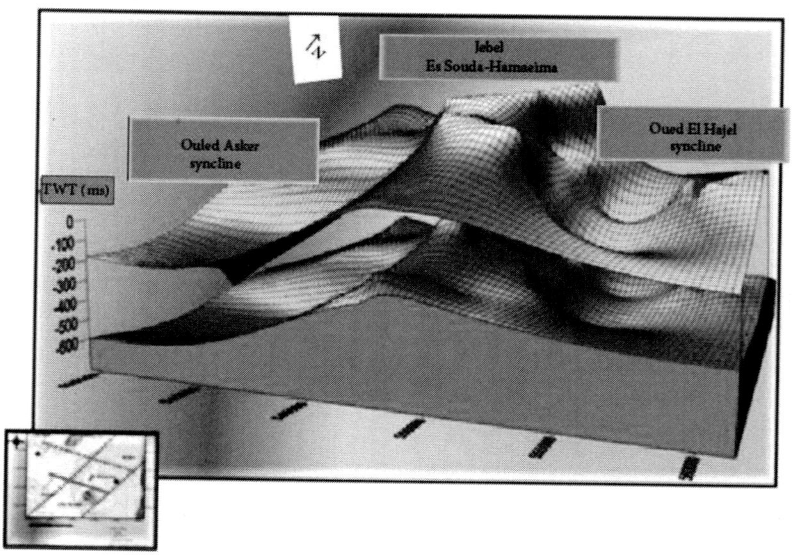

Figure 15: 3D visualization of the Langhian-Serravallian (Middle Miocene) reservoir.

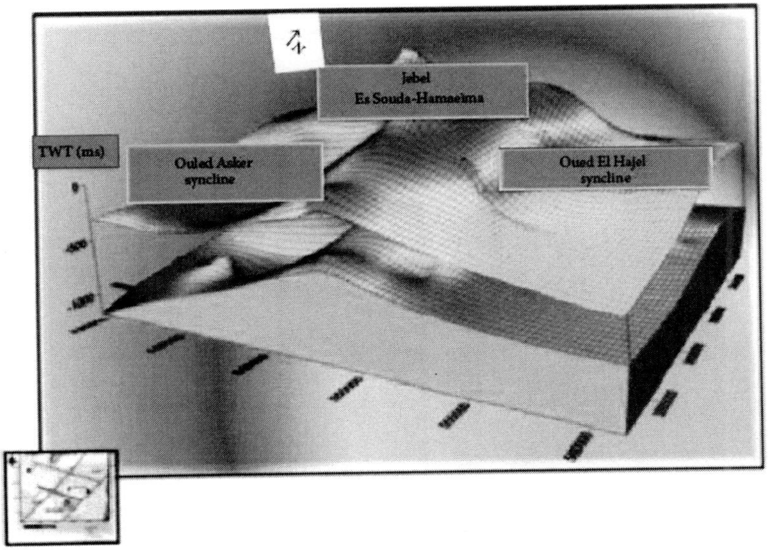

Figure 16: 3D visualization of the Oligocene-Aquitanian (Lower Miocene) reservoir.

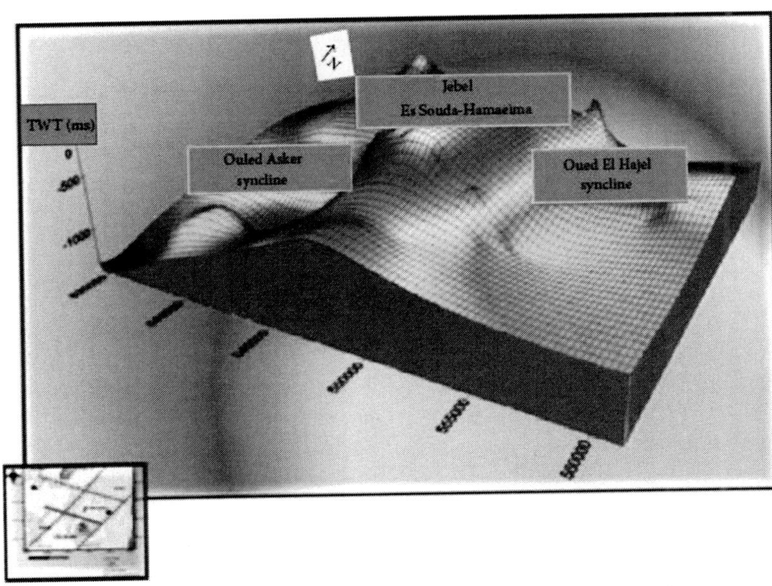

Figure 17: 3D visualization of the Upper Cretaceous reservoir.

5. CONCLUSION

The integration of the geophysical, geologic, stratigraphic, and well data leads to the following.

(i) The geometrical characterization of the Oued El Hajel and Ouled Asker water table with NE-SW, N-S, and E-W lineaments.

(ii) The identification of target aquifers for the hydrogeological prospection which was based on the seismic analysis and interpretation. The obtained results helped in identifying the different units forming the aquifers: limestone levels (Upper Cretaceous to Lower Eocene)—sandstones and sand levels from the Oligo-Miocene to the Mio-Plio-Quaternary. In fact, the study area is mainly formed of(a)large gutter structures characterized by a differential encasement from Oued El Hajel to Ouled Asker,(b)strait anticlines and an overlapping tendency along the NE-SW à NNE-SSW fault corridors; based on the integration of different geophysical methods, it was possible to conclude that the study area is mainly formed by horst and graben structures, box folds and, also several flower structures with fan-shaped configuration. The latter present an important hydrogeological interest. The study area is also affected by an important faulting system with deep rooted faults that implicated the individualization of depressed area presenting potential zone in which it was possible to suggest a number of fictive wells.

CONFLICT OF INTERESTS

The authors declare that there is no conflict of interests regarding the publication of this paper.

ACKNOWLEDGMENTS

The research was supported by Sahel-Kairouanais project (CERTE 2010–2013, resp., Pr. M. Bedir). The authors are very grateful to ONM and ETAP for the scientific supports.

REFERENCES

1. H. Smida, Apports des Systèmes d'Informations Géographiques (SIG) pour une approche intégrée dans l'étude et la gestion des ressources en eau des systèmes

aquifères de la région de Sidi Bouzid (Tunisie centrale) [Ph.D. thesis], Université de Sfax, 2008.

2. P. F. Burollet, Contribution à l'Étude Stratigraphique de la Tunisie Centrale, Annales des Mines et de la Géologie de Tunisie no. 18, 1956.
3. D. Tanfous, H. Gabtni, H. Azaiez, M. Soussi, and M. Bedir, "Integrated gravity and seismic investigations over the Jebel Es Souda-Hmaeima structure: implication for basement configuration of the eastern frontal fold-and-thrust belt of Tunisian Atlasic Mountains," Arabian Journal of Geosciences, vol. 5, no. 3, pp. 513–524, 2012. r · ·
4. A. Kadri, Evolution Tectonosedimentaire (Aptien-Quaternaire) des DJ. Koumine, Hamra et Lessouda (Tunisie Centrale), Université de Paris-Sud Centre D'Orsay, 1988.
5. S. Bouaziz, E. Barrier, M. Soussi, M. M. Turki, and H. Zouari, "Tectonic evolution of the northern African margin in Tunisia from paleostress data and sedimentary record," Tectonophysics, vol. 357, no. 1–4, pp. 227–253, 2002. r · ·
6. H. Buness, P. Giese, C. Eva et al., "The EGT'85 seismic experiment in Tunisia: a reconnaissance of the deep structures," in Proceedings of the 6th Workshop on the European Geotraverse Project, Data compilations and synoptic interpretation, R. Freeman and S. Muller, Eds., pp. 197–210, European Science Foundation, Strasbourg, France, 1989.
7. R. J. Blakely and R. W. Simpson, "Approximating edges of source bodies from magnetic or gravity anomalies," Geophysics, vol. 51, no. 7, pp. 1494–1498, 1986. r · ·
8. A. B. Reid, J. M. Allsop, H. Granser, A. J. Millett, and I. W. Somerton, "Magnetic interpretation in three dimensions using Euler deconvolution," Geophysics, vol. 55, no. 1, pp. 80–91, 1990. r · ·
9. N. Boukadi, Structuration de l'Atlas de Tunisie: signification géométrique et cinématique des noeuds et des zones d'interférences structurales au contact de grands couloirs tectoniques [Thèse d'Etat], Université de Tunis II, Tunis, Tunisia, 1994.
10. H. Azaïez, H. Gabtni, I. Bouyahya, D. Tanfous, and M. Bédir, "Lineaments extraction from gravity data by automatic lineament tracing method in Sidi Bouzid Basin (Central Tunisia): structural framework inference and hydrogeological implication," International Journal of Geosciences, vol. 2, no. 3, pp. 373–383, 2011. r ·
11. H. Gabtni, Caractérisation profonde et modélisation géophysique des zones de transition entre les différents blocs structuraux de la Tunisie centro-méridionale [Thèse de Doctorat], Université de Tunis El Manar, 2006.
12. C. Doglioni, A. Bosellini, M. C. Frare, F. Dhaha, and E. A. Ben Said, "Aspects tectoniques de la region au sud-ouest de Kairouan (Tunisie centrale)," Estratto da: Annali dell' Università di Ferrara (Nuova Serie). Science Terra, vol. 2, no. 5, pp. 77–94, 1990.
13. M. Bédir, Mécanismes géodynamiques des bassins associés aux couloirs de coulissement de la marge atlasique de la Tunisie: sismo-stratigraphie, sismo-tectonique et implications pétrolières [Thèse de Doctorat es Sciences], Université de Tunis, 1995.
14. A. Hlaiem, "Halokinesis and structural evolution of the major features in eastern and southern Tunisian Atlas," Tectonophysics, vol. 306, no. 1, pp. 79–95, 1999. r · ·
15. T. Zouaghi, Distribution des séquences de dépôt du Crétacé (Aptien-Maastrichtien) en subsurface: rôle de la déformation tectonique, l'halocinèse et évolution géodynamique (Atlas central de Tunisie) [Thèse Doctorat], Université de Tunis El Manar, 2008.
16. M. Ben Haj Ali, M. Jedoui, T. Dali, H. Ben Salem, and L. Memmi, "Geology map of Tunisia, Office National des Mines (ONM publication)," 3 Sheets, Scale 1: 500.000, Département de Géologie, Tunis, Tunisie, 1985.
17. G. Creuzot and J. Ouali, "Extension, diapirisme et compression en Tunisie Centrale. Le Jebel Es Souda," Géodynamique, vol. 4, no. 1, pp. 39–48, 1989.

CHAPTER 8

Inexpensive Geophysical Instruments Supporting Groundwater Exploration in Developing Nations

James A. Clark[1], Richard Page[2]

[1]*Department of Geology and Environmental Science, Wheaton College, Wheaton, Illinois, USA*

[2]*Design Engineer, Wheaton, Illinois, USA*

ABSTRACT

Geophysical methods are often used to aid in exploration for safe and abundant groundwater. In particular resistivity and seismic refraction methods are helpful in determining depth to bedrock and zones of saturation in the subsurface. However the expense of these instruments ($5000 to $20,000) has resulted in their limited use in developing countries. This paper describes how to construct these devices for less than $250 each. The instruments are small, light and robust and are as useful for groundwater exploration as the commercial models for shallow aquifers (less than 35 m deep) where wells can be hand dug, augured or drilled with small portable drill rigs. Data interpretation can be accomplished quickly in the field with free software implemented on a laptop computer. A suite of geophysical instruments and software can therefore be assembled for less than $850. This paper gives the design for these instruments and essential information needed to use them. It is hoped that these inexpensive geophysical instruments can be widely distributed among drillers and aid workers in developing countries, improving the success rate of water wells.

Keywords: Groundwater Geophysics, Inexpensive, Developing World, Resistivity, Seismic Refraction, Groundwater Exploration

1. INTRODUCTION

Because the need for safe water in much of Africa and many developing nations is so great and sustainable water production is desired, training of a host of indigenous workers with skills to drill, develop, complete and maintain good wells is essential. To this end a team from LifeWater International trained an effective team of Tanzanians to drill and complete shallow wells (<35 m deep) with a small portable drill rig (LS-100). This mud rotary rig can drill quickly through soft and unconsolidated sediments, but cannot penetrate the hard crystalline bedrock common in East Africa. During the following year the highly productive indigenous team drilled 50 boreholes but, because there was little geological rationale for siting each well other than avoiding proximity to sources of contamination, 15 of those wells were dry. The aquifer in the southern highlands of Tanzania is slightly weathered granite or gneiss found immediately above unweathered bedrock, but below 5 to 40 m of clay. Dry boreholes typically occurred where bedrock was close to the surface (<10 m) so that the water table was below the permeable weathered zone, or where bedrock was so deep (>35 m) that the aquifer was beyond the depth limit of the small drill rig. Similar depth limits constrain hand auger drilling methods and hand-dug wells. Prior knowledge of the approximate depth to bedrock would be extremely valuable information aiding the selection of drilling sites in this region.

In a different geological regime villagers in eastern Chad hand-dig shallow wells, 1.5 to 3 meters deep, in the sands of wadis to provide their families with water for 6 months of the year. During the dry season the water table drops below the bottom of the wells and the women must travel several kilometers to obtain water from large dry sandy river beds. Hard crystalline bedrock limits the depth of the wells, but its surface is highly irregular so that overlying sand thicknesses differ dramatically over short distances. Knowledge of the approximate depth to bedrock would enable wells to be dug in thick sand sequences which remain saturated throughout much of the year.

In each of these cases the striking contrast in physical properties between the overlying unconsolidated material, clay or sand, and the hard bedrock makes geophysical methods particularly attractive as a means of detecting depth to bedrock. In particular, the well known electrical resistivity and seismic refraction techniques are very well suited to determination of depth to bedrock under these geological situations [1]. In the resistivity method a slowly alternating electrical current flows through the ground, such that, through analogy to Ohm's law (resistance = voltage/current), measurement of voltage and current at the ground surface gives an estimate of the electrical resistance in the subsurface. Because the electrical resistance is related to lithology and saturation, the subsurface geological conditions can be estimated. In the seismic refraction method sound waves produced by an explosion or sudden shock at the surface travel through the subsurface. Shock waves travel much faster in bedrock than in unconsolidated material and, in analogy to Snell's law of optics, eventually refract back towards the surface. Accurate measurement of initial arrival times of these refracted waves allow inference of depth to bedrock layers.

These techniques using commercially available instruments have been shown to be very helpful in finding water in Africa. In one study of 370 wells drilled in Zimbabwe the use of resistivity surveys improved the success rate for wells from 50% to 85 % [2]. Kirsch [3] provides an overview of geophysical methods in support of groundwater exploration, and in that volume Rabbel [4] considers seismic and resistivity methods as second only to drilling for usefulness in determining the subsurface structure. His paper also explains that both methods should be used for maximum information because they complement each other.

This paper will add little to the immense literature about the theory, use, and interpretation of resistivity and seismic refraction methods developed over the past 70 years in support of oil exploration and, more recently, groundwater exploration [3,5]. The objective is to demonstrate that instruments costing tens of thousands of dollars if purchased commercially can be constructed for less than $250. Such inexpensive tools could be disseminated widely among drillers who could use them to determine depth to bedrock, and therefore increase the chance of completing successful water wells. Thus the instruments would be operated not by highly-paid consultants or a few geologists employed by governments but by those who are actually drilling wells. Although these instruments are much less elaborate than the commercial models, they are capable of similar accuracies where prospecting targets are shallow. They are much less cumbersome to transport and easier to repair under field conditions than commercial units. Furthermore, if an inexpensive laptop computer is available, free software can be used to interpret the geophysical data while in the field. However, commercial models typically allow more efficient data collection and commercial software provides more refined data interpretation.

Others have reported developing inexpensive resistivity [6] and seismic instruments [7] for educational purposes, but the use of similar instruments for water prospecting in developing countries has not been addressed. Olowofela et al. [8] working in Nigeria have recently outlined the plans for a resistivity meter in the context of physics research, but their plans are more elaborate than those proposed by us and have not been applied specifically to water prospecting.

The following sections briefly describe the theory of each technique in turn and then outline construction and interpretation procedures. Each section concludes with an application of the technique to the geological situations outlined above for Tanzania and Chad.

2. RESISTIVITY METHOD

It is not necessary to review in detail the theoretical basis for "vertical electrical sounding (VES)" because, for all but the simplest homogeneous earth model, the theory is difficult and adds little to meeting the objectives of this paper. A brief discussion of the theory and field methods, however, is important to understand the rationale for features included in the instrument and data interpretation method.

2.1. Theory

A DC voltage across two current electrodes (A, B) inserted into soil causes an electrical current to flow through the ground (**Figure 1**). Potential electrodes (M, N) are inserted between and in line with the current electrodes. Measurements of the electrode spacing, the current (I) and the voltage drop between the potential electrodes (V_m) are made and then the electrode spacing is increased and the measurement repeated. As the electrode spread increases, the depth of penetration of the current increases so the surface readings become more affected by electrical properties at depth. The electrical property of interest is ground resistivity with units given in ohm-meters.

Two electrode configurations are common (**Figure 2**). In the Wenner arrangement the spacing between each electrode, "a", is constant. When the current electrodes are moved farther apart to induce a deeper current, the inner potential electrodes must also be moved. With the Schlumberger method the potential electrodes remain fixed and only the outer current electrode spread increases.

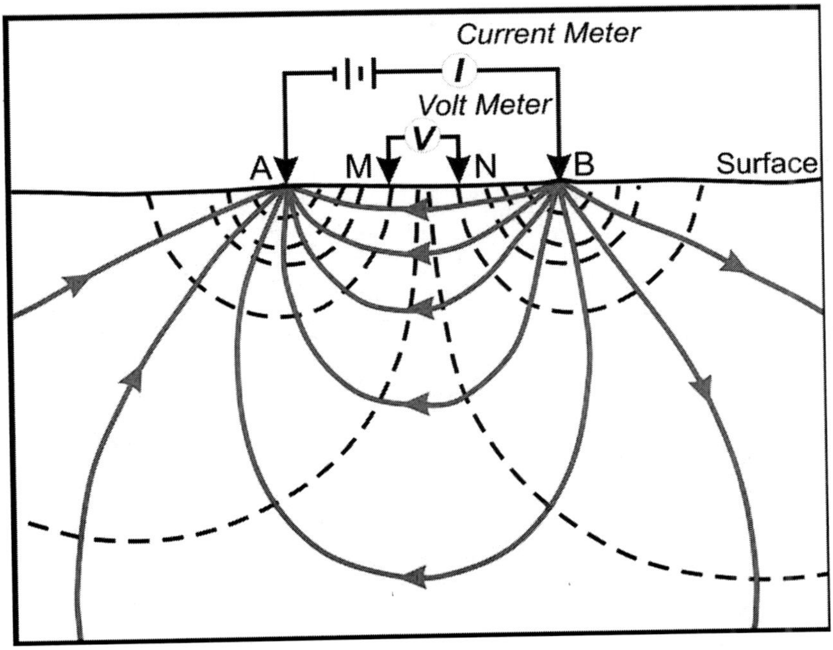

Figure 1. Schematic illustration of the resistivity method. Solid red lines represent current flow. Dashed black lines contour electrical potential (voltage).

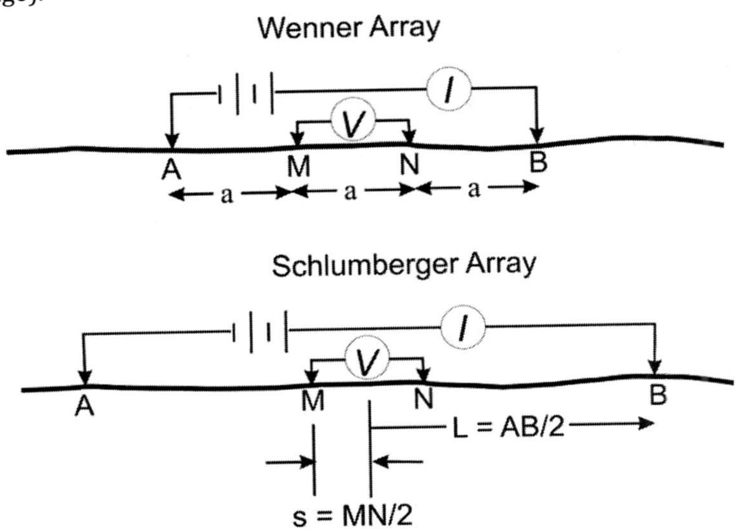

Figure 2. The two most common electrode configurations.

The measurements of I and V_m are used to calculate the "apparent resistivity," which is the resistivity for the equivalent homogeneous earth that produces the same I and V_m. For the Wenner electrode configuration the apparent resistivity is

$$\rho_a = 2\pi a \frac{V_m}{I} \tag{1}$$

For the Schlumberger arrangement the relationship is

$$\rho_a = \frac{\pi V_m \left(L^2 - s^2\right)}{2sI} \tag{2}$$

with the L and s distances defined in **Figure 2**. For a homogeneous earth structure, as the current electrodes are spread farther apart the apparent resistivity will not change. If the actual earth resistivity decreases with depth, as often occurs in Tanzania, the calculated apparent resistivity will decrease as the electrode spread increases. This occurs because at large spread a greater proportion of current is flowing in the lower resistivity region at depth, and that is reflected in the measured current and voltage at the surface. It is this relationship, coupled with theory of current flow in layered materials, that allows inference of resistivity properties at depth.

Two problems can arise in that 1) the earth may have a natural voltage or "self potential" that is unrelated to earth resistivity at depth, and 2) the DC current source itself can gradually induce an electrical potential around the electrodes from polarization. Both the self potential and the induced polarization effects can be minimized by periodically reversing the current at a rate of about 1 Hertz. A slowly alternating current reduces polarization because a charge concentration cannot accumulate. If V_{SP} is the natural ambient earth self-potential, assumed to be constant, and V_p is the voltage induced by the current electrodes, then the measured voltage, V_m, across the potential electrodes is

$$V_m^+ = V_{SP} + V_p \tag{3}$$

If the current is reversed only the V_p voltage is affected by the sign change so that the reading is then

$$V_m^- = V_{SP} - V_p \tag{4}$$

The difference in the pair of readings gives $2V_p$ and half this difference is then the desired V_p. In the field it is convenient to use reversing switches for both the current and the potential reading. If both are switched simultaneously, the measured voltage values only need to be added rather than subtracted, and the average of the measurements is then V_p. At the African sites described above the average of four values of voltage and current is adequate and reproducible. More readings can be averaged, but there must be an even number of measurements to ensure that the V_{SP} term is eliminated.

2.2. Construction

Commercial resistivity devices are excellent, and they now contain microprocessors that control current reversal and averaging procedures and even calculate apparent resistivity for each standard electrode configuration. They are also expensive ($5000 to $10,000) and repair is costly and time-consuming. The design goal was to build a resistivity device for less than $250 because then it could remain with the LS-100 drill rig for national drill team use. Further requirements were simplicity, so that repair in the field is possible, and robustness so that repair would not be needed even under harsh African conditions. Furthermore, because of airline transport issues, it had to be light, compact, and with no large batteries, which are now prohibited on many flights. Because of ready availability in almost any country, a 12-volt car battery was used as the power source. A small commercially available 400 watt inverter that produces an AC "modified" sine wave from a 12 volt DC source was used to increase the voltage to the required value of several hundred volts. A full-wave bridge rectifier in parallel with a 1 μF capacitor converts the AC voltage to a high steady DC voltage. The inverter used in Tanzania had a 120 volt AC output producing a DC voltage output from the resistivity device under no load of 170 volts. A high voltage across the current electrodes is desired because the maximum spread, and hence the depth of current penetration in the ground, is limited by the voltage magnitude across the current electrodes. In Tanzania resistivities at depths exceeding 35 m were barely detected. An even higher DC voltage is needed for the Schlumberger electrode array because the potential electrodes are not widely spaced and so have a lower voltage drop across them than the equivalent Wenner array. An identical device built upon return from Tanzania uses a commercial inverter that produces 240 volts AC, resulting in a DC no-load voltage of 330 volts. An added benefit of the inverter is that it can provide power in the field for a laptop computer, useful in data interpretation.

The electrical schematic (**Figure 3**) includes two reversing switches (DPDT)—one for the ammeter and one for the voltmeter. To measure the current and voltage two small inexpensive digital multimeters worked well. For U.S. applications filtering AC noise is important, but in Tanzania that problem did not exist as no electricity was close to the drill sites. The isolation from electrical noise in developing countries is a great advantage. The four electrodes were 10-inch lag screws, and were attached with large alligator clips to stranded single conductor 14-gauge house wire which was cheaper than the lighter 18 gauge wire used in commercial devices. Two 152 m (500 feet) spools of house wire will allow a maximum current electrode spread with the Wenner array of 228 m (750 feet), and would be adequate for almost any survey. From each spool cut off 38 m (125 feet) for the potential electrodes. Two tape measures, preferably more than 60 m long, to extend in both directions from the array center are the only additional equipment needed for a survey.

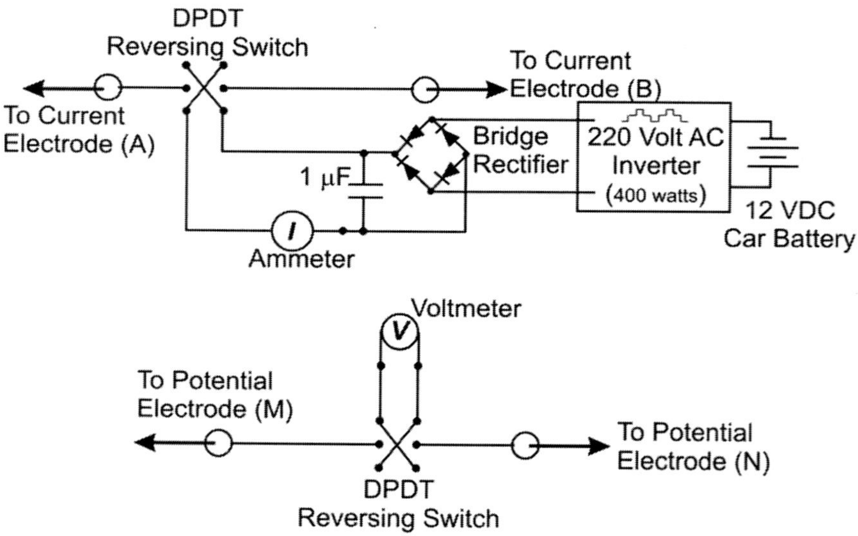

Figure 3. Electrical circuit used in the inexpensive resistivity device.

The total cost (**Table 1**) of $206 is well within the design budget. Most of the materials can be obtained from electronic supply and hardware stores. The 220 VAC inverter and the inexpensive multimeters were purchased from internet sources. Pictures of the unit and the field arrangement (Figures 4 (a), (b)) illustrate how small and simple it is. Tests of the device at Northern Illinois University confirmed that results were virtually identical to those from a commercial $10,000 resistivity instrument (Sting R1), though the commercial instrument provided apparent resistivity values instantaneously whereas the values for ρ_a for our instrument had to be calculated from the current and voltage readings using Equation (1). A test in Nigeria with another Sting R1 resistivity instrument validated the accuracy of the inexpensive instrument (**Figure 5**).

(a)

(b)

Figure 4. (a) Resistivity instrument exterior and (b) field setup at Mgongo, Tanzania.

Table 1. Itemized list of resistivity device materials and cost.

Quantity	Cost/Item US Dollars	Item	Cost US Dollars
2	22	500 ft roll 14 gauge stranded wire	44
1	2	1 microFd capacitor	2
1	1	Bridge Rectifier	1
2	2	DPDT Switch	4
2	15	Digital Multimeter	30
2	13	200-ft Tape Measures	26
4	1	Alligator clips	4
1	60	12 VDC to 220 VAC 400 Watt Inverter	60
1	35	Miscellaneous electronic supply store parts (box, banana plugs, extension wire etc)	35
		TOTAL COST	206

2.3. Field Method

For efficient surveying at least 3 people are needed to operate the instrument and insert the electrodes, but conceivably only one operator is necessary. Once a site is selected the two tape measures are extended outward and the electrodes inserted symmetrically about the center. When the battery, inverter, and multimeters are connected to the resistivity device and the wires are attached to the electrodes, readings can commence. In the Tanzanian field area annual rainfall is 91 cm, but during 6 months of the year only 6 cm is typical and the soil is very dry. A serious error can occur if any electrode does not have a good electrical contact with the soil and so water was poured into a small shallow hole, made with a hammer, to saturate the ground near each electrode. The electrode was then pounded into the saturated ground. The reversing switches are thrown manually, and the ammeter and voltmeter readings taken after a brief delay to allow values to stabilize. The switches were thrown four times and an average calculated for voltage and current.

The Schlumberger configuration is much faster and easier to implement in the field, but when the current electrodes are very far apart (e.g. 100 m) the voltage V_m may be too small to detect reliably. In that case the potential electrodes must be spread farther apart without moving the current electrodes and another reading taken. Normally some slight offset in ρ_a values occurs, and this requires adjustment during interpretation. It is best to keep the spacing of the current electrodes more than 5 times the potential electrode spacing. The Wenner array always has maximum potential electrode spacing, and therefore would be more appropriate where the depths investigated are near the limit of the instrument. It is helpful to have a tabulated chart of the location of the electrodes for different Wenner "a" spacing. This is not necessary for the Schlumberger spread. After each reading the spread is increased until the instrument readings exceed the accuracy of the multimeters. This typically occurred at a 90 m spread with the 120 volt AC inverter.

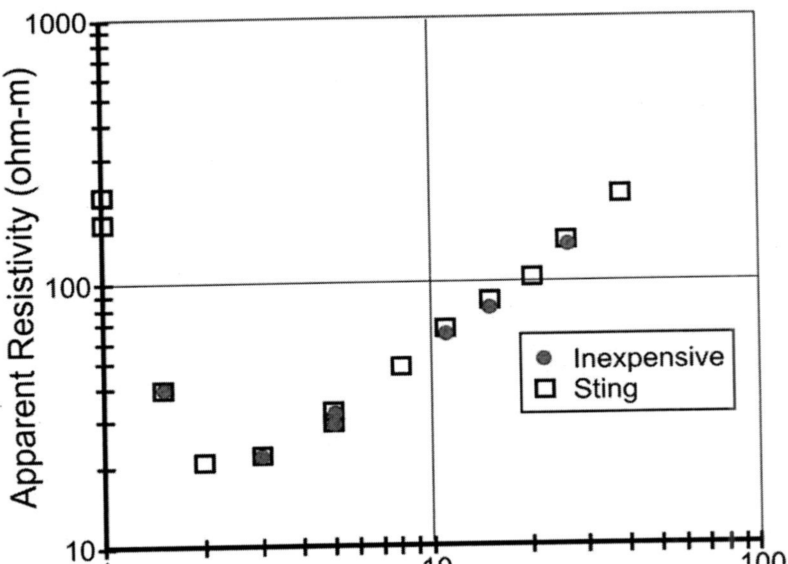

Figure 5. Apparent resistivity comparison between the commercial resistivity meter Sting R1 and the inexpensive resistivity instrument. Results are very similar.

2.4. Data and Results

Using the inexpensive resistivity instrument and a Wenner electrode spacing geometry, the Tanzanian drill team collected data at three sites (**Figure 6**). Wells already existed at two of the sites and they planned to drill a borehole at the third (Mafinga). Data from all VES indicate that apparent resistivity decreases as the current penetrates deeper into the ground with increasing Wenner "a-spacing." At an a-spacing that differs among all surveys, apparent resistivity starts to increase, suggesting that at depth there is a resistivity increase. It is clear that the a-spacing at the minimum apparent resistivity value is related to bedrock depth.

Computer software for resistivity interpretation is readily available commercially. These codes can be very elaborate and often include the possibility of predictions of both vertical and horizontal changes in resistivity. This software is expensive. If the goal is to find only vertical variation in resistivity and one is content to assume that the layer interface is horizontal (or at least dipping less than 15 degrees) then free software can be used. In Tanzania freeware provided by Dr. Philip Carpenter of Northern Illinois University was used. It followed the theory of Ghosh [9,10] and required the DOS operating system. The simplest resistivity model that fits the Tanzanian data has three layers and gives an acceptable estimate of the depth to bedrock. It is possible to estimate geological materials from resistivity values (**Table 2**), but the large range in the values makes such estimates approximate. In this region surface clay is dry to a depth of a few meters. Below that the clay is saturated and of very low resistivity, but its extremely low permeability precludes it as an aquifer. Below the clay is bedrock that is essentially an electrical insulator halfspace of several thousand ohm-m. The fit to the data at the Mafinga site (**Figure 6**) resulted in the interpretation that depth to bedrock is approximately 30 meters. The upper layer is 1.5 m thick with resistivity of 110 ohm-m, and overlies a 29 m layer with 6.5 ohm-m resistivity. The substrate halfspace is at least 2000 ohm-m, which is essentially a perfect insulator. The curves in **Figure 7** represent models with these resistivities where only the thickness of the middle layer is changed. Where the geological materials are similar, the depth can be estimated by the a-spacing value at the minimum point of the curve. Comparison of predictions to the data clearly rules out depths to bedrock, and the overlying aquifer, less than 15 m, which would be too shallow and wells would likely dry up during part of the year. So the well was drilled and reached bedrock at 28 meters. Water from the aquifer of slightly weathered bedrock underlying the clay rose in the well to within 5 m of the surface, and a high yield well was completed.

Table 2. Typical Geophysical Properties (from [7,13-15] and [1] pp. 452-456).

Geological Material	Resistivity (ohm-m)	Velocity (meters/sec)
Soil	20 - 300	100 - 600
Clay (wet)	1 - 50	1500 - 2500
Clay (dry)	10 - 100	500 - 900
Sand (dry)	1000 - 10,000	200 - 1000
Sand (wet)	50 - 500	600 - 2200
Sandstone	50 - 1000	3000 - 5000
Shale/Siltstone	10 - 400	4000 - 5300
Limestone/Dolomite	1000 – 10,000	4800 - 6200
Granite	1000 – 10,000	5000 - 6000
Metamorphic rocks	300 - 3000	3500 - 5600
Air	very large	331.5
Water (fresh)	50 - 100	1400 - 1600
Water (salty)	0.2	

2.5. DISCUSSION

The Tanzanian drill team quickly learned how to perform the field operations, but navigating the DOS operating system was daunting. We subsequently developed user friendly software for the Windows operating system to calculate apparent resistivities. It also provides a Monte Carlo method to invert the observed apparent resistivities for the likely lithologic thicknesses. The Schlumberger geometry is more efficient in the field than the Wenner configuration, but a greater input voltage is needed to penetrate to the same depth because the potential electrodes are closer together, and so are sampling a smaller range of voltage. Both a Schlumberger and a Wenner array were used at one locality (Mgongo), and the final interpretation from the two methods was similar.

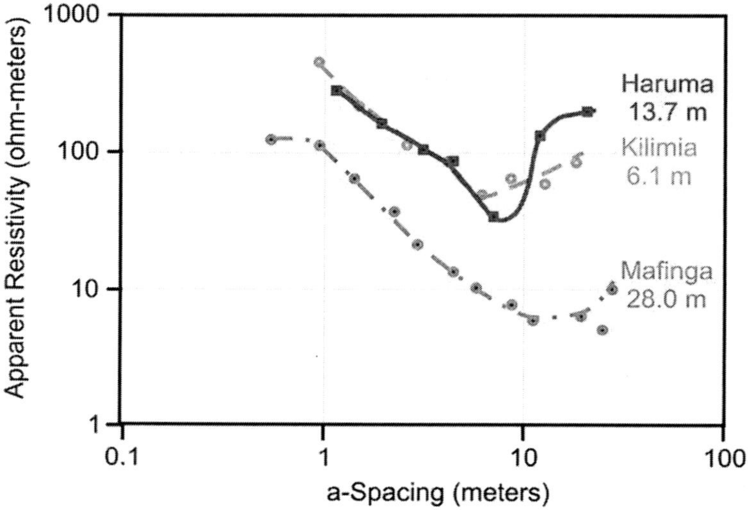

Figure 6. Resistivity data at three sites with known depth to bedrock given.

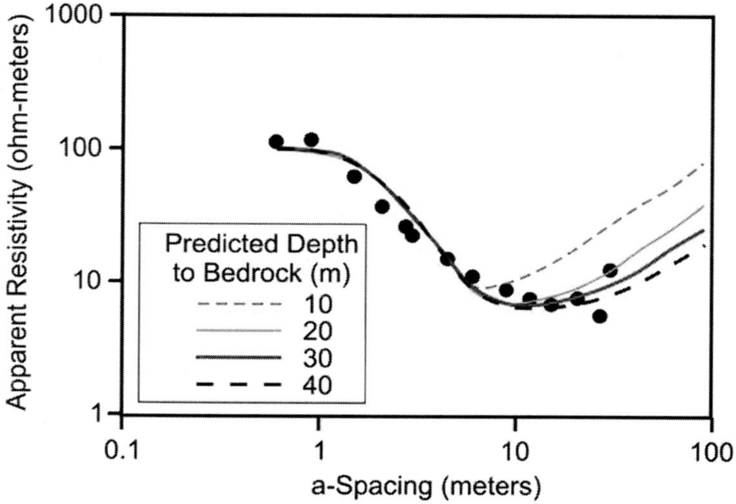

Figure 7. Mafinga resistivity data compared to predictions at differing predicted depths to bedrock. Depths of 20 to 30 m fit best. Actual bedrock was at 28 m depth.

Data collection typically took about two hours, but learning and training was occurring simultaneously. It is estimated that three trained workers can do a survey in about one hour, including the interpretation, if a laptop computer is available. This provides quick and useful screening to detect if basement rock is too close to the surface. Most of the field time was spent in measuring and moving the electrodes, so, relative to the Wenner approach used in the Tanzanian study, the Schlumberger method would halve the field time as only the two current electrodes are moved for each reading. For bedrock depths approaching 30 m the voltage source (120 VAC or 170 VDC) was near its limit. Use of the 220 VAC (400 watt) inverter would double the DC voltage output and make interpretations down to 35 m depth more reliable. A higher voltage is necessary for the Schlumberger approach because the closer spacing of the potential electrodes results in a smaller voltage drop. It would be easy and cheap (<$10) to add a small instrument amplifier chip (up to 50× voltage amplification) to allow more precise readings when the voltages are in the 5 to 10 millivolt range. It is not clear, however, whether noise in the environment and from the inverter would mask the signal.

The software allows rapid fitting of the resistivity model to the data, but, as with most geophysical techniques, the resulting resistivity model is not unique. Because of errors in the data and physical trade-offs (e.g. thicker layer but lower resistivity vs thinner layer but higher resistivity), it is possible to fit the data within its errors with different geologic models. With experience in a given geological regime an operator will likely be able to recognize and eliminate unlikely models.

If no computer is available it is still possible to interpret the data using the "Master Curve" method. This method was used in the 1960's to interpret resistivity data, and is still used by government hydrogeologists in Tanzania. It is described in most textbooks, but sets of Master Curves are needed [11].

3. SEISMIC REFRACTION METHOD

Whereas resistivity methods give a general estimate of depth, seismic refraction is capable of revealing a seismic section showing undulations of the subsurface contacts. It can be used where dry coarse sediment at the surface limits electrical penetration into the subsurface. Again numerous papers and texts describe the theory of seismic refraction, so the method is only briefly described below.

3.1. Theory

When a sudden disturbance, such as a dynamite explosion or a blow from a sledgehammer, occurs on the earth's surface a shock wave propagates outward in wavefronts and travels through rocks and sediments at different wave velocities (**Table 2**). In general saturated sediments have higher velocities than unsaturated sediments and weathered rocks have lower velocities than unweathered rocks of the same lithology. Where there is a velocity contrast across a contact the shock waves refract with critically refracted waves eventually propagating back to the surface where they are detected by a geophone (**Figure 8**). The elapsed time of travel for the shock wave between the source and the geophone receiver is a function of the velocity of the surface and subsurface materials and the depth of the layers. Although commercial refraction seismographs utilize many geophones (12 to 24) and few "shot points," the instrument proposed here has one fixed geophone and multiple shot locations. Using one geophone reduces the number of costly geophones and simplifies the instrument. From Snell's law and geometrical considerations for a horizontal contact, it can be shown that if the arrival time of the shock wave is plotted against the horizontal distance (**Figure 9**) the depth, h, to a horizontal contact is:

$$h = \frac{t_i}{2} \frac{V_2 V_1}{\sqrt{V_2^2 - V_1^2}} \qquad (5)$$

with V_1 and V_2 the seismic wave velocity of each layer determined from the reciprocal of the slope of the linear segments of the curve (**Figure 9**) and t_i the intercept time. A similar equation applies to a three-layer case where three linear segments are observed in the travel-time curve.

Because the subsurface geology is rarely horizontal, a completely different approach, the delay-time method, can be used to determine variable relief on the contact of the refracting layer. This method requires data from both forward and reverse seismic lines by placing the geophone at each end of the line, and collecting data from identical shot points for each geophone location. **Figure 10** gives the geometrical arrangement showing the forward and reverse paths for a given source location. It can be shown [5] that the delay-time at point C, T_C, is related to the depth to the refracting surface directly below C by the expression

$$z(C) = T_C \frac{V_1 V_2}{\sqrt{V_2^2 - V_1^2}} \qquad (6)$$

The subsurface profile is determined through calculation of z at successive locations. The value of T_C can be determined by the simple expression

$$T_C = \frac{t_{AC} + t_{BC} - t_{rec}}{2} \tag{7}$$

where t_{AC} and t_{BC} are the times required for the seismic wave to travel from the source at C to the geophones at A and B respectively ([5] p. 123). Only waves that travel along the lower layer for part of their path can be used in Equation (7). The reciprocal time, t_{rec}, is the travel time from A to B. From geometrical considerations t_{rec} for a wave going from A to B must equal the time for the wave to go from B to A. It is therefore possible to determine T_C at each intermediate shot point between A and B. All that is needed to determine the actual depth, z, are V_1 and V_2. V_1 is found immediately from the travel-time curve. Burger et al. ([5] p. 124) show that V_2 can be determined from the slope of the line defined by

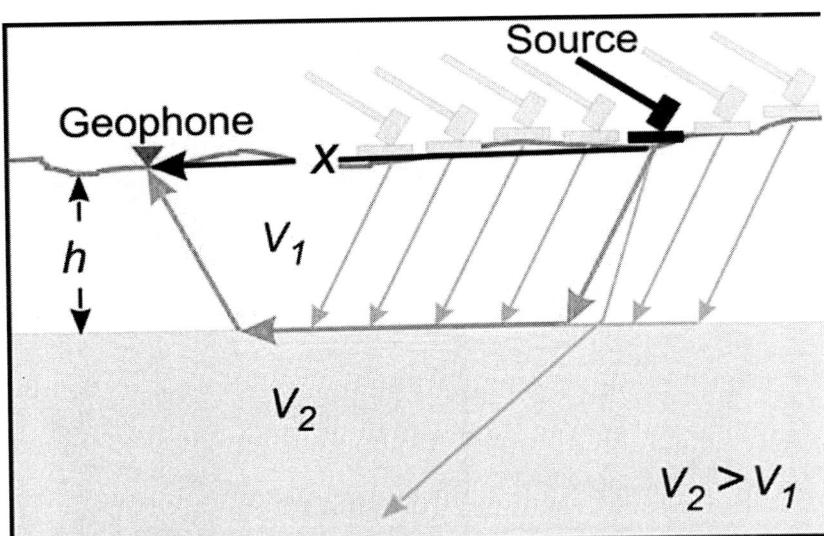

Figure 8. Illustration of the propagation of shock waves at the ground surface (black line) and subsurface (blue line) for a layered case with velocity in the upper layer V_1 less than V_2 the velocity of the shock wave in the lower layer. The wave that refracts at the critical angle will eventually propagate back to the surface and be detected by the geophone. For the inexpensive instrument the geophone is fixed and the source moves repeatedly.

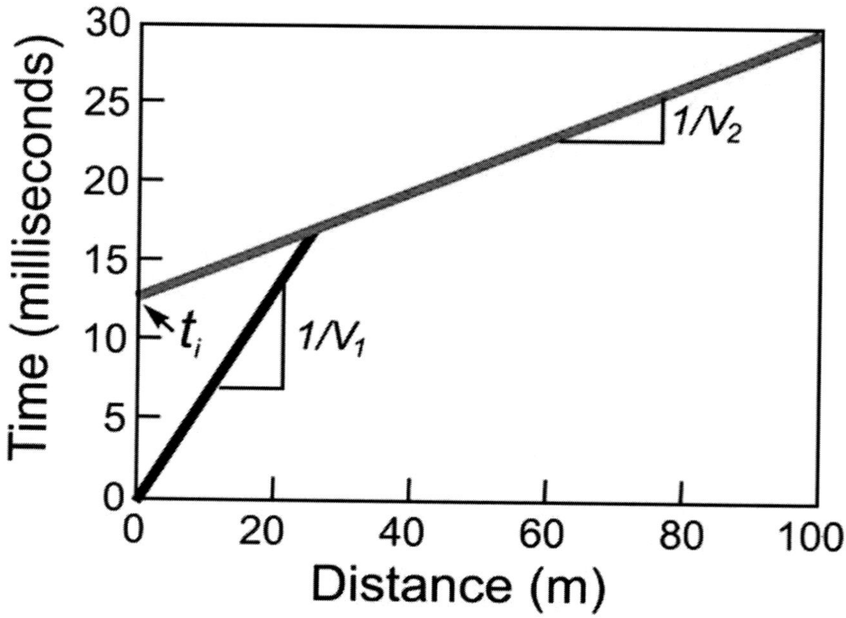

Figure 9. Typical Travel-Time curve from a seismic refraction survey. Velocities of the surface (black line) and subsurface (red line) layers ($V_1$and V_2) can be determined from the inverse of the slopes. The intercept time, t_i, together with the velocities can be used to determine the depth to the subsurface layer if the layer is horizontal.

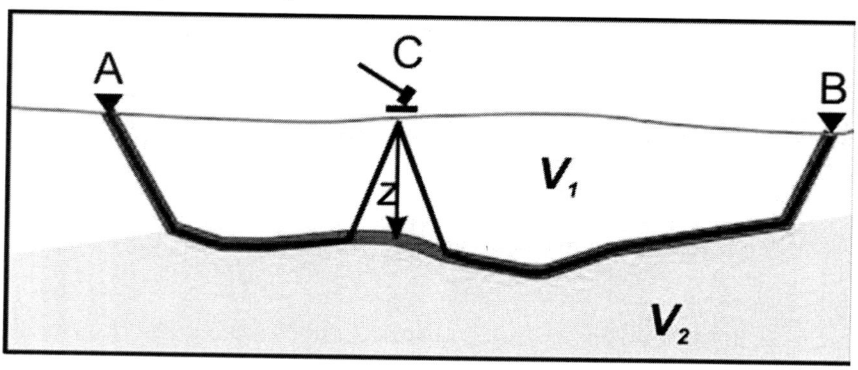

Figure 10. Illustration of the shock wave propagation for the Delay-Time method. The path for the reciprocal time is indicated in red. The depth z can be determined below each sledge hammer location to give the subsurface relief.

$$\left(t_{AC} - t_{BC}\right) = k + \left(\frac{2}{V_2}\right)x \tag{8}$$

where x is the distance from A to C and k is a constant.

Although the delay-time method offers a simple means of determining subsurface relief, a much more accurate but complex procedure uses the delay time profile initially but modifies it through ray tracing until the observations are fit [12]. This procedure is coded in FORTRAN as public domain software SIPT developed by the United States Geological Survey. We developed free Windows software to make execution of SIPT more user friendly.

3.2. Construction

The design goal is to construct a reliable, easily transported seismic device for less than $250. A seismic refraction survey requires three important items: seismic source, geophone, and a seismograph. For shallow surveys a sledgehammer striking a 1-inch thick aluminum plate is usually an adequate source. A geophone is a transducer that converts slight vertical ground motion into a voltage signal. Usually this is accomplished through suspending a magnet from a spring within a coil of wire. Another approach is to use a MEMS accelerometer integrated circuit chip, which is much cheaper and smaller than a mechanical geophone and can be implemented in the same manner. Initial experiments suggest these chips have adequate sensitivity, but for the seismic survey in Chad a used geophone was purchased for $40. The seismograph records the voltage change through time, and must be able to record this signal with millisecond accuracy so that the time of first arrival of the seismic wave at the geophone can be determined. If a laptop computer is available an analog to digital converter manufactured by Measurement Computing (Personal Measurement Device "PMD") can be used to transfer the signal to the computer through its USB port. The PMD-1208LS model can convert analog voltages into a 12-bit digital signal at a rate of 8000 samples/sec. A computer code, written in Visual Basic Net using Measurement Computing subroutines, displays the waveform on a laptop computer and permits the operator to determine elapsed time between impact of the seismic source and arrival of the first seismic wave at the geophone (**Figure 11**). Timing of the record must begin when the sledgehammer first strikes the plate. Upon contact, wires attached to both the plate and the sledgehammer complete an electric circuit, triggering the A/D converter. The voltage signal generated by the geophone is very small so it must be amplified before input into the PMD. A simple op-amp differential amplifier circuit which provides variable gain up to 200×

(**Figure 12**) is adequate. The PMD must be configured properly to per mit triggering, which requires a few jumper wires and a 1000 ohm resistor (**Figure 13**). To power or recharge a laptop a small inverter, such as the one used for a resistivity survey, should be included in the equipment list. A 60 m tape measure is also required. The instrument and geophone are shown in **Figure 14**. The approximate cost of each item is in **Table 3** and the $245 total cost is within the design budget.

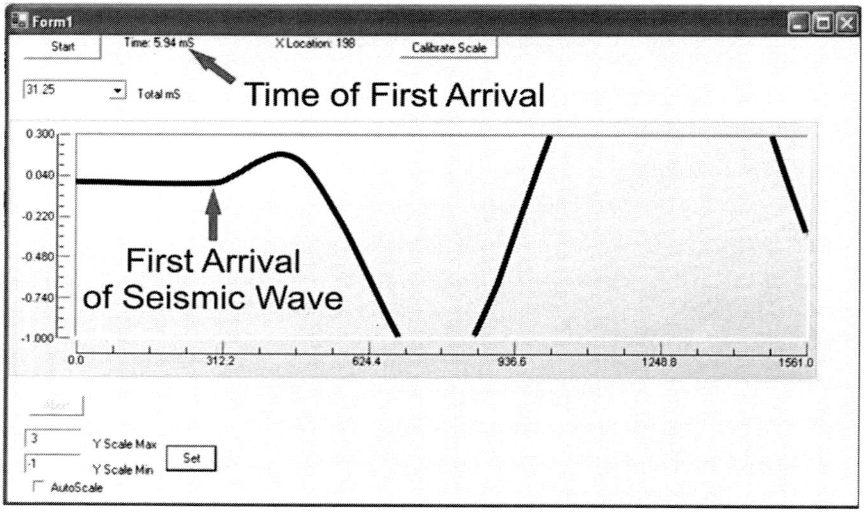

Figure 11. Laptop screen capture of a seismic wave from the PMD1208-LS Visual Basic program. The time of arrival of the first wave for known distance between the sledge and the geophone is the data required for the analysis.

3.3. Field Method

Two workers are required to conduct a seismic survey. Once a possible well site is selected, insert the geophone into the ground. If there is wind, burying the geophone a few inches greatly decreases noise and enhances the signal. Plug the geophone wires into the amplifier circuit, the trigger wires from the sledgehammer into the PMD, and then plug the PMD into the USB port of the laptop (**Figure 15**). Extend a tape measure about 40 m from the geophone with the tape origin at the geophone. If forward and reverse surveys are desired it is most convenient to have a constant interval for the source locations or to insert small surveying flags at each location. Launch

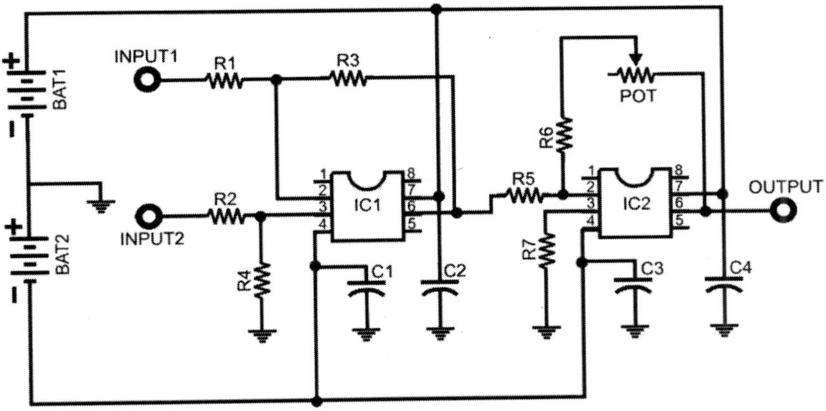

R1, R2, R3, R4	10 K ohm resistors
R5	510 ohm resistor
R6, R7	1.8 K ohm resistor
C1, C2, C3, C4	0.01 μF capacitors
POT	100 K ohm variable resistor
BAT1, BAT2	4 x 1.2 volt rechargeable AA batteries
IC1, IC2	741 operational amplifiers

Figure 12. Electronic schematic for amplification of the seismic signal.

the free seismic data collection program on the laptop and strike the plate with a mighty sledgehammer blow. It is good to involve the local people as seismic "sources" (**Figure 16**) because there is often healthy competition to see who can create the largest shock wave. Upon completion of the forward seismic line repeat the method for the reverse seismic line. Move the geophone to the other end of the line and bury it at the exact location of the last shot. As a simple check, the arrival time for the most distant point for the forward seismic line should equal the arrival time for the distant point in the reverse direc tion. This is the reciprocal time, t_{rec}, used in Equation (7). The calculations for depth and velocity can be done on a laptop in the field using our free version of the SIPT software. Output is a seismic section showing a profile of the subsurface lithologies.

3.4. Data and Results

To check the accuracy of the inexpensive seismograph, data collected with the inexpensive instrument were compared to data from an EG & G SmartSeis 12 Channel seismograph. The data from the inexpensive instrument compares favorably with the data from the $15,000 SmartSeis seismograph. The differences of only 0.4 m in depth determination (**Figure**

17) can easily be due to errors in assigning the time of the first arrival of the seismic wave.

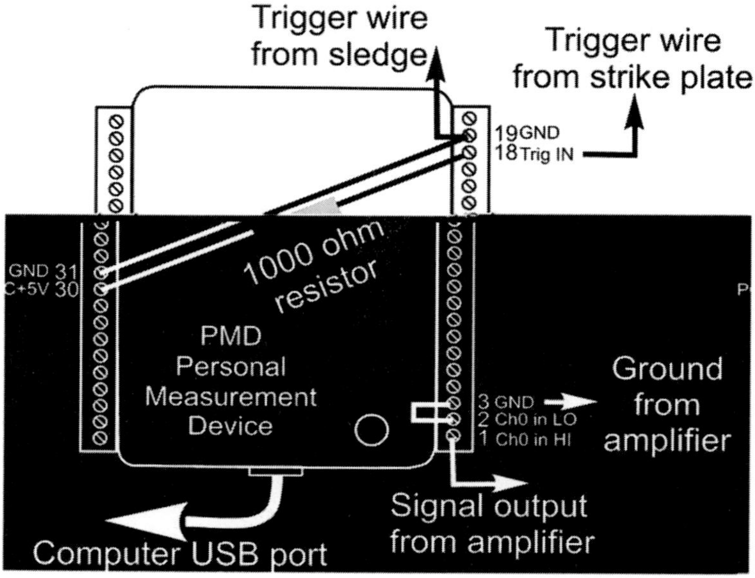

Figure 13. Electrical hookup of the Measurement Computing "Personal Measurement Device" (PMD) 1208LS.

Figure 14. Instrument and geophone.

Table 3. Itemized list of seismograph materials and cost.

Item	Cost in US Dollars
PMD-1208LS	110
Electronics Parts	10
Case (Pelican)	20
Strike Plate	15
Geophone (used)	40
Sledge Hammer	15
Speaker Wire	10
8 Rechargeable AA Batteries	25
TOTAL	245

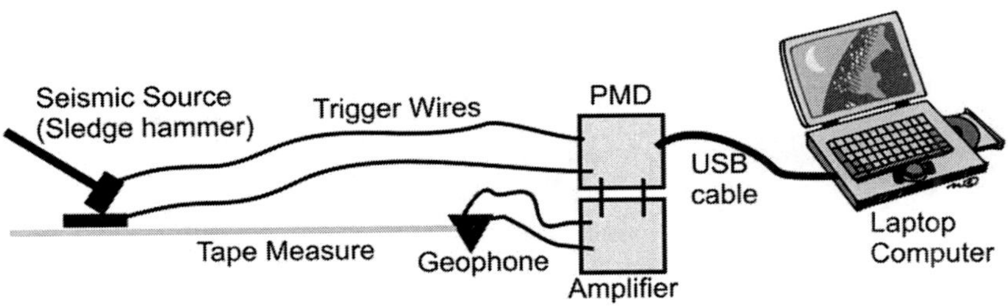

Figure 15. Schematic of field setup.

The equipment was first used in Chad at the Faguire site, where the three segments in the travel-time curve (**Figure 18**) indicated three subsurface layers with velocities of 373, 1202 and 5356 m/sec. It was known that dry sand is the surface material and, from nearby outcrops, it was evident that granite is the bedrock at depth. Referring to **Table 2** the surface velocity of 373 m/s is exactly what would be expected for dry sand and 5356 m/s is appropriate for solid granite. The intermediate layer, with a velocity of 1202 m/s could be saturated sand, but it is more likely that this is a weathered granite layer, because the existing well reached rock close to the surface. The thickness of the upper layer (Equation (5)) is then 3.4 m and the thickness of the intermediate layer is 6.4 m, so the total depth to solid granite is expected to be 9.8 m.

(a)

(b)

Figure 16. Use of the instrument in Chad. Local residents provided the seismic source power at Faguire (a) and Berkoman (b).

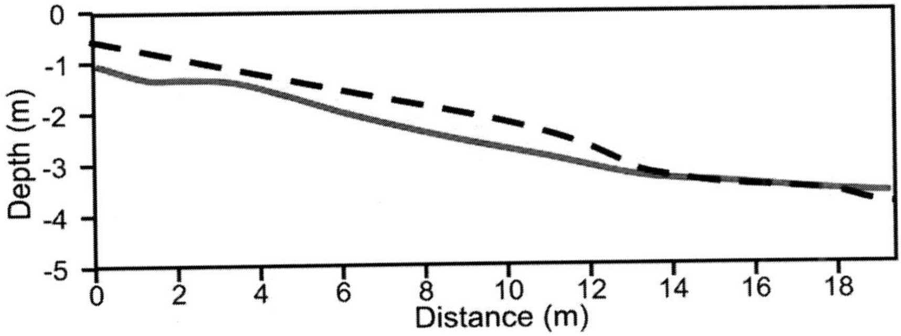

Figure 17. Comparison of the subsurface profile for the inexpensive device (solid red line) to the EG & G SmartSeis 12-channel seismograph (dashed black line). The WindowsSIPTwater program was used to calculate each profile.

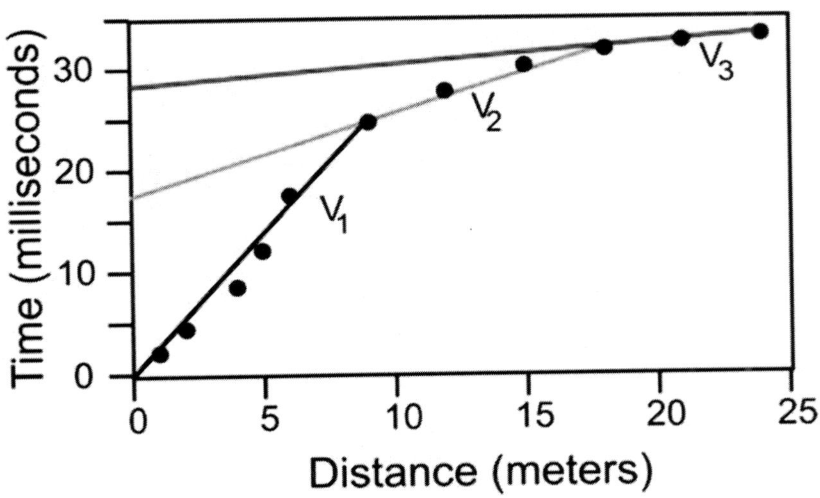

Figure 18. Indication of three layers at Faguire in a traveltime plot. Each successive change in slope represents another layer.

At Berkoman in southwestern Chad both forward and reverse seismic lines were completed. The data shown in the travel-time plot (**Figure 19**) indicate a 2-layer earth. Some scatter about the lines suggests that the dipping contact has some subtle relief. These data were used in our modified SIPT program to produce a seismic section (**Figure 20**). Where the time-delay method is used, the predicted curve shows depth to the contact between the upper sediment (374 m/s velocity) and the underlying substrate (1796 m/s). When the ray tracing algorithm is included using SIPT the contact depth is only slightly changed and smoothed.

3.5. DISCUSSION

The inexpensive seismograph is accurate and capable of detecting shallow strata under developing world conditions. In addition to the huge cost savings, this instrument is very small, light, and portable. The sledge hammer has the greatest bulk and mass. All but the sledge can be packed into a small suitcase. The modified SIPT software (called WindowsSIPTwater.exe) provides excellent interpretations of subsurface relief and is readily used while in the field. Although the instrument was tested where layers are shallow, it would be ideal to penetrate at least 20 m, which would typically require a spread about 80 m long. A stronger seismic source such as an explosive or a "Betsy gun" shotgun source would be required, but would certainly be out of the question in the developing world. Transporting explosives on aircraft would also be problematic. A solution is to stack sledge hammer blows so that a weak signal is added to previous signals, enhancing the signal while simultaneously canceling any random noise. This stacking ability is now included in our seismic recording software. Another feature common on commercial units is the ability to filter certain frequencies. The physical properties of the geophone do this to some extent, and this feature is needed less for refraction surveying than for reflection surveying. An improvement to the device would be a simple active low cut or high cut filter.

Two workers can measure and record each seismic signal in 3 to 4 minutes. A seismic survey with 12 shot locations would take about 1.5 hours to complete, since both forward and reverse lines are needed. Importing the data and running WindowsSIPTwater takes about 30 minutes.

The problems that can arise with seismic refraction are clearly described in most texts. The most important assumption is that deeper strata must have progressively higher velocities. For example where low velocity sediment (e.g. sand) underlies a higher velocity layer (e.g. well-packed clay) the seismic wave refracts downward, not upward, and the sand will be invisible in the interpretation. Very thin layers may not be observed.

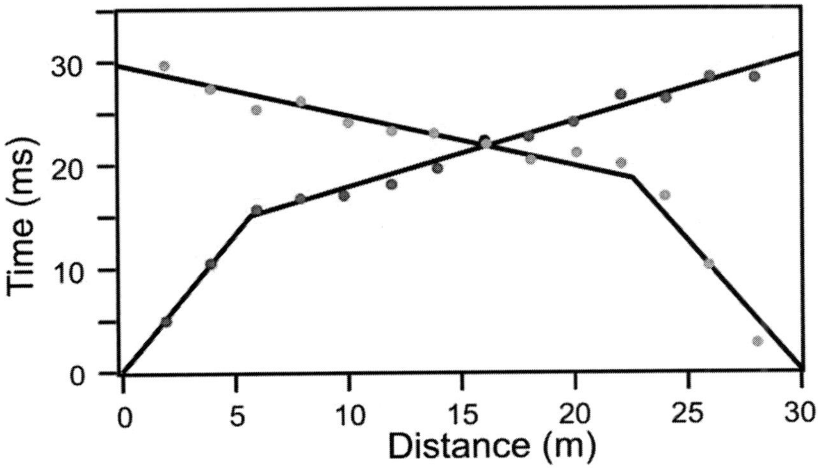

Figure 19. Travel-Time data at Berkoman, Chad. The forward (red data points) and reverse (blue data points) surveys are not symmetrical indicating the subsurface layer is tilted.

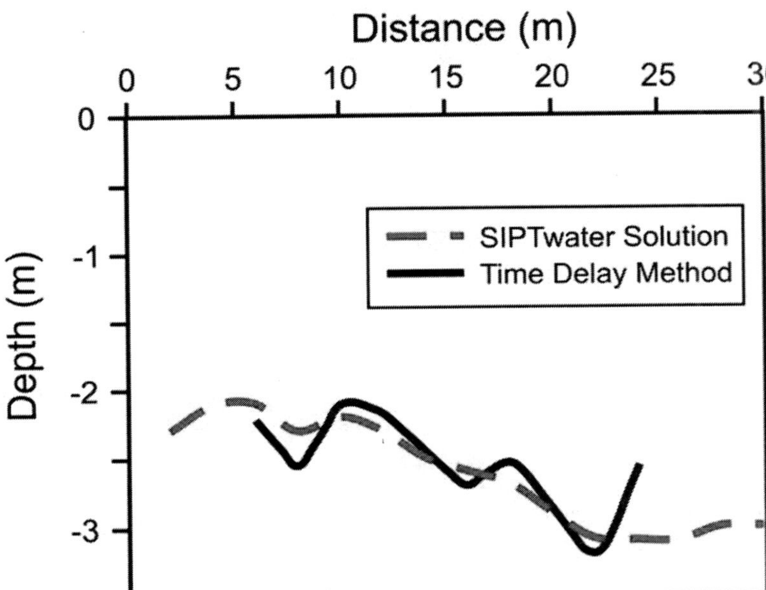

Figure 20. Comparison of the Time-Delay method to the ray tracing method of WindowsSIPTwater at Berkomann, Chad.

4. FINAL DISCUSSION

The indigenous men and women who we trained were eager to learn the methods and quick to apply them. Some may argue that extensive training is necessary to insure interpretation errors are avoided, which would be true if elaborate instruments are used or highly varied geology is encountered. More likely these inexpensive instruments will be used in localized regions by drillers and aid workers who will not travel great distances. In Tanzania the geologic stratigraphy consists of clay overlying bedrock and in Chad sand or silt overlies bedrock with depth the only unknown. Once a geophysical survey is finished the well that is subsequently drilled will provide excellent ground truth to compare to predictions. Such comparisons will undoubtedly lead to revisions and improvement of future geophysical interpretations in the region.

The seismic refraction method requires a laptop computer, and interpretation of earth resistivity is greatly aided with one. It is the laptop that is most likely to fail under field conditions. Fortunately even the cheapest of modern laptops has ample CPU and storage capacity to execute the software needed for the surveys. Acceptable laptop computers are now available for less than $400. Furthermore the computer can perform other useful tasks in support of drilling. For example, the free OpenOffice suite provides spreadsheet, word processing and presentation capability useful in report preparation. QuantumGIS is a public domain Geographic Information System (GIS) with good remote sensing capability in its GRASS module. Satellite images are often free and can be used with GIS to search for natural fracture trends, lineaments or moist regions [16]. An inexpensive GPS (i.e. less than $120) can be used to record well locations for input into the GIS as well as for navigation. If a geophysical team travels first and records the locations of good well sites, then a GPS will allow accurate re-establishment of the sites for subsequent drilling. In Tanzania original survey locations recorded with GPS by government geologists were successfully re-established by us within 2 meters.

Finally, free groundwater models, such as PMWIN, can give insight into groundwater flow in a region and the effects of well pumping in low yield areas. A laptop computer may serve many useful purposes in support of groundwater exploration, so its expense is readily justified.

The suite of geophysical resistivity and seismic refraction instruments including a laptop computer can be assembled for less than $850, the cost of a hand auger. More than half of that cost is the laptop, which can be used for much more than just geophysical surveys.

5. CONCLUSIONS

With the cost so reasonable (less than $850) this equipment should become common among those desiring to provide water in developing countries. The instruments are small, light, robust, inexpensive and easy to construct. With the free software available the interpretation of data is not difficult, although care still is needed. The free software described in this paper can be downloaded from http://cs.wheaton.edu/~jclark.

6. ACKNOWLEDGEMENTS

We wish to thank the Wheaton College Alumni Association for support both in the development of the resistivity device and for funding the field work. St. Paul Partners paid for drilling the wells, materials used in the resistivity meter, salaries of the Tanzanian drilling team, and general logistics within Tanzania. Keith Olson led the LifeWater International team and greatly encouraged the development of the instruments. It is to his memory that this paper is dedicated. The Evangelical Lutheran Church of Tanzania and Rev. Lambert Mtatifikolo provided much needed support while in Tanzania. Christy Page assisted with the Tanzanian field work and Daniel Clark helped in the development of the resistivity instrument. Dr. Phil Carpenter helped test the resistivity device.

James A. Clark, Richard Page (EEMET) in Chad, provided advice, hospitality, and friendship. Sambim Kosdonlengar (EEMET), Stephen Moss (Wheaton College) and Tanya Thomas (World Relief) provided essential help during all phases of the field work. The EEMET and the United Nations High Commission on Refugees (UNHCR) provided travel while in Chad. Chris Gregory tested the seismograph with the EG&G SmartSeis instrument. We would especially like to thank Don Church for his encouragement in promoting all phases of these projects. Publication of this paper was possible through funds provided by Will Chester and Emma Sarbu at the time of their marriage engagement. May their life together be God glorifying.

REFERENCES

1. W. M. Telford, L. P. Geldart, R. E. Sheriff and D. A. Keys, "Applied Geophysics," Cambridge University Press, Cambridge, 1982.
2. R. D. Barker, C. C. White and J. F. T. Houston, "Borehole Siting in an African Accelerated Drought Relief Project," In: E. P. Wright and W. G. Burgess, Eds.,

Hydrogeology of Crystalline Basement Aquifers in Africa, Geological Society Special Publication, No. 66, 1992, pp. 183-201.

3. R. Kirsch (Ed.), "Groundwater Geophysics: A Tool for Hydrogeology," Springer, New York, 2006.
4. W. Rabbel, "Seismic Methods," In: R. Kirsch, Ed., "Groundwater Geophysics: A Tool for Hydrogeology," Springer, New York, 2006, pp. 23-83.
5. H. R. Burger, A. F. Sheehan and C. H. Jones, "Introduction to Applied Geophysics: Exploring the Shallow Subsurface," W. W. Norton & Company, New York, 2006
6. R. Herman, "An Introduction to Electrical Resistivity in Geophysics," American Journal of Physics, Vol. 69, 2001, pp. 943-952.
7. M. J. Hornbach, "Development and Implementation of a Portable Low Cost Seismic Data Acquisition System for Classroom Experiments and Independent Studies," Journal of Geoscience Education, Vol. 52, 2004, pp. 386-390.
8. J. A. Olowofela, V. O. Jolaosho and B. S. Badmus, "Measuring the Electrical Resisitivity of the Earth Using a Fabricated Resisitivity Meter," European Journal of Physics, Vol. 26, 2005, pp. 501-515.
9. D. P. Ghosh, "The Application of Linear Filter Theory to the Direct Interpretation of Geoelectrical Resistivity Sounding Measurements," Geophysical Prospecting, Vol. 19, 1971, pp. 192-217.
10. D. P. Ghosh, "Inverse Filter Coefficients for the Computation of Apparent Resistivity Standard Curves for a Horizontally Stratified Earth," Geophysical Prospecting, Vol. 19, 1971, pp. 769-775.
11. E. Orellana and H. M. Mooney, "Master Tables and Curves for Vertical Electrical Sounding over Layered Structures," Interciencia, Madrid, 1966.
12. J. H. Scott, "SIPT—A Seismic Refraction Inverse Modeling Program for Timeshare Terminal Computer Systems," USGS Open-File Report, 1977, pp. 77-365.
13. S. H. Ward, "Resistivity and Induced Polarization Methods," In: S. Ward, Ed., Geotechnical and Environmental Geophysics: Review and Tutorial, Vol. 1, Society of Exploration Geophysicists Investigations in Geophysics, 1990, pp. 147-189.
14. D. W. Steeples, "Shallow Seismic Methods," In: Y. Rubin and S. S. Hubbard, Eds., Hydrogeophysics, Vol. 50, Part. 2, 2005, pp. 215-251.
15. R. H. Brown, P. A. Konoplyantsev, J. Ineson and V. S. Kovalevsky (Eds.), "Groundwater Studies: An International Guide for Research and Practice," UNESCO Studies and Report in Hydrology, No. 7, Sect. 9.1, 1997, p. 2.
16. S. Mutiti, J. Levy, C. Mutiti and N. S. Gaturu, "Assessing Ground Water Development Potential Using Landsat Imagery," Ground Water, Vol. 48, 2010, pp. 295-305.

CHAPTER 9

The Application of Joint Inversion in Geophysical Exploration

Ákos Gyulai, Mátyás Krisztián Baracza, Éva Eszter Tolnai

Department of Geophysics, Faculty of Earth Science and Engineering, University of Miskolc, Miskolc, Hungary

ABSTRACT

The paper presents a short overview about the application of joint inversion in geophysics. It gives also an alternative explanation for the term of "different data sets" and discusses what types of inversion procedures can be considered as joint inversion. Nowadays there are no standard standpoints using the appellation joint inversion. What is joint inversion? Based on the information matrix an answer could be given for this question what could be regarded as various types of data sets that are inverted simultaneously. We would like to expand the explanation—that is professed by many researchers—of the method that regards only the simultaneous inversion of data sets based on different physical parameters as joint inversion.

Keywords: Joint Inversion; Simultaneous Inversion; Geophysical; Serious Expansion

1. INTRODUCTION

The adaptation of joint inversion is the straightforward consequence of the applied complex interpretation methods in geophysics together with the development of inversion methods. The first phrasing and application of joint inversion were given by Vozoff in 1975 as "inverting several different kinds of geophysical measurements". He realized the simultaneous inversion of DC (Direct Current) resistivity and magnetotelluric measured data sets [1]. This method could be regarded as the basis of joint inversion methods. Vozoff as the first researcher and author of joint inversion method developed the simultaneous inversion of different measured electric conductivity data. We could put a question about why do we consider two data sets as different data sets? Is it because one is gained from DC measurement and the other one is from electromagnetic measurement or because the difference is in the deeper physical content through the data about the geological structure? Perhaps it is not so difficult question to answer.

Furthermore the term of joint inversion was applied in literature for the inversion of data measured by different physical principles. For instance, in near-surface exploration it generally represents the joint inversion of seismic and geoelectric data. Distinguishing inversion methods based on different data combinations researchers have tried to introduce different appellation e.g. cooperative inversion, simultaneous inversion, combined inversion etc. However, most of them remained with the term "joint inversion" for naming the method that gives the solution of various types of data sets inverted simultaneously. In most cases the complex geophysical exploration means the application of physically different methods hence these methods appeared more frequently as joint inversion.

2. THE METHOD OF JOINT INVERSION

The scheme overview of joint inversion by [2] demonstrates the joint inversion of diverse subsurface electrical soundings and VSP (Vertical Seismic Profiling) measurements. The seismic model parameter vector is (the first and the fifth layers are considered to be half-spaces)

$$\boldsymbol{P}_s = \left[h_2, h_3, h_4, v_1, v_2, v_3, v_4, v_5\right]^{\mathrm{T}} \quad ,(1)$$

where h_i denotes the layer-thicknesses and v_i denotes the seismic propagation velocities. The geoelectric model parameter vector is

$$\boldsymbol{P}_e = \left[h_2, h_3, h_4, \rho_1, \rho_2, \rho_3, \rho_4, \rho_5\right]^{\mathrm{T}} \quad ,(2)$$

where ρ_i denotes the resistivities. The joint parameter vector $(\boldsymbol{P})$ is the combination of vectors $\boldsymbol{P}_s$ and $\boldsymbol{P}_e$

$$\boldsymbol{P} = \left[h_2, h_3, h_4, v_1, \cdots, v_5, \rho_1, \cdots, \rho_5\right]^{\mathrm{T}} . \quad (3)$$

The common elements in the joint parameter vector in this example are the layer thicknesses as Equation (3) shows. The direct problem is

$$Y_i^{\text{calculated}} = Y\left(\boldsymbol{P}, S_i\right) , \quad (4)$$

where S is a geometrical parameter S = z (depth) in seismic case and S = r (electrode spacing applied in underground electric measurement) in electric case. Joint inversion arises from the joint solution of non-linear functions

$$Y\left(\boldsymbol{P}, S_i\right)^{\text{calculated}}, Y^{\text{observed}}$$
$$\Leftrightarrow \begin{cases} \Delta t\left(\boldsymbol{P}_s, z_i\right), \Delta t_i & i = 1, \cdots, N_i \\ \rho_a^{(\text{roof})}\left(\boldsymbol{P}_e, r_i\right), \rho_{ai}^{(\text{roof})} & i = N_1 + 1, \cdots, n_2 \\ \rho_a^{(\text{floor})}\left(\boldsymbol{P}_e, r_i\right), \rho_{ai}^{(\text{floor})} & i = n_2 + 1, \cdots, n_3 \\ R_a^{(\text{seam})}\left(\boldsymbol{P}_e, r_i\right), R_{ai}^{(\text{seam})} & i = n_3 + 1, \cdots, N \end{cases} \quad (5)$$

where

$$n_2 = N_1 + N_2, n_3 = n_2 + N_3, N = n_3 + N_4,$$

and N_1, N_2, N_3, N_4 are the total numbers of seismic, roof, floor and seam sounding data, respectively, Δt is the travel-time difference between the travel-times measured by the seismic receivers (geophones), $\rho_a^{(\text{roof})}$, $\rho_a^{(\text{floor})}$ denotes the apparent resistivities in roof and floor sounding measurements, and $R_a^{(\text{seam})}$ is the apparent resistance in the case of seam sounding measurement. The joint inversion involves the calculation of an appropriate parameter vector $\boldsymbol{P}$ where quantities $Y^{calculated}$ and $Y^{observed}$ fit the best. Roof and floor sounding means soundings measured on different formation boundaries with identical arrays of the surface measurements. Seam sounding means a sounding measured with vertical dipole array in mine tunnels. The basic equation for the solution of the linearized inverse problem is

$$\underline{\underline{G}}^{\mathrm{T}}\underline{\underline{G}}\boldsymbol{p} = \underline{\underline{G}}^{\mathrm{T}}\boldsymbol{y} \quad ,(6)$$

where $\boldsymbol{p}$ denotes the normalized model parameter vector dP/P, $\boldsymbol{y}$ is the normalized data vector

$(Y^{observed} - Y^{calculated})/Y^{calculated}$ and $\underline{\underline{G}}$ is the Jacobi's (sensitivity) matrix. Equation (5) presents that joint inversion makes use of one seismic and three types of geoelectric data sets. Roof and floor soundings provide different measured data sets because of the diversity of parameter sensitivities. Further geoelectric data measured in different arrays are involved in the joint inversion. One more seismic data set is attached to the previous data sets in the entire joint inversion. The data sets is different in three ways, i.e. in measurement array, in measurement (reference) level and physically different data sets are processed simultaneously in the inversion procedure, hence multiple joint inversion is realized.

2.1. The Role of Parameter Sensitivity in Joint Inversion

The definition of parameter sensitivity in geoelectric practice was given by Gyulai [3,4].

$$G_{kj} = \frac{P_j^{(0)}}{Y_k^{calc}}\left(\frac{\partial Y_k^{calc}}{\partial P_j}\right)_{P=P^{(0)}} , \quad (7)$$

where G_{kj} denotes the sensitivity of the k-th data with respect to the j-th parameter. The form of the Fischer Information Matrix [5] with the nominations of Equations (6) and (7) is

$$I \cong \frac{1}{\sigma^2}\left(\underline{\underline{G}}^{\mathrm{T}}\underline{\underline{G}}\right) , \quad (8)$$

where σ^2 is the uncertainty (standard deviation of geophysical data) and $\underline{\underline{G}}$ is the Jacobi's matrix [6,7]. The Fischer Information matrix gives the possibility to quantify the information content of geophysical data related to the geological structure, which is used to decide whether the data sets are different. The diversity of data sets are joined to physical content that substantially affects the result of the inversion and explains the meaning of joint inversion.

It is questionable whether the joint inversion of similar measurements data results in any improvement in the parameter estimation or in the covariance matrix. The improvement is affected strongly by the data noise and the parameter sensitivities including the correlation between the parameter sensitivities as the biggest problem of the inversion. Generally we can state that if more complex the model then the application of the joint inversion is more required. The researches of our Department prove that the adequate joint inversion causes significant improvement of accuracy and reliability in case of 1D (1 Dimensional) model approximations. The selection of the measurement methods is based on the parameter sensitivities and noise tests. The next chapter presents a short overview about some typical case of joint inversion without striving for completeness.

2.2. Typical Cases of Joint Inversion

A common fact in geophysical exploration is that several measured data are connected to one station, e.g. VES (Vertical Electric Sounding) and magnetotelluric data. The simultaneous inversion of two different types of sounding data sets measured in the same station is joint inversion because of the diversity of their parameter sensitivities. If the data is duplicated according to the regular sampling then it does not signify joint inversion. The inversion of the average of VES data measured in dip and strike direction is a single inversion. In case of complex geological structures the symmetrization of the measurement array causes information loss depending on the approximation applied in the model dimension during the inversion.

The VES measurements are made with close spacing in both cases (29 VES stations in each section, 25 m away from each other). The difference between the two figures is due to measurements made by electrode spacings of various directions in the same station. One of the electrode spacing is made in "dip direction" (in direction of the shorter structural element in cross direction), the other is in "strike direction" (in direction of the longer structural element in straight direction). The joint inversion of dip and strike direction measurements has an advantage only in case of 2D (2 Dimensional) or 3D (3 Dimensional) structures. The information content of measurements in dip or strike direction significantly differs in this case as Figures 1 and 2 show. The complex model was built up of 2D and 3D elements and the distribution of resistivity sensitivities was calculated from VES data measured in Schlumberger array. Synthetic data were calculated with Spitzer's algorithm [8]. The difference between the spatial and "along the profile" parameter sensitivities refers to the advantage that is provided by the simultaneous (joint) inversion. It is worth to look forward to the joint inversion of regional and "along the profile" measurements that are proposed to realize with series expansion based inversion.

The different orientations of the measurements eventuates differing data in case of anisotropic structures. Data derived from measurements with different directions open the door to simultaneous inversion. The previously mentioned simultaneous inversion can also be named as joint inversion.

In the recent the simultaneous inversion of data measured in different orientation (potential, gradient, dipoledipole array) is also joint inversion and it is especially referenced in cases of 2D and 3D structures.

Other various types of data sets can be derived from subsurface measurements where it is possible to measure the vertical derivates either in different depths besides the horizontal derivates or potentials (based on mine surveys measured in different levels and arrays by the Department of Geophysics). The simultaneous analysis of these data is also joint inversion.

Few decades ago it was a frequently used technique in geoelectric exploration that the data sets of some measuring array, was transformed into another array [9,10]. The inversion results of the transformed data set show no difference from the inversion results of the original data sets. The reason is that the transformation does not increase the information content of the data because it also affects the noise rate and it possibly increases the parameter sensitivities together with the noises. According to this the simultaneous inversion of the original and transformed data sets is not a joint inversion.

Joint inversion can also be the inversion of only gravitational or only magnetic data measured in different reference levels [11]. In this case the simultaneous inversion of the original data and the one transformed into other levels cannot be accepted as joint inversion.

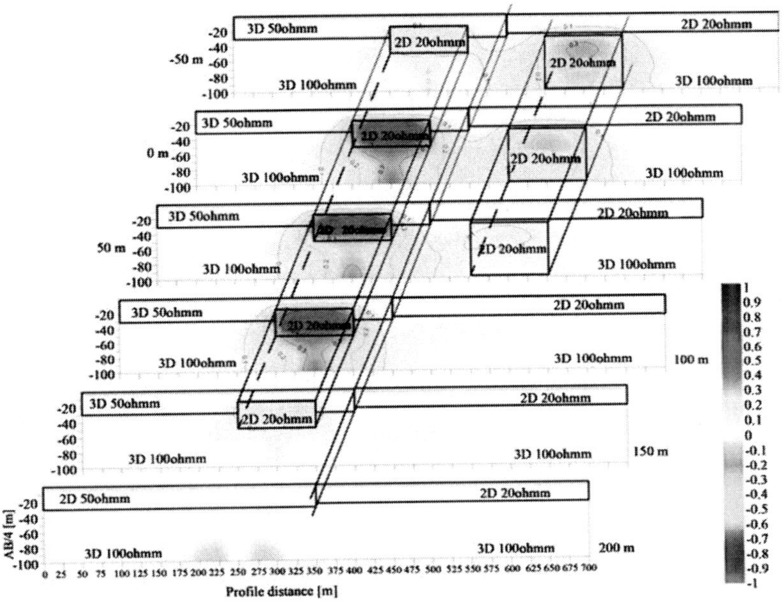

Figure 1. 3D parameter sensitivities of VES data measured in Schlumberger array in dip direction.

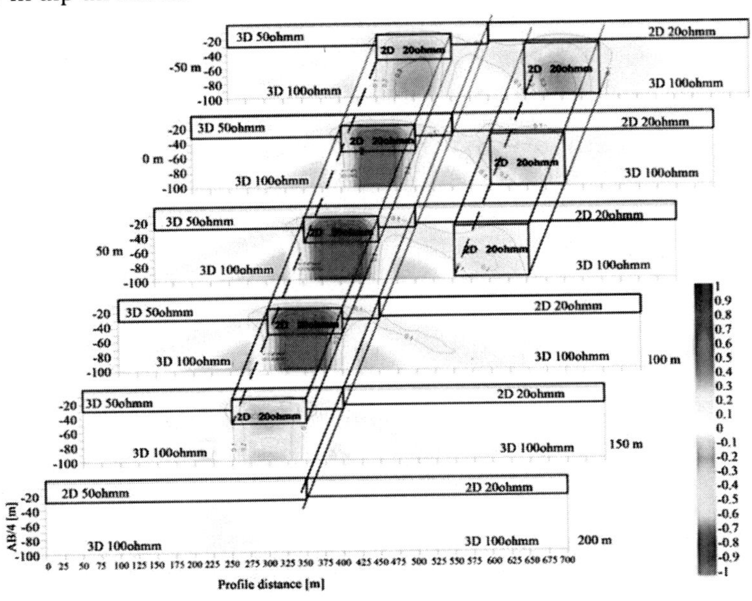

Figure 2. 3D parameter sensitivities of VES data measured in Schlumberger array in strike direction.

It is worthwhile to mention the time-lapse measurements used for the monitoring investigations in environmental geophysics where time dependence of physical properties of the structure is observed. The parameter sensitivity of time dependence means diverse information matrix which provides various type of data sets for the basis of a new type of joint inversion.

3. SPECIAL CASE: SERIES EXPANSION BASED JOINT INVERSION

The series expansion based inversion method can be considered as a simultaneous inversion technique, where each data measured along a profile (or over an area) assists in the determination of the series expansion coefficients describing the geological structure [12-14]. The inversion technique allows to couple further data measured by different physical principles in one inversion procedure, which can improve the reliability of the interpretation results [15,16].

The principle of the series expansion based inversion method is that variations of layer boundaries and physical parameters along the profile are described by continuous functions. The discretization of model parameters is based on series expansion [17]

$$p_k(x)=\sum_{q=1}^{Q_k} C_q^{(k)}\Phi_q(x), \tag{9}$$

where p_k denotes the k-th physical or structural parameter $(k=1,2,\cdots,K)$, C_q is the q-th expansion coefficient and Φ_q is the q-th basis function (up to Q number of additive terms), which is the function of the independent variable x. Basis functions are known quantities, which may be chosen arbitrarily for the environmental geological setting. In earlier studies, it was demonstrated that geoelectric structures can be described properly by periodic basis functions [4,13].

In a simpler way a set of power basis functions can be used in Equation (9)

$$\Phi_q(x)=x^{q-1}. \tag{10}$$

By this formulation all data measured along the profile are inverted simultaneously to determine the series expansion coefficients. The advantage of this technique is that the variation of structural and physical parameters can be described with significantly less unknowns (i.e. series expansion coefficients) than data, thus a highly overdetermined inverse problem is formulated, which is more favourable to solve than a marginally overdetermined or an underdetermined inverse problem. The strategy of choosing the number of expansion coefficients is detailed in Gyulai et al. [13].

In the inversion method the data of each VES station participate simultaneously in the determination of each coefficient hereby in the approximation of the parameters because the parameters of the structure are calculated from the coefficients. The determination of the structure with inversion method using series expansion means the simultaneous inversion of spatial data where the parameter sensitivities are different in each station. By our interpretation we consider this methodology a joint inversion approach. The simultaneous inversion of different data sets measured with different physical methods along the profile means further joint inversion based on the different measurements of orientation, lay-out of the array and reference levels and the different geophysical parameters. The further advantage of series expansion based inversion methods is the possibility of using joint inversion in case of not relevant boundaries. The unknowns of the inversion are the coefficients of the series expansion in Equation (9). In case of multi-layered models the condition of joint inversion is realized even if not all the layer-thicknesses fit to each other vice versa in every method. This term could also be fulfilled even if only a few coefficients are the same. For instance, the displacement of parallel boundaries can be described with the deviation of the constant coefficient while all the other coefficients fit to each other vice versa, thus still possible to perform "strong joint inversion". Some common coefficients could be given often to different physical parameters that improves the result of the inversion. This possibility of inversion using series expansion was presented earlier [18] and the results were published [19].

4. DISCUSSION AND CONCLUSION

Consider the problem again what we could call various types of data sets that are inverted simultaneously. The general answer to this question could be given by referring to the information matrix. Those measurements could be called diverse where the Fischer Information Matrix is different to the data.

Wide range of the researches shows that there is intention for the application of joint inversion in the fields of geophysics [20-33].

We do not consider it a matter of principle in joint inversion that the various types of data sets are inverted simultaneously, receive nominations separately. Our opinion is that the best is to leave it to the practice. The posterior shows that standardized system evolves hardly and there are huge variances between the uses of the nominations. The definition of joint inversion often is limited to simultaneous inversion based on different physical principles. We prefer more the comprehensive application of joint inversion as far as possible numerical qualification based on physical principles and regarding to the accuracy of the approximation. We consider it especially relevant in environmental geophysics.

5. ACKNOWLEDGEMENTS

Ákos Gyulai, Mátyás Krisztián Baracza, Éva Eszter Tolnai This described work was carried out as part of the TÁMOP-4.2.1.B-10/2/KONV-2010-0001 project in the framework of the New Hungarian Development Plan. The realization of this project is supplied by the European Union, co-financed by the European Social Fund. The authors are also grateful for the support of the research team of the Department of Geophysics, University of Miskolc.

REFERENCES

1. K. Vozoff and D. L. B. Jupp, "Joint Inversion of Geophysical Data," Geophysical Journal of the Royal Astronomical Society, Vol. 42, No. 3, 1975, pp. 977-991.
2. M. Dobróka, Á. Gyulai, T. Ormos, J. Csókás and L. Dresen, "Joint Inversion of Seismic and Geoelectric Data Recorded in an Underground Coal Mine," Geophysical Prospecting, Vol. 39, No. 5, 1991, pp. 643-665.
3. Á. Gyulai, "Parameter Sensitivity of Underground DC Measurements," Geophysical Transactions, Vol. 35, No. 3, 1989, pp. 209-225.
4. Á. Gyulai and T. Ormos, "A New Procedure for the Interpretation of VES Data: 1.5-D Simultaneous Inversion Method," Journal of Applied Geophysics, Vol. 41, No. 1, 1999, pp. 1-17.
5. P. Salát, Gy. Tarcsai, L. Cserepes, M. Vermes and D. Drahos, "Information-Statistical Methods of Geophysical Interpretation (in Hungarian)," Tankönyvkiadó, 1982.
6. W. Menke, "Geophysical Data Analysis-Discrete Inverse Theory," Academic Press, Inc., London, 1984.
7. M. Dobróka, "Introduction to Geophysical Inversion (in Hungarian)," Miskolci Egyetemi Kiadó, Miskolc, 2001, 209 p.
8. K. Spitzer, "A 3-D Finite Difference Algorithm for DC Resistivity Modelling Using Conjugate Gradient Methods," Geophysical Journal International, Vol. 123, No. 3, 1995, pp. 902-914.
9. S. P. Dasgupta, "A Note on the Conversion of DC-Dipole Sounding Curves to Schlumberger Curves," Geoexploration, Vol. 22, No. 1, 1984, pp. 43-45.
10. R. Kumar and U. C. Das, "Transformation of Schlumberger Apparent Resistivity to Dipole Apparent Resistivity over Layered Earth by the Application of Digital Linear Filters," Geophysical Prospecting, Vol. 26, No. 2, 1978, pp. 352-358.
11. Y. Li and D. W. Oldenburg, "Joint Inversion of Surface and Three Component Borehole Magnetic Data," Geophysics, Vol. 65, No. 2, 2000, pp. 540-552.

12. M. Dobróka and L. Völgyesi, "Inversion Reconstruction of Gravity Potential Based on Gravity Gradients," Mathematical Geosciences, Vol. 40, No. 3, 2008, pp. 299-311.
13. Á. Gyulai, T. Ormos and M. Dobróka, "A Quick 2-D Geoelectric Inversion Method Using Series Expansion," Journal of Applied Geophysics, Vol. 72, No. 4, 2010, pp. 232- 241.
14. M. Kis, "Investigation of near Surface Structures with Joint Inversion of Seismic and Direct Current Geoelectric Data (in Hungarian)," Ph.D. Thesis, University of Miskolc, Miskolc, 1998.
15. M. Dobróka and N. P. Szabó, "Combined Global/Linear Inversion of Well-Logging Data in Layer Wise Homogeneous and Inhomogeneous Media," Acta Geodaetica et Geophysica Hungarica, Vol. 40, No. 2, 2005, pp. 203-214.
16. D. Drahos, "Inversion of Engineering Geophysical Penetration Sounding Logs Measured along a Profile," Acta Geodaetica et Geophysica Hungarica, Vol. 40, No. 2, 2005, pp. 193-202.
17. M. Dobróka, "The Establishment of Joint Inversion Algorithms in the Well-Logging Interpretation (in Hungarian)," Scientific Report for the Hungarian Oil and Gas Company, University of Miskolc, Miskolc, 1993.
18. Á. Gyulai, T. Ormos and L. Dresen, "A Joint Inversion Method to Solve Problems of Layer Boundaries, Differently Defined by Seismics and Geoelectric," 6th Meeting of Environmental and Engineering Geophysical Society— European Section, 3-7 September 2000, Bochum.
19. Á. Gyulai and T. Ormos, "New Geoelectric-Seismic Joint Inversion Method to Determine 2-D Structures for Different Layer Thickness and Boundaries," Geophysical Transactions, Vol. 44, No. 3-4, 2004, pp. 273-300.
20. M. Breitzke, L. Dresen, J. Csókás, Á. Gyulai and T. Ormos, "Paratmeter Estimation and Fault Detection by ThreeComponent Seismic and Geoelectrical Surveys in a Coal Mine," Geophysical Prospecting, Vol. 35, No. 3, 1987, pp. 832-863.
21. P. Dell'Aversana, "Joint Inversion of Seismic, Gravity and Magnetotelluric Data Combined with Depth Seismic Imaging," EMG International Workshop, Capri, 2007.
22. J. Doetsch, N. Linde, I. Coscia, S. A. Greenhalgh and A. G. Green, "Zonation for 3D Aquifer Characterization Based on Joint Inversions of Multimethod Crosshole Geophysical Data," Geophysics, Vol. 75, No. 6, 2010, pp. G53-G64.
23. M. Dobróka, N. P. Szabó, E. Cardelli and P. Vass, "2D Inversion Borehole Logging Data for Simultaneos Fetermination of Rock Interfaces and Petrophysical Parameters," Acta Geodaetica and Geophysica Hungarica, Vol. 44, No. 4, 2009, pp. 459-482.
24. L. A. Gallardo and M. A. Meju, "Joint Two Dimensional DC Resistivity and Seismic Traveltime Inversion with Cross-Gradients Contrains," Journal of Geophysical Research, Vol. 109, No. B3, 2004, pp. B03311-B03321.
25. L. A. Gallardo-Delagdo, M. A. Perez-Flores and E. Gomez-Trevino, "A Versatile Algorithm for Joint 3D Inversion of Gravity and Magnetic Data," Geophysics, Vol. 68, No. 3, 2003, pp. 949-959.
26. E. Haber and D. Oldenburg, "Joint Inversion: A Structural Approach," Inverse Problems, Vol. 13, No. 1, 1997, pp. 63-77.
27. A. Hering, R. Misiek, Á. Gyulai, T. Ormos, M. Dobróka and L. Dresen, "A Joint Inversion Algorithm to Process Geoelectric and Surface Wave Data, Part. I.," Geophysical Prospecting, Vol. 43, No. 2, 1995, pp. 153-156.
28. M. Jegen, R. W. Hobbs, P. Tartis and A. Chave, "Joint Inversion of Marine Magnetotelluric and Gravity Data Incorporating Seismic Constrains. Preliminary Results of Sub-Basalt Imaging off the Farve Shelf," Earth and Planetary Science Letters, Vol. 282, No. 1-4, 2009, pp. 47-95.
29. N. Linde, A. Tryggvason, J. E. Peterson and S. S. Hubbard, "Joint Inversion of Crosshole Radar and Seismic Travel Times Acquired at the South Oyster Bacterial Transport Site," Geophysics, Vol. 73, No. 4, 2008, pp. G29- G37.
30. G. F. Margrave, R. R. Steward and J. A. Larsen, "Joint PP and PS Seismic Inversion," The Leading Edge, Vol. 20, No. 9, 2001, pp. 1048-1052.

31. R. Misiek, A. Liebig, Á. Gyulai, T. Ormos, M. Dobróka and L. Dresen, "A Joint Inversion Algorithm to Process Geoelectric and Surface Wave Seismic Data Part II," Geophysical Prospecting, Vol. 45, No. 1, 1997, pp. 65-85.
32. S. P. Sharma and S. K. Verma, "Solutions of the Inherent Problem of the Equivalence in Direct Current Resistivity and Electromagnetic Methods through Global Optimalisation and Joint Inversion by Successive Refinement of Model Space," Geophysical Prospecting, Vol. 59, No. 4, 2011, pp. 760-776.
33. N. P. Szabó, "Global Inversion of Well-Logging Data," Geophysical Transactions, Vol. 44, No. 3-4, 2004, pp. 313- 329.

CHAPTER 10

The Impact of Frequency in Surveying Engineering Slopes Using Ground Penetrating Radar

Angelo Indelicato

Dragages Hong Kong Ltd., Hong Kong, China

ABSTRACT

Ground Penetrating Radar (GPR) is one of the non-invasive techniques commonly used to identify "anomalies" in the ground. It has been proven very effective in different fields ranging from the location of pipes and other underground services to the identification of archaeological sites. After the 1994 Kwun Lung Lau accident in Hong Kong, the Government has been commissioning the feasibility of different geophysics techniques to identify any issues related to engineering slopes and retaining walls. Among the different techniques tested during phase I, Electrical Imaging (EI) and Ground Penetrating Radar (GPR) were the most applicable in the study of old masonry walls. This paper aims to stress the importance of using the appropriate frequencies during the GPR survey of engineering slopes. In order to do that, two independent contractors who used different frequencies to carry out the GPR survey on the same area will be compared.

Keywords: Ground Penetrating Radar, Slope, Survey, Hong Kong

1. INTRODUCTION

Ground Penetrating Radar, or GPR, is a non-invasive technology used to study subsurface stratigraphy. It employs electromagnetic waves with frequencies between 10 MHz and 4 GHz to detect changes in electrical properties [1] .

The electromagnetic waves are essentially the same as the radio waves that a car radio antenna receives from a broadcasting radio station. The reflection of the waves is due to the different dielectric properties of the material under the surface. The dielectric difference of an object within the ground produces a reflected wave. Using the reflections, a cross sectional reflection profile is created.

Since its first use, GPR has gained popularity as a shallow subsurface geophysical imaging tool due to its ease of use, portable recording and high-resolu- tion section [2] . It has been successfully applied to very diverse tasks ranging from the geological mapping of structures to the identification of defects in concrete [3] .

The use of this technique in the survey of old retaining walls was first considered in 1994 after the collapse of the old retaining wall in Kwun Lung Lau on Hong Kong Island. During this event, 1000 m^3 of debris ended up in the pathway, killing 5 people and injuring 3 others. For this reason, the Hong Kong Government was adamant in evaluating the potential of a geophysical survey to define the thickness of the retaining wall.

Retaining structures (Figure 1) in Hong Kong have been classified mainly into two groups: Pre-war and post-war [4] .

Post-war structures, which are mainly cantilever, counterfort and gravity structures, are generally designed and built in accordance with engineering principles [5] . However, pre-war walls are largely gravity structures of masonry or mass concrete construction. The condition and the geometry of these walls require new investigations to test their stability and integrity [5] .

When, in 1995, the Geotechnical Engineering Office (GEO) invited six contracting firms to take part in a site characterisation study to determine the possible uses of geophysical techniques for the assessment of subsurface stability indicator [4] , GPR was one of the non-invasive methodologies considered to determine possible slope instabilities. Geophysical techniques are more effective when studying old masonry walls than slopes because walls have a more clearly defined structure [5] .

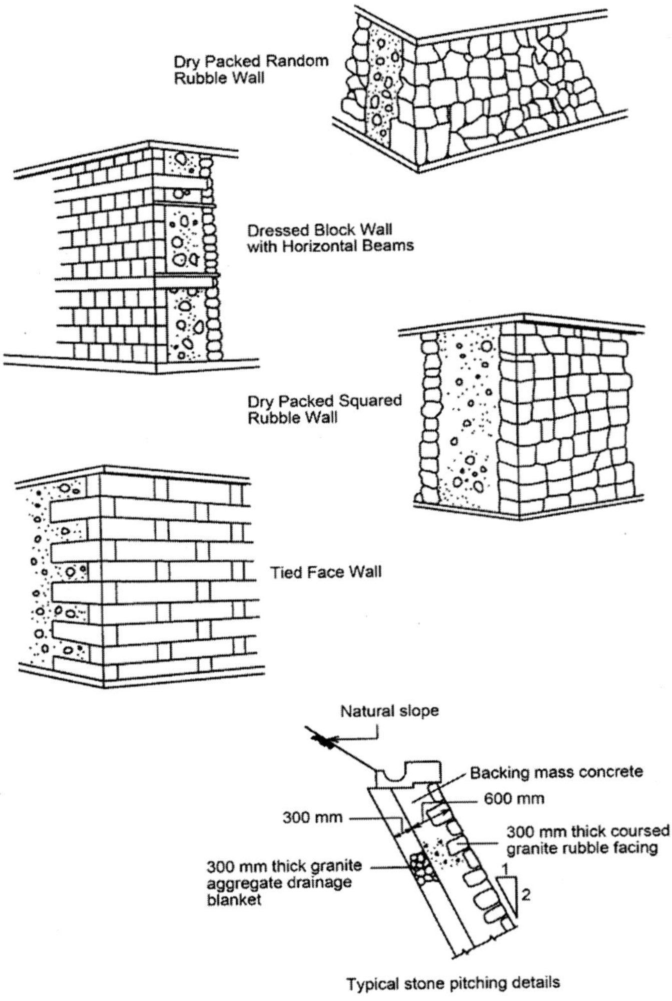

Figure 1. Types of retaining walls used in Hong Kong [5] .

This paper focuses on the GPR survey carried out by two of those subcontractors on a specific site using different frequencies (200 MHz and 35 MHz for Fugro and 500 MHz and 100 MHz for Golder). The results of the surveys will be compared to determine how the choice of the right frequencies might affect the interpretation of the data.

2. LOCATION

The location chosen for this paper is the masonry wall near the Police Quarters on Hollywood Road. This area was built in 1951 and originally served as the first dormitory for Chinese rank and file police officers (Site F, Figure 2).

Site F consisted of a mortared random rubble masonry retaining wall on the southeast to northwest end. A staircase at the southeast end connected a concrete covered platform at the top of the wall to a play area at the bottom [6] . This area was investigated using four horizontal drill holes (Figure 3(a)) and two trial pits located on the western part of the terrace above the retaining wall (Figure 3(b)). GPR and Resistivity Imaging survey were carried out in October 1996.

3. METHODOLOGY

The centre frequencies of commercial GPR antennas typically range from 25 MHz to 4 GHz [3] . Generally there is a direct relationship between the transmitter frequency and the resolution that can be obtained. At the same time, there is an inverse relationship between frequencies and penetration depth. For civil engineering purposes, most of the GPR frequencies range between 1.5 GHz and 400 MHz, for which the penetration is between 0.3 and 2 m.

While metallic targets are very good reflectors and PVC pipes generally produce much weaker signals, the presence of moisture and clay can also limit the signal penetration. Other external factors that might affect the performance are site topography and vegetation, as well as weather conditions [3] .

The presence of kaolinite clays, which are most commonly derived from weathered granites, gives reasonable results during the survey, whereas montmorillonite clays derived from weathered basalts have much higher signal at tenuation rates [3] .

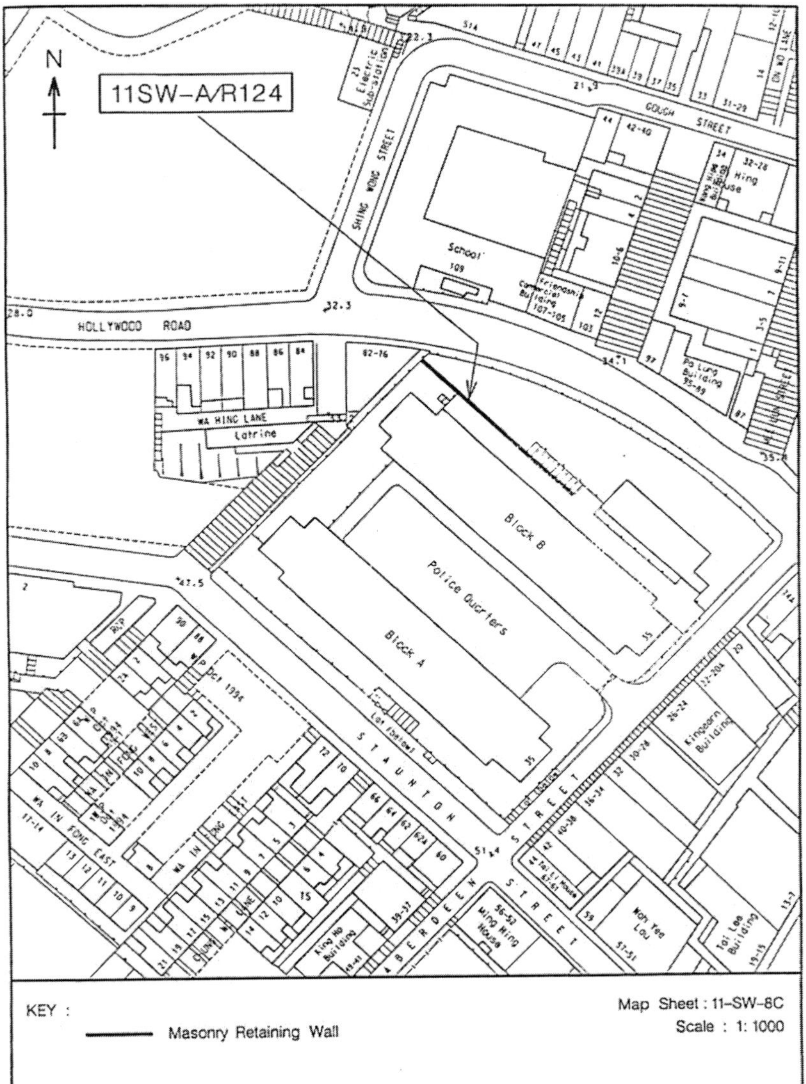

Figure 2. Site F (©Slope Information System 2016).

The ground resolution also depends upon a series of other factors such as:

- Amplitude and wave length of the transmitted pulse;

- The electrical properties and electromagnetic propagation characteristics of the host material;
- The complexity of the geology;
- Noise;
- Depth shape and size of the target;
- Electrical impendance of the target [7] .

Attenuation of the GPR signal can be caused by conductivity losses and dielectric or magnetic absorption. Conductivity losses are the most common within geological materials [8] .

The Geotechnical Engineering Office (GEO) determined that on Site F, the location of two traverse lines (TF-01 and 02) was to be surveyed using GPR. Other than that, the contractor could decide on the frequency to be used, as well as whether or not to survey the other traverses.

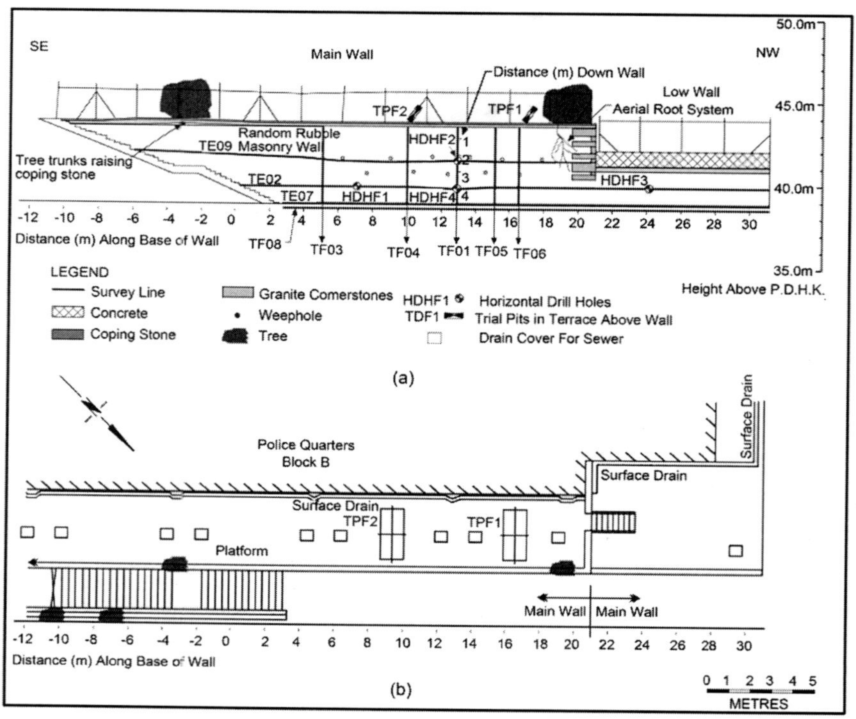

Figure 3. (a) Front elevation of the wall at Hollywood Road Police Quarters. (b) Plan of the wall and platform at Hollywood Road Police Quarters [5] .

For their survey, Fugro used both 35 MHz and 200 MHz frequencies for all the traverses (TF01 to 08). Golder, on the other hand, used two different frequencies for the GPR survey at Site F: a 500 MHz antenna on TF1 to TF6 and a lower frequency (100 MHz) one on TF02 [5] .

4. RESULTS

The GPR surveys carried out by the contractors produced very different results. Fugro's high frequency (200 MHz) accurately identified features near the surface (Figure 4). Granite and volcanic boulders were easily recognised, as well as sub- vertical features interpreted as drainage pipes.

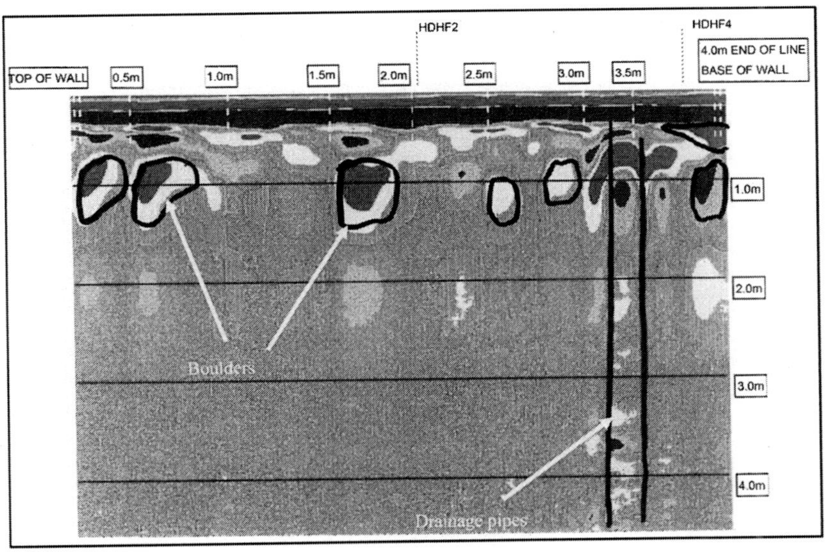

Figure 4. Results with 200 MHz antenna (©Fugro Report 1997).

Horizontal boreholes were drilled on the Site F wall and the cores showed the heterogeneity of the fill material, which contained large boulders also visible through the GPR survey. The trench pits excavated also confirmed the nature of the fill deposit, containing bricks, fragments of concrete slabs and metal objects easily picked up by the 200 MHz antenna. However, it is important to note that the thickness of the retaining wall was not so clearly marked.

The survey carried out by the same contractor at 35 MHz aimed, instead, to identify deeper features that would be out of the reach of higher frequencies. The results show the disappearance of the shallow features and the presence of a few parallel to sub-parallel linear features located between 2.5 to 10 m away from the surface (Figure 5). Initially they were interpreted as the interface between the fill deposit and the completely decomposed granite (CDG). However, none of the boreholes or the trench pits confirmed this interpretation.

The results obtained by the GPR survey highlight that the antenna at 200 MHz works very well to define superficial features present in the slope. It does not, however, define the true thickness of the masonry wall. In cases such as the collapse of the retaining wall in Kwun Lung Lau, the earlier identification of the type and size of masonry may have prevented such an event.

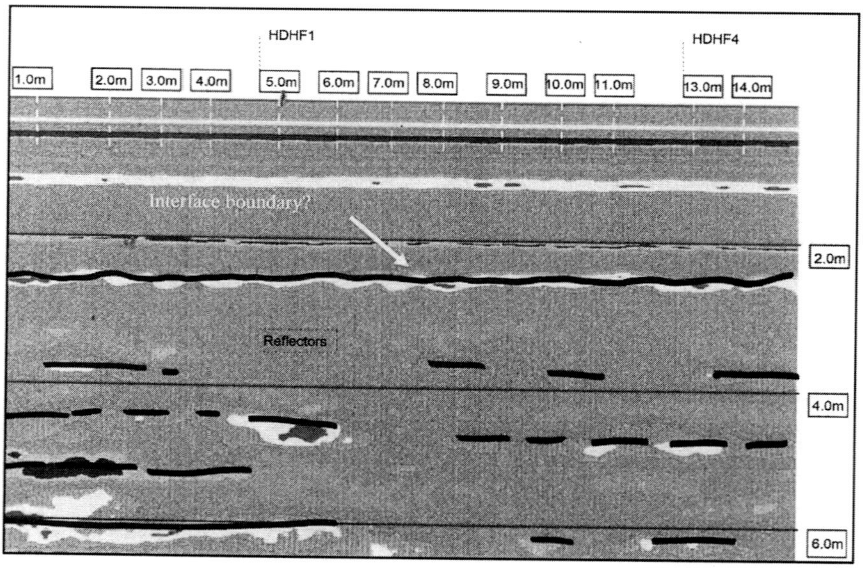

Figure 5. Results with 35 MHz antenna (©Fugro Report 1997).

Golder obtained better results at the same location using two frequencies: 100 and 500 MHz [5] . The 500 MHz frequency allowed the radargram to identify the thickness of the masonry wall. Other important features were also picked up with this frequency, features such as a granite-lined sewer system and loose cobble and boulder fill with concrete, brick fragments and voids in the trench dug for the new sewer [5] .

The 100 MHz frequency identified few coherent reflectors at a depth of 2 - 3 meters into the wall. The radar energy appears to be severely attenuated by the material present in this area. The correlation with electrical imaging indicates the presence of conductive material such as clay. The result from the 35 Mhz survey shares some similarities in identifying such linear features.

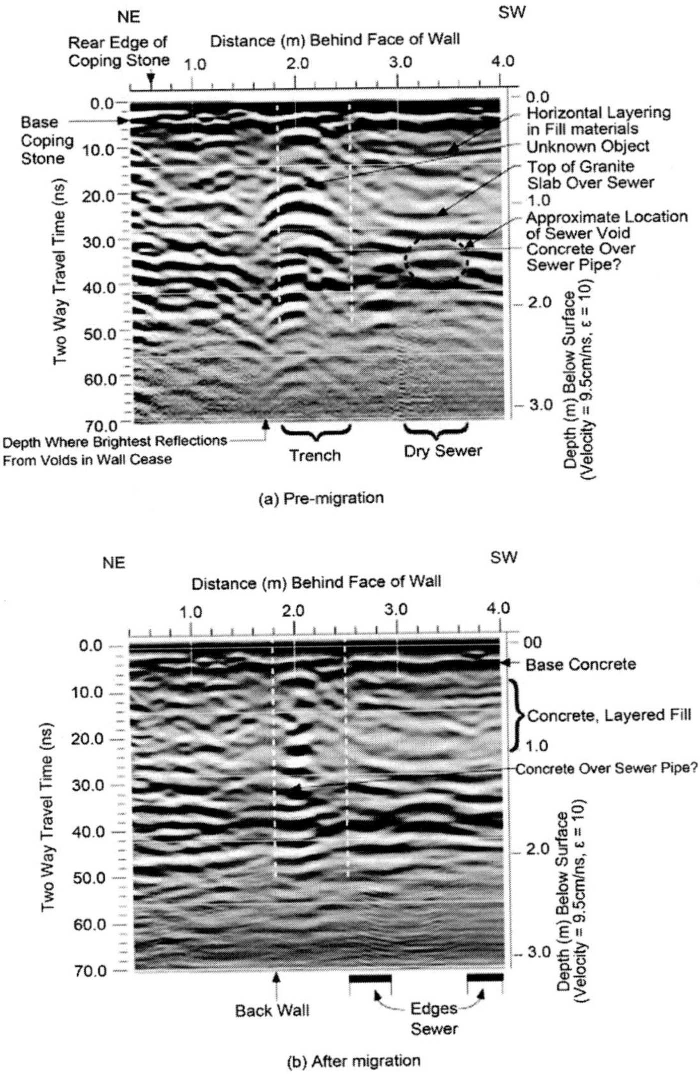

Figure 6. Comparison between pre-migration and after migration signal on Golder's survey [5] .

5. CONCLUSIONS

This geophysical technique has potential in the assessment of the stability of the pre-war retaining walls but it is important to choose the correct frequency. The choice must be related to the aim of the survey itself. A 500 MHz antenna provides better results than a 200 MHz one if we want to identify the thickness of a retaining wall and the presence of any shallow features present within the fill. For deeper surveys, the 35 MHz and 100 MHz antennas seem inadequate to identify major features. However, these frequencies have spotted a few common reflectors that can be interpreted as air void or accumulation of clay rather than a clear interface between fill and CDG.

Other factors have probably played an important role in identifying the main features behind the retaining wall and improving the quality of the results. Golder, for example, repeated the traverses more than once and made adjustments to the gain levels, ranges, sampling intervals and filter readings [4] . We should also consider the importance of migration, which aside from allowing us to correctly interpret the GPR data from structurally complex areas and determine the geometry and correct location of reflectors (or scatters) in the subsurface (Figure 6), it also provides quantitative information on the reflectivity coefficients at those surfaces or interfaces [8] . In the survey carried out by Golder, the migration process was used to successfully collapse the diffraction from an unknown piece of material such as a steel bar buried in the trench behind the wall [5] .

The frequency in surveying engineering slopes has a significant impact on Ground Penetrating Radar results. Nonetheless, other elements should be taken into account so as to have better quality results, with less room for misinterpretation.

ACKNOWLEDGEMENTS

The author wishes to thank Hong Kong University Professor L. S. Chan for his help in retrieving the unpublished reports and Miss Judy Wu for the editing work.

CITE THIS PAPER

Indelicato, A. (2017) The Impact of Frequency in Surveying Engineering Slopes Using Ground Penetrating Radar. International Journal of Geosciences, 8, 296-304.

REFERENCES

1. Milsom, J. and Eriksen, A. (2015) Field Geophysics, Wiley Blackwell, Chapter 10, 185-205.
2. Fisher, S.C., Stewart, R.R. and Jol, H.M. (1992) Processing Ground Penetrating Radar (GPR) Data. CREWES Research Report, 4, 11-1 to 22.
3. Yelf, R.J. (2007) Application of Ground Penetrating Radar to Civil and Geotechnical Engineering. Electromagnetic Phenomena, 7, 102-117.
4. Koor, N.P. (1996) A Review of Non-Invasive Engineering Geophysical Techniques for Site Characterization. Hong Kong Interim Report, No. 2 (2nd Draft).
5. Bishop, I. and Koor, N.P. (2000) Integrated Geophysical and Geotechnical Investigations of Old Masonry Retaining Wall in Hong Kong. Quarterly Journal of Engineering Geology and Hydrogeology, 33, 335-349.
6. Fugro Report (1997) Site "F" Masonry Wall near Police Quarters, Block B, Hollywood Road, Volume 2.
7. Lau, K.C. (1998) A Review of Downhole Geophysical Methods for Ground Investigation. Geo Report No. 99, Geotechnical Engineering Office, Civil Engineering Department, The Government of the Hong Kong Special Administrative Region.
8. Sena, A.R., Stoffa, P.L. and Sen, M.K. (2005) Migration of Ground Penetrating Radar Data in Heterogeneous and Dispersive Media. New Strategies for European Remote Sensing, Rotterdam, 711-719.

CHAPTER 11

Characteristic Seismic Waves Associated with Cryosphere Dynamics in Eastern Dronning Maud Land, East Antarctica

Masaki Kanao,[1] Akira Yamada,[2] and Mikiya Yamashita[3]

[1]*National Institute of Polar Research, Research Organization of Information and Systems, 10-3 Midori-cho, Tachikawa-shi, Tokyo 190-8518, Japan*

[2]*Geodynamics Research Center, Ehime University, Bunkyo-cho 2-5, Matsuyama, Ehime 790-8577, Japan*

[3]*Institute for Research on Earth Evolution, Japan Agency for Marine-Earth Science and Technology, Showa-machi 3175-25, Kanazawa-ku, Yokohama 236-0001, Japan*

ABSTRACT

Several kinds of natural source signals are recorded by seismic exploration stations on the continental ice sheet in Eastern Dronning Maud Land, East Antarctica, during 2002 austral summer. They include not only tectonic earthquakes, but also ice-related phenomena possibly involving recent global climate change. The recorded signals are classified into (1) teleseismic events, (2) local ice quakes, and (3) unidentified events (X-phases). The teleseismic waves show the high signal-to-noise ratio in spite of the small magnitude of the event; this indicates that it is highly feasible to study not only the local shallow structure but also the deep structure of the earth by using teleseismic events. Frequency spectra of the all waveforms represent discordances along the observation seismic profile. The abrupt change of topography in the valley along the seismic profile might cause both the anomalous frequency content and travel times. Finally, an origin of the X-phases is speculated as the intraplate earthquakes or possibly large ice-quakes (glacial earthquakes) around Antarctica, involving global warming appeared in polar region.

1. INTRODUCTION

It was generally understood by a majority of seismologists in the IGY era that no extreme earthquakes occurred in the Antarctic. Though the Antarctic was known as an aseismic region, some significant earthquakes have occurred both on the Antarctic continent and in the surrounding oceans. Although the seismic stations in Antarctica have been operated as a part of global network, no detailed studies of local events have been made until recently [1, 2]. The density of station distribution of both permanent stations belonging to the global network, as well as the temporary deployed stations at specified regions, has improved over time, and has recently permitted several studies of seismicity in the Antarctic [3–5].

Several kinds of signals associated with environmental changes in the atmosphere-ocean-solid earth system have been detected in Antarctica and the surroundings. Ice-related seismic motions for small-magnitude events are generally named "ice quakes" (or "ice shocks" for smaller events) and can be generated by glacially related dynamics [6–8]. Such cryoseismic sources include the movements of ice sheets, sea ice, oceanic tide cracks, oceanic gravity waves, icebergs and the calving fronts of ice caps. Sometimes, we can hardly distinguish between the waveforms generated by local tectonic earthquakes and those of ice-related phenomena; however, the cryoseismic sources are likely to be influenced by surface environmental conditions, and the study of their temporal variation may provide indirect evidence of climate change.

Temporary deep seismic exploration was performed by the Japanese Antarctic Research Expedition (JARE) on the Mizuho Plateau, Eastern Dronning Maud Land, East Antarctica, during the 2002 austral summer. The exploration was conducted as a part of the project on "Structure and Evolution of the East Antarctic Lithosphere (SEAL)" [9, 10] in order to reveal the structure of the crust and upper mantle beneath the investigated region (Figure 1). The SEAL-2002 experiment recorded chiefly the artificial waveforms originated by seven large explosions with 161 temporary stations along the transect profile, which is almost parallel with the coast of the Lützow-Holm Bay [11]. They determined the P-wave velocity structure of the crust and upper mantle as well as the overlying ice sheet by seismic travel time analysis.

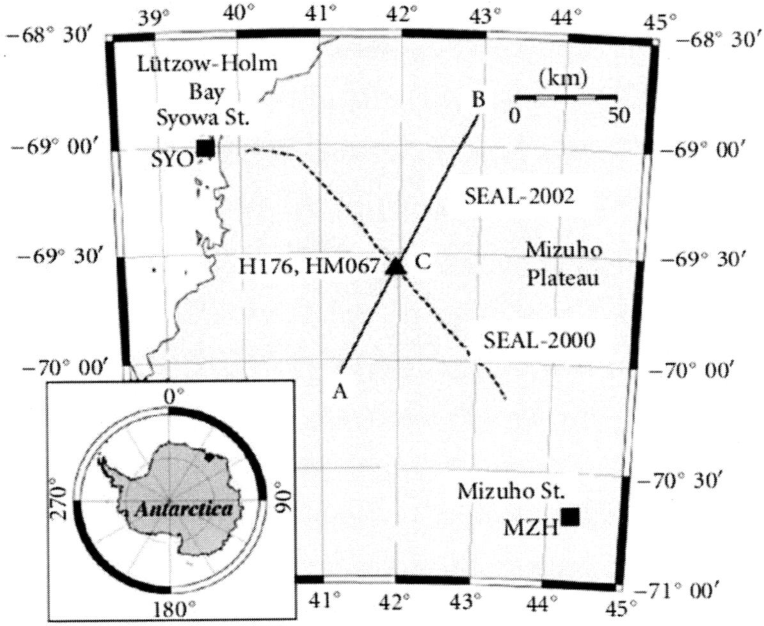

Figure 1: Location map of the seismic profile for SEAL-2002 exploration (solid line; between A and B), together with SEAL-2000 profile (broken line) on the Mizuho Plateau, Eastern Dronning Maud Land, East Antarctica. The data retrieved by SEAL-2002 exploration were used in this study. Names of nearby Japanese stations are as follows: Syowa St. (SYO) at the coast of Lützow-Holm Bay and Mizuho St. (MZH) on ice-sheet plateau. The intersection station of the two seismic profiles (C) is indicated by a solid triangle (H176 and HM067 are station codes for SEAL-2000 and SEAL-2002, resp.). The distance measured from the southwestern terminal (A) along the SEAL-2002 profile is utilized as the abscissa in several record sections in this paper (Figures 3, 4, 6, 7, 9, 10, 11, 12, 13, 14, and 15). The extended area in Antarctic continent is indicated by solid rectangle as shown at the lower left.

An interesting feature of the obtained structure presents a complex topography of the boundary between ice sheet and crystalline crust. Several geophysical data measured along the SEAL-2002 profile are compiled in Figure 2. Elevation of ice-sheet surface, free-air gravity anomaly, depths of the bedrock inferred from radioecho soundings [13], together with the bedrock from seismic travel time analysis [11], is shown in the same figure. The bedrock topography is attributed by a remarkable valley structure having about 10 km width beneath the middle part of the seismic profile. The P-wave velocities beneath this valley area are estimated to be slower (5.9–6.0 km/s) than those of the other area in the profile (6.1–6.2 km/s).

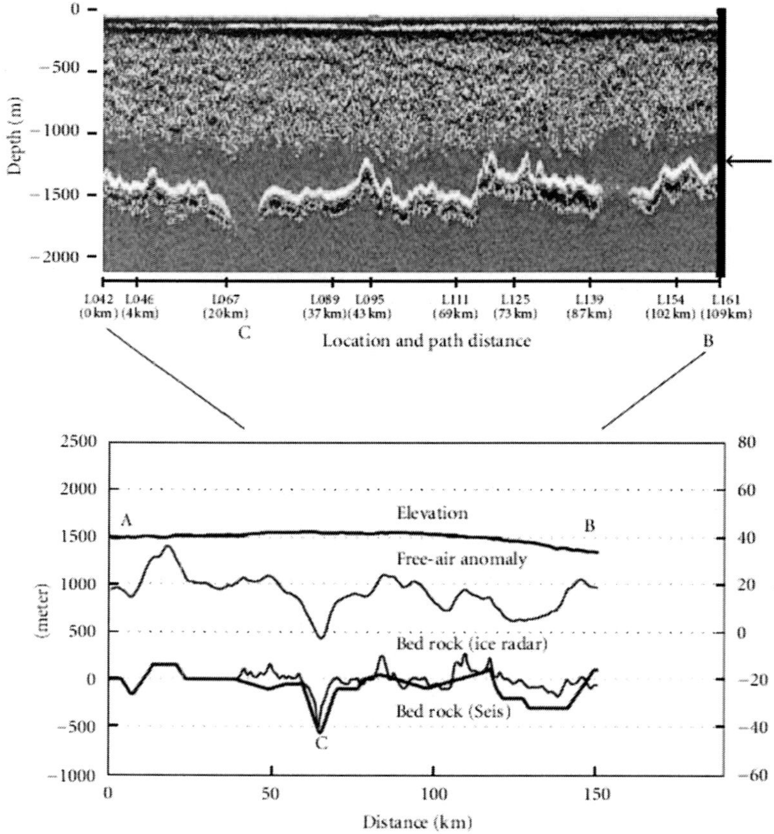

Figure 2: Comparison of several geophysical data measured along the SEAL-2002 profile (A-C-B) in Figure 1; (lower) elevation of the ice-sheet surface (scale on the left side), free-air gravity anomaly (scale on the right side, modified after [11]), depths of the bedrock inferred from radioecho soundings [13], together with those from seismic travel-time analysis [11]. (upper) detail result from ground-traverse based radioecho soundings between the stations (L042)-C-B (modified after [13]). Several refection layers of the inner ice-sheet together with strong reflections from bedrock are identified. It is also noticed that there are areas in the profile without

any reflections from the bedrock (around C, and the area between L139 and L154). In the areas, there is a possibility of water-filled layer spreading over the bedrock surface just as the valley structure (see the text for details).

In addition to the bedrock topography, detailed information of reflection images from radioecho sounding within the ice sheet, as well as strong reflections from bedrock surface, is represented in upper of Figure 2(modified after [13]). It is noticed that there are specific areas in the profile without any reflections from the bedrock (around C, and the area between the stations of L139 and L154). A plausible possibility of low P-wave velocity appeared in the valley is explained by the presence of water infilling the area [14], which also produces less reflections by radioecho sounding. These phenomena are reported as the existence of a significant number of subglacial lakes in wide area of inland Antarctic continent [15, 16].

In addition to artificial seismic sources by deep explosion exploration, several kinds of characteristic waves associated with natural environmental variations are clearly recorded in the SEAL-2002 data. In spite of the recording time lengths limited to only several minutes in a day during the one month of the experiment, a total of sixteen time windows clearly identified the contamination of characteristic natural source events (regional and teleseismic events, local events including ice quakes, and unknown phases; Table 1).

In this paper, remarkable features of the anomalies in the travel-times, amplitudes, and frequency contents are demonstrated in several examples of the recorded waveforms from natural sources recorded by the SEAL-2002 data. In discussion, particularly, the origin of unknown characteristic phases is treated by focusing on the relationship with recent dynamics of ice-related phenomena (including glacial earthquakes), climate changes appearing in polar region, as well as the hidden (uninvestigated) microseismic activities around the Antarctic.

2. DATA AND ANALYSIS

A total of 161 seismic stations by vertical component with a natural period of 0.5 s are aligned in the SEAL-2002 exploration profile (Figure 1) with an interval of approximately 1 km (0.5 km interval for a 10 km span in the middle part of the line). Digital recordings with duration of seven and half minutes were performed twice a day (14:00 and 16:00 in UTC) during the experiments from January 8 to January 27, 2002. More detailed information on the observation specification is given by [11]. In spite of these short recording times, several kinds of seismic phases are identified and classified in Table 1.

Table 1: Recordable event list during the SEAL-2002 exploration. A total of sixteen data windows are listed with a raveling of the sequenced numbers for "event#". Source parameters for teleseismic events, detected seismic phases, local tectonic event, ice-quakes, and X-phases (unknown phases) are notified for individual data window(with italic and bold italic).Some of the phases are not clearly identified; however, the most considerable candidates for each phase are raveled. The teleseismic events, whose waves are recorded, are indicated inbold.

Data time window (UTC)		Recordable events in data window and event information								Phases	Event#
		hh:mm:ss	Latitude	Longitude	Depth	Mb	Location name	Delta	Back azimuth		
2002/1/7	14:00	13:26:26	18.957	144.958	599.4	5.7	Mariana Islands	112.155	95.33	***ice-quake?***	eq001
2002/1/7	16:00	15:49:07	−25.049	−176.839	33	4.7	**South of Fiji Islands**	81.6	215.06		eq002
		15:49:26	−25.1239	−177.1569	200	4.7	**South of Fiji Islands**	81.463	215.327	*P,ice-quake?*	eq002
		15:54:44	−49.8259	110.806	10	4.2	**Southeast Indian Ridge**	37.299	94.61		eq002
2002/1/9	16:00	15:41:01	34.126	139.456	128.8	4.9	Near S. Coast of Honshu	124.119	83.436	*X-phase?*	eq003
2002/1/13	16:00	15:49:11	−19.67	−174.774	63.4	4.7	**Tonga Islands**	87.24	214.347	***P,pP,ice-quake?***	eq004
2002/1/14	14:00	14:03:20					(local event)			***ice-quake?***	eq005
2002/1/14	16:00	15:26:38	−55.547	−25.91	33	4.7	South Sandwich Islands Region	32.262	260.473		eq006
		15:49:16	−8.3799	127.789	33	4.0	Timor Region, Indonesia	80.747	89.044	*X-phase?*	eq006
2002/1/15	16:00	15:51:54	8.165	127.213	10	4.0	Philippine Islands Region	95.934	82.713		eq007
		15:52:33	−5.8581	105.7325	0	4.1	Sunda Strait	75.567	67.154	?	eq007
2002/1/16	14:00	13:33:06	23.901	125.415	10	5.3	Southwestern Ryukyu Islands	109.916	75.271	Local	eq008
2002/1/16	16:00	15:26:52	−6.157	105.244	10	4.9	Sunda Strait	75.128	66.779	***ice-quake?***	eq009
2002/1/18	14:00	13:54:17	−41.89	172.71	90	3.8	South Island, New Zealand	63.144	140.657	—	eq010
2002/1/20	14:00	13:07:11	−45.923	34.891	10	5.3	Prince Edward Islands Region	23.98	347.807		eq011
		13:45:27	-9.809	159.254	33	4.6	Solomon Islands	90.012	118.86	*Shot,X-phase?*	eq011
2002/1/21	14:00	13:47:11	26.721	103.139	10	4.0	Yunnan, China	105.511	54.418		eq012
		13:50:58	−31.255	179.633	436.6	4.5	**Kermadec Islands Region**	74.862	143.321	P	eq012
2002/1/21	16:00	15:42:29	−15.255	167.371	100	4.6	Vanuatu Islands	87.222	128.044		eq013
		15:42:35	−15.384	167.483	115.5	5.3	Vanuatu Islands	87.13	128.185	***ice-quake?***	eq013
		15:47:23	20.2	169.4	10	6.9	North Pacific Ocean	121.459	119.051		eq013
2002/1/24	16:00	15:24:05	3.513	95.607	33	5.3	Off W. Coast of Northern Sumatera	81.333	54.413	Local	eq014
2002/1/27	14:00	13:42:43	0.775	29.716	10	4.7	Zaire	70.702	347.022		eq015
		13:42:44	51.725	177.014	50.9	4.8	Rat Islands, Aleutian Islands	152.591	107.361	*X-phase?*	eq015
		13:50:23	−7.3999	116.858	160	4.4	Bali Sea	77.849	78.378		eq015
2002/1/28	16:00	15:49:32	−15.17	−173.472	33	4.8	Tonga Islands	91.876	214.057	—	eq016

Characteristic seismic waves from natural sources recorded in the experiments are evaluated by the following process. These obtained waves have a high signal-to-noise ratio in spite of the small magnitudes. These waves were classified into the following three categories: (1) teleseismic events; for example, those occurring in the Kermadec Is. Region, (2) local events extremely attributed by ice-quakes around the Lützow-Holm Bay Region, near the Mizuho Plateau, and (3) the source unidentified events by present information; which appeared to have the origin around Antarctic (here we call these phases as the "X-phases"). Regarding the teleseismic events, some phases are not clearly identified, however; the most considerable candidates for each phase are picked up in Table 1.

Figure 3 is an example of waveforms for a record section of artificial active source by dynamite explosive by the SEAL-2002 exploration (shot no. SP3 (2002/01/20/14:01 UTC); event#; eq011 in Table 1) together with band-pass filtered (1.0–4.0 Hz) waveforms for the same record section. A few local events that considered as the ice-quakes are recognized on the seismograms. The latest arrival phases in the lower have relatively long duration time in 20–30 s, which might be considered as the large ice-quakes, otherwise the source unknown events (X-phases). Another waveform example containing teleseismic event (event#; eq004; Tonga Islands) is presented in Figure 4. Many local events which appeared to be ice-quakes are clearly recognized with apparent slowness determined by different injection back-azimuths into the seismic profile.

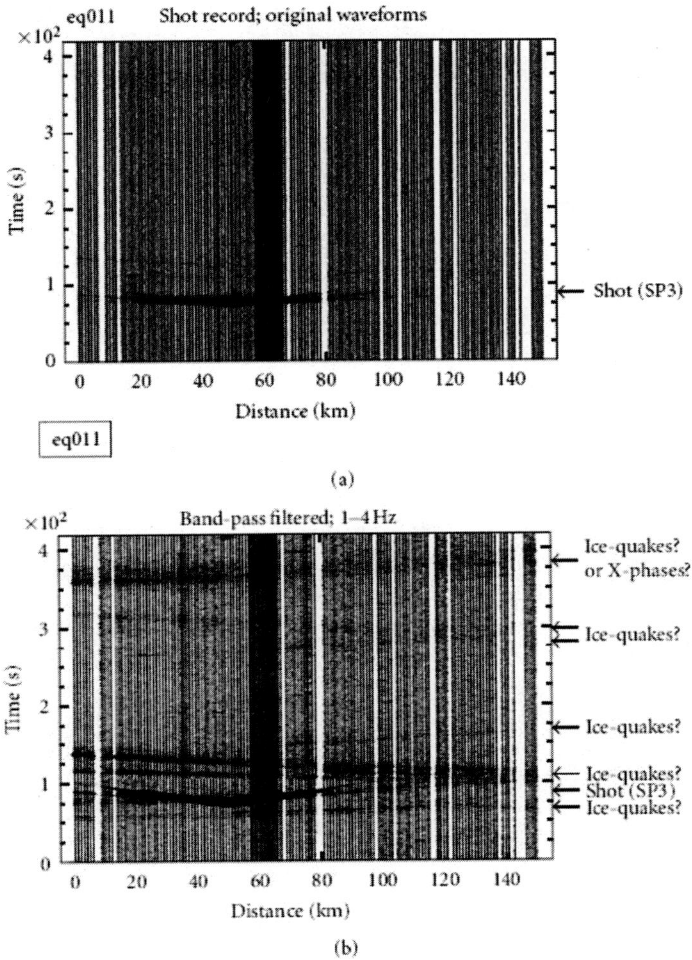

Figure 3: (upper) Original waveforms for record section of artificial active source by SEAL-2002 exploration (shot No. SP3 (2002/01/20/14:01 (UTC)); event#; eq011). (lower) Band-pass filtered (1.0–4.0 Hz) waveforms of the same record section. Many local events which appeared to be the ice-

quakes are recognized in the record. Starting time of the vertical axis is 14:00 (UTC) on January 20, 2002.

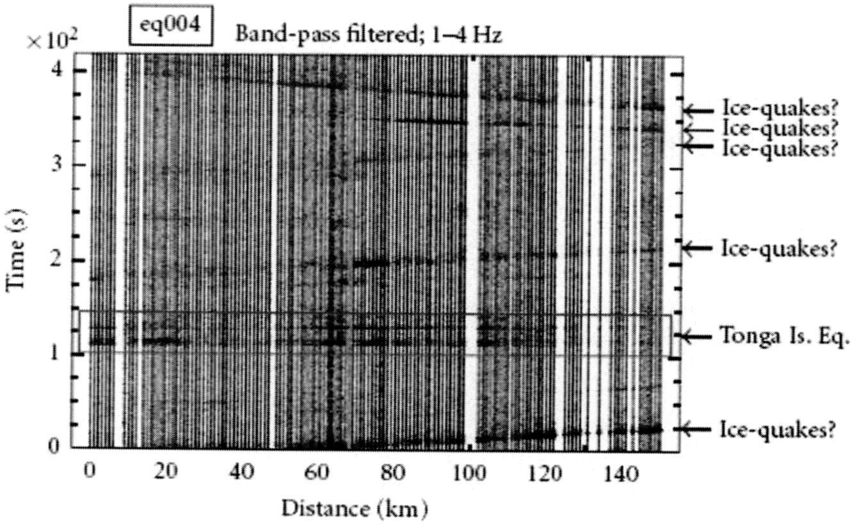

Figure 4: Band-pass filtered (1.0–4.0 Hz) waveforms of the record section including teleseismic event (event#; eq004; Tonga Islands, indicated by red squared area). Many local events which appeared to be ice-quakes are clearly recognized. The vertical axis starts from 16:00 (UTC) on January 13, 2002. Seven minutes of the data length.

An overview of the observed waveform signals indicates the difference in (a) coherency of signals and (b) amplitude fluctuations along the seismic profile. Here, we chiefly performed two analyses by a stacking method and a running spectrum in order to figure out the above two differences. First, we applied to several waveform dataset by a phase-weighted stacking (PWS) approach [17]. The PWS is a kind of nonlinear stacking procedure, and we use it instead of an ordinary linear stacking method. Each data point of a linearly stacked trace is weighted with an absolute value on average, as a sum of instantaneous phases for each trace. Thus, coherent signals throughout the seismic profiles can be constructively stacked and incoherent signals, even if some waveforms have larger amplitudes, result in a stacked signal with small amplitude due to the destructive sum of the phase components. The stacking procedure is performed by adopting variation in the slowness with a scanning interval of 0.001 s/km. We measure the slowness along the SEAL-2002 profile.

Second, against the several dataset, a running spectraum is taken by applying the FFT to a 2 s window (both sides of the window are hanning-tapered) with a time interval of 1 s. Before calculating the spectrum, instrumental response is deconvolved from the original waveforms. By using these methods, we can achieve the discordances in both frequency and arrival times of the waveforms. Significant anomalous features are revealed in the waveforms of the stations just above the valley structure, as the topographic boundary between the ice sheet and the topmost crust. A difference in the PWS stacked waveforms, as well as the running spectra, is clearly obtained by applying to three kinds of waveforms of teleseismic, local, and unknown X-phase events.

3. TELESEISMIC EVENTS

A hypocentral map of the plausibly detected teleseismic events during the SEAL-2002 exploration is shown in Figure 5. The teleseismic events shown in the map are the same listed in Table 1, and an expanded local map of the studied area on the Mizuho Plateau, East Antarctica, is inserted on the same figure. Five teleseismic events are successfully recorded (indicated in red in Table 1). Some events are located at larger epicentral distances over 90 degrees. Seismic phases expected to be recorded at the profile from such events are Pdiff and core phases of PKP family which have usually small amplitudes due to those longer wave paths within the earth. Therefore, it is expected that the events of smaller magnitude cannot be detected clearly on the seismographs in Antarctic stations.

Figure 6 indicates a teleseismic event from the Kermadec Is. Region (January 21, 2002, 13:50:58.8 (UTC), 4.5 m_b; event#; eq012). The epicentral distance of the event to the SEAL-2002 profile is about 75°. The direct P waves can be evidently and coherently seen throughout the recorded stations of the seismic profile, regardless of the degree of magnitude for the earthquake. The high signal-to-noise ratio of P-waves for this event is achieved by high-sensitivity recording deployed at the Mizuho Plateau, which could give rise to the benefits of investigation not only to local structure of Antarctica but also to deep interiors of the Earth.

The teleseismic signals include a sufficient correlation of waveforms between the neighborhood stations and have frequency components lower than 2.0 Hz due to long-path propagation in the Earth's mantle. It is also identified clearly that the phase after 30 s from the first arrival appeared. This phase has almost the same slowness as that of the direct P-waves. Therefore, the origin of this later phase is presumably converted or reflected waves from discontinuity or heterogeneity located near the event source. A possible explanation of the later phase is the S-P converted waves at the 660 km discontinuity (S660P) beneath the earthquake. A travel time difference in stacked waveforms between the direct P and the later wave leads to a preliminary estimation of a converted depth at about 760 km based on a velocity model of IASP91 model [18]. The depth is deeper than those reported in the previous studies (~720 km [19, 20]) for the event region. Thus interpretations as seismic scatterers [21] or multiple discontinuities in the uppermost lower mantle [22] might be possible for the waves observed here. However, at the present, it is difficult to constrain the origin of the wave because the direction and length of the observation line are unfortunately unsuitable for slowness analysis which gives us additional and useful information of the origin.

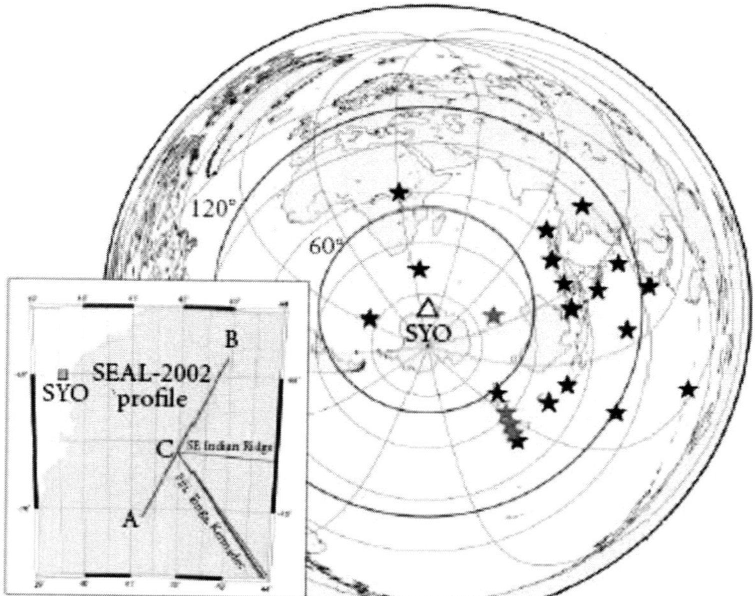

Figure 5: Hypocentral map of the teleseismic events during the SEAL-2002 exploration. Red/black stars indicate events whose waves can be identified/unidentified in the observed records, respectively (corresponding to Table 1). A black triangle indicates SYO. Blue circles show great circle distances of 60 and 120 degrees from SYO. (inset) An extended map of the studied area on the Mizuho Plateau (red line: SEAL-2002 profile, blue lines, incoming ray paths from observed teleseismic events).

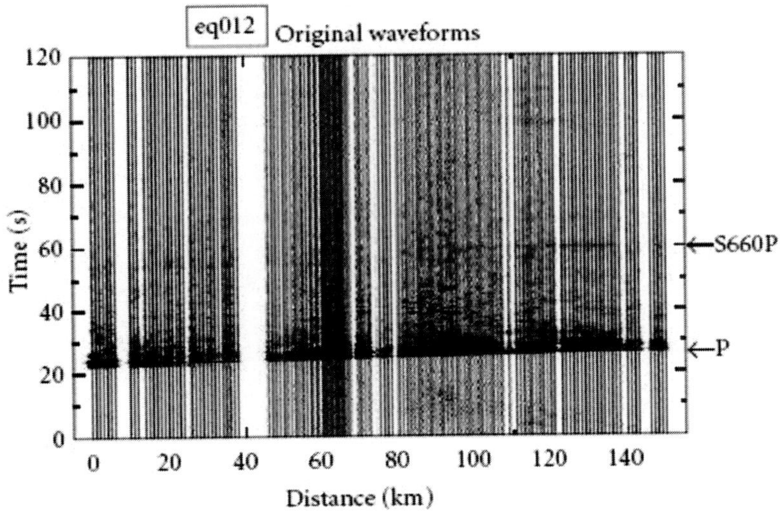

Figure 6: Record section indicating a typical teleseismic event occurred at Kermadec Is. (; event#; eq012; modified after [23]). The vertical axis is the relative time measured from 14:01:30.0 (UTC) on January 21, 2002. Original (no filtered) traces for all station along the profile represent high signal-to-noise ratio. Very strong signals during 20–30 s are composed of direct P and possibly reverberations from the ice-sheet. Faint signals found in 50–60 s may be an S-P converted wave at the 660 km discontinuity just beneath the event (S660P).

Figure 7 represents the PWS images applied to teleseismic waves of the Kermadec Is. event (modified after [23]). The individual traces in the upper of Figure 7 are the low-pass filtered ones with a cutoff of 2.0 Hz. A linearly stacked trace is calculated by searching the slowness that maximizes the amplitude of stacked direct P-waves. All the traces are aligned on the direct P-waves by applying the cross-correlation method between each trace and the stacked P trace. The alignment of direct P-waves represents a line of zero slowness, and also identified in the upper of Figure 7. The term of "slowness" here means a relative value compared with that of the direct P. Then we apply the PWS to the aligned traces by changing the slowness. The stacked traces are converted into envelopes by using the Hilbert transform and are plotted as a contour map in the lower of Figure 7.

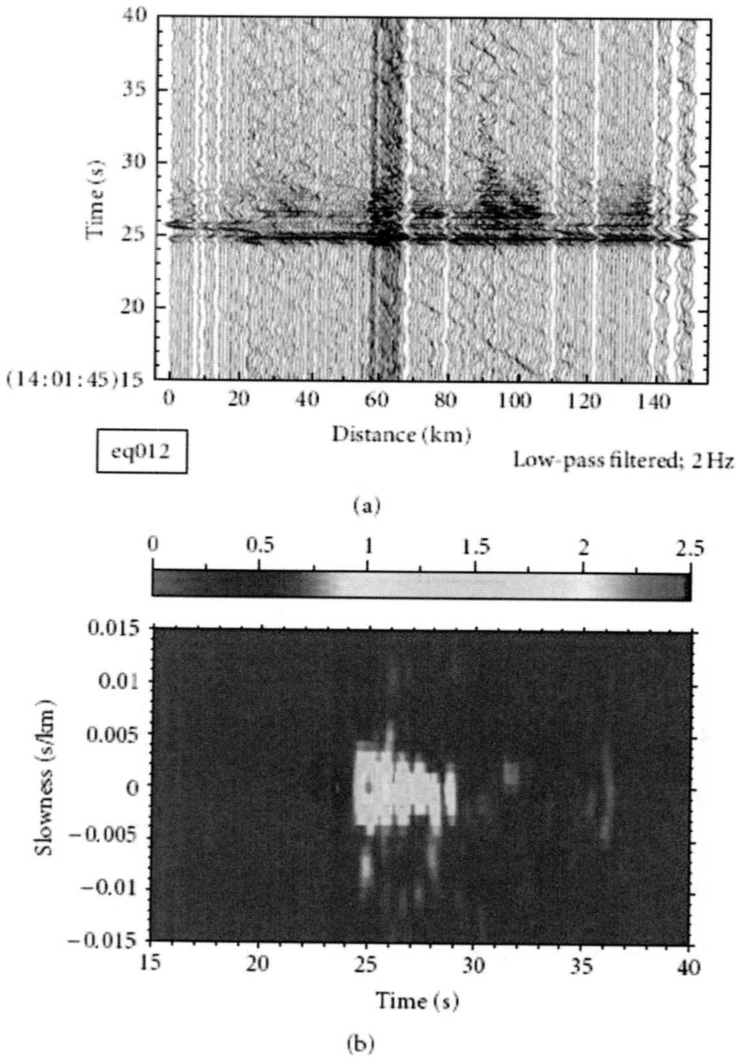

Figure 7: (upper) Time-expanded record section of the Kermadec Is. event (event#; eq012), which are aligned on the onset of the direct P and are 2.0 Hz low-pass filtered. The pivot station is located at 74 km. The vertical time-window contains the direct P and the core-reflected PcP. (lower) Contour map of the phase-weighted-stacking (PWS) applied to all traces of upper in the same time-window (modified after [23]).

Successive peaks with an interval of about 0.8 s between 25–29 s in Figure 7 are the direct P-waves. Since an incident angle of the direct P-waves is close to almost vertical direction in case of this event, the interval of about 0.8 s can be interpreted as a two-way travel time within the ice-sheet layer beneath the seismic profile on the Plateau. Therefore, the five peaks following the direct P-waves could be the reverberations within the ice-sheet. The thickness of the ice-sheet beneath the observation stations is approximately 1.5 km (Figure 2), which can explain the time intervals of the five peaks in later phases. The seismic source time function of this event is not identified, however, a deep focus event with small magnitude has assumed to be a simple source time function, and these peaks can be caused by the reverberations rather than complexity of the source. All the peaks have almost the same slowness as that of the direct P-waves.

Figure 8 represents the running spectra of three different time windows (from upper to lower; 24–26 s, 25–27 s, and 26–28 s) applied to the Kermadec Is. Earthquake (event#; eq012; modified after [23]). A dominant frequency is identified at about 2.0 Hz in these windows. It is clear that there is a significant gap in high energy volume around 2.0 Hz at about 65 km, at where the valley structure is underlying beneath the ice-sheet (Figure 2). The observation stations on both sides of this gap (at around 59 km and 72 km) appeared to receive the higher energies. It is also identified that the peaks at frequency about 3.0 Hz are located in the middle part of the seismic profile.

Another example of the record section containing simultaneously occurred teleseismic events is shown in Figure 9 (event#; eq002). Several seismic phases from both the "Southeast Indian Ridge Earthquake" and "South of Fiji Island Earthquake" are duplicated recorded on the seismogram about the same times on January 7, 2002. Major teleseismic phases are clearly recognized at the time window around 100 s. Such overlapping arrivals of waves can be disentangled by applying the PWS stacking procedure. Stacking the waveforms by changing values of slowness yields separation and identification of these events because of expected large difference in slowness between the events. In addition to the teleseismic events, several local events plausibly associated with ice-related dynamics are recorded in later phases.

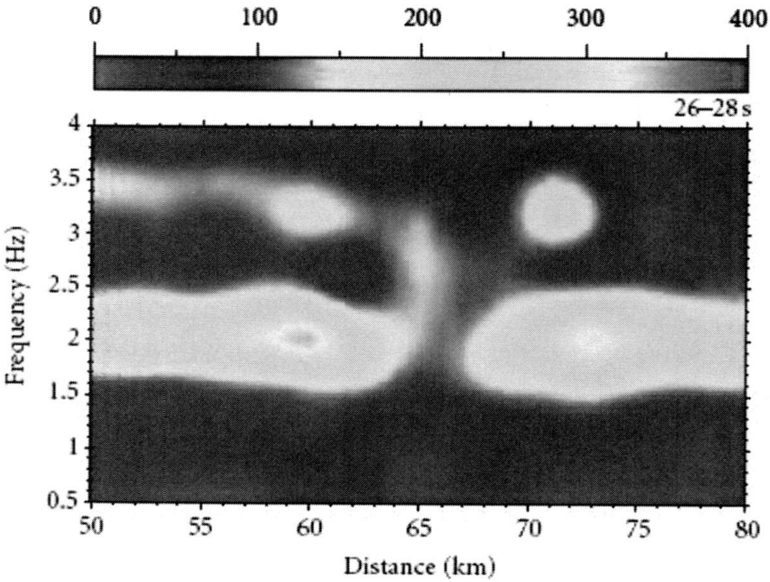

(c)

Figure 8: Contour maps showing amplitudes of seismic energies plotted in the frequency-distance domain for traces of the Kermadec Is. Earthquake (event#; eq012; after [23]). Spectra are taken from the traces aligned on the direct P. Time windows are shown at the upper right of each plot (from upper to lower; 24–26 s, 25–27 s, and 26–28 s, resp.).

In spite of the short period of the SEAL-2002 exploration (three weeks) and short recording duration (7.5 minutes twice a day), clear observations from teleseismic events were obtained. As is listed in Table 1, all the events recorded here have rather small magnitude (smaller than 5.0) that is generally difficult to observe. These waves also deliver scientifically interesting information, that is, not only near-receiver structural complexity but also global Earth's deep structures like a possibility of core phases, secondary waves from the mantle discontinuity/scatterer. The examples here suggesting further experiments of densely deployment of the short-period seismographs in Antarctica will provide us with fruitful data for investigations of both shallow structure of the Antarctica and deeper Earth structures.

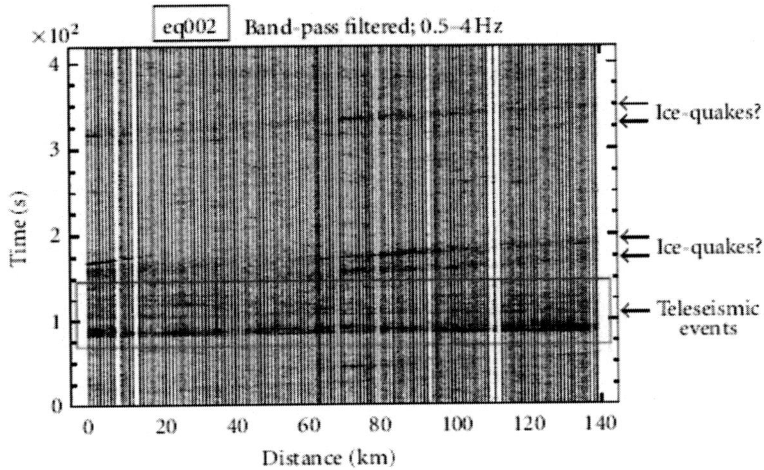

Figure 9: Band-pass filtered (0.5–4.0 Hz) waveforms of the record section containing multiple teleseismic events (event#; eq002; Southeast Indian Ridge Earthquake, and South of Fiji Island Earthquake, January 7, 2002). Major phases are clearly identified at the time-window around 100s (red colored square area). Several local (ice-related?) events are recorded as the later phases, which are raveled as "ice-quakes?". Starting time of the slowness stacking analyses is indicated by red colored broken line (see the caption in Figure 10).

4. ICE-QUAKES

Seismic signals involving ice-related phenomena are so-called "ice-quakes" ("ice-shocks" for smaller ones) and are most frequently reported in association with glacially related mass movements of ice sheets, sea ice, tide cracks, and icebergs [7, 20]. In West Antarctica, for instance, recent seismic and geodetic experiments have led to new discoveries related to the deformation of ice sheets environment. For example, daily tidally modulated, rapid slip episodes of the ice stream are discovered by microseismicity studies, and subsequently confirmed by GPS measurements [6, 24, 25]. The recent slip events also radiate long-period surface waves, which allows additional constraints to be placed on the source mechanism [24].

Local seismicity in 1987–2003 around the Lützow-Holm Bay Region is reported by [7]. The seventeen events were detected by local network deployed around the bay area. Almost hypocenters were located along the coast, apart from a few on the northern edge of the continental shelf. Several events may be large ice-quakes associated with sea-ice dynamics around the bay or in the southern ocean. The broadband seismographs at Syowa Station (SYO; 69.0°S, 39.6°E), moreover, significantly recorded the characteristic waveforms associated with discharge events of large sea-ice volume from the Lützow-Holm Bay [8]. The long-duration sea ice tremors represent spectral characteristics that distinguished them clearly from ordinary teleseismic and/or local tectonic events. A few sequences of harmonic overtoned signals, presumably associated with the merging of multiple ice volume at the bay.

As an example of local events relating to the ice-quakes, Figure 10 shows the original waveforms and the band-pass filtered (4.0–8.0 Hz) ones applying to the event#; eq001. A significant number of local events are identified on the waveforms. It is noted that relatively strong and long duration wave packets recognized in 150–180 s do not match with the traveltimes of teleseismic phases. Moreover, when taking into account the positive slowness of these phases, the two phases with large amplitude may be a set of injected waves from the southern direction against the SEAL-2002 profile on the Mizuho Plateau.

From these evidences, several signals traversing the observation profile may have common features: (1) including high-frequency components that are comparable with those of background noise, (2) containing relatively large slowness, and (3) large fluctuation of amplitudes and weak correlation of the waveforms along the seismic profile. These characteristics support an inference that local events such as ice-quakes occurred near the observation stations. According to as the above speculation, the locally originated phases are presumably involved with long-duration ice-dynamics occurred at the vicinity of studied area in Antarctica.

For another example, the upper of Figure 11 represents large ice-quakes detected on the recording stations of the SEAL-2002 profile (event#; eq005; after [23]). Two obviously identical phases between 10 and 70 s arrive from the northern side of the observation profile (right of Figure 11), although the accurate direction of the incoming waves cannot be determined due to the line setting of the seismographs in the SEAL profile. Considering similarity of the lengths of wave trains and the difference in slowness, these two large phases may be a set of either P and S waves or direct and reflected waves from an ice-quake. In the figure, another two later phases with small amplitudes can be seen in time windows between 80 and 120 s. When considering the positive slowness, these two phases may be a set of waves from another pair of ice-quakes from the southern side of the observation profile.

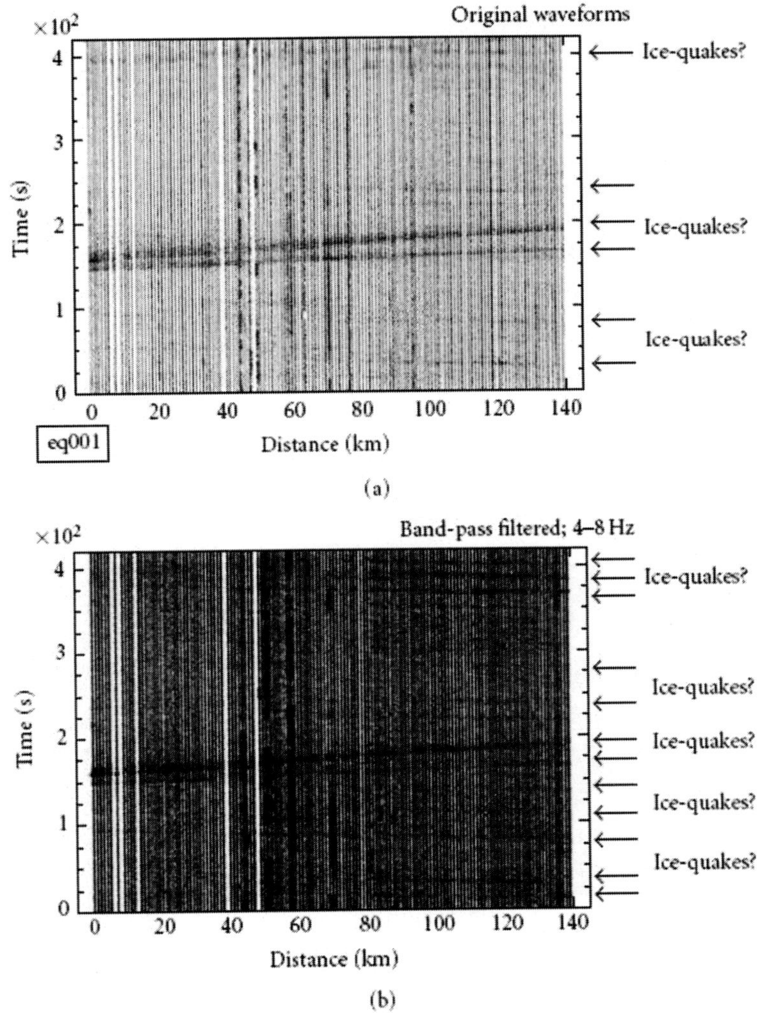

Figure 10: (upper) original waveforms for record section of local events, which appeared to be ice related events (ice-quakes) (event#; eq001). (lower) Band-pass filtered (4.0–8.0 Hz) waveforms of the same record section. A significant number of local (ice-related) events are raveled as "ice-quakes?".

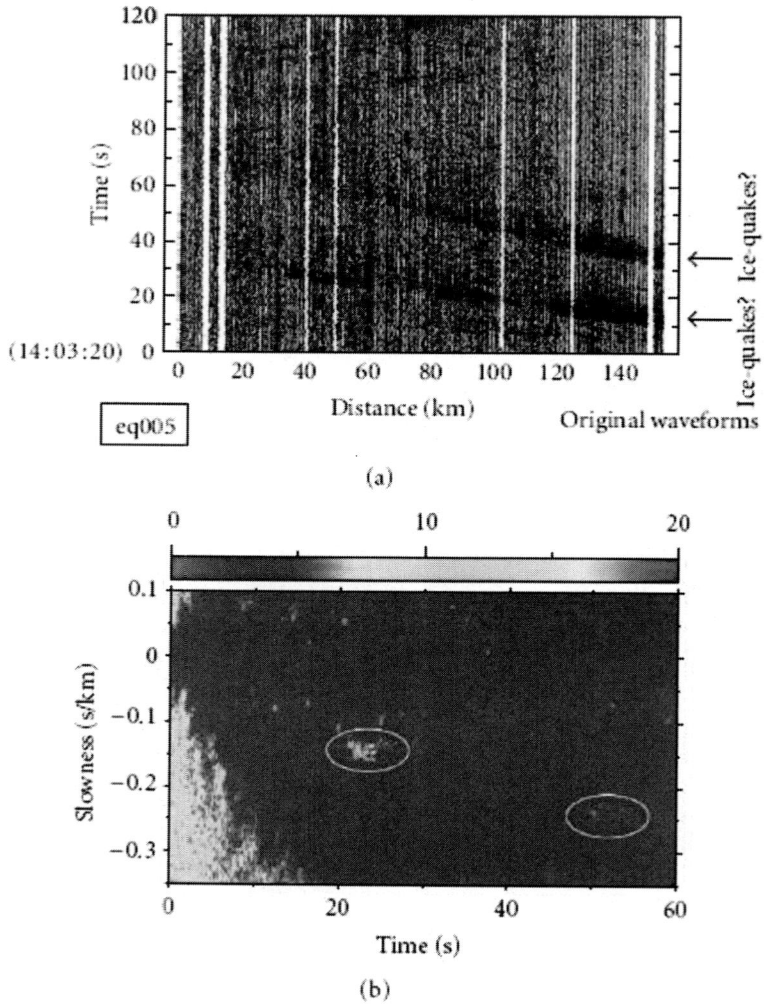

Figure 11: (upper) Recorded section for all the traces containing two large ice-quakes (event#; eq005; after [23]). Two identical signals can be clearly traced. The vertical axis starts from January 14, 2002, 14:03:20 (UTC). (lower) Contour map of PWS applied to these ice-quakes in the same time-window (modified after [23]). Peaks corresponding to two obvious events in the upper are indicated by white ellipses: peaks at (23 s, −0.135 s/km) and (51 s, −0.235 s/km). It is noted that high amplitudes found at the upper and lower left are artifacts caused by the PWS applied to zero-amplitude portions: since the zero amplitudes have the same value of phase components, a sum of the phase components should have a large value.

The PWS contours of the same ice-quakes (event#; eq005) are shown in the lower of Figure 11. Because of the incoherency of these ice-quakes, alignment on arrival cannot be done as above. Thus, the vertical axis in the lower of Figure 11 indicates the absolute slowness. Two peaks corresponding to the ice-quakes in the upper are seen at (23 s, −0.135 s/km) and (51 s, −0.235 s/km), respectively. As compared with the peaks in the lower of Figure 7, those in Figure 11 represent weak and scattered features. Since the PWS results enhance the signal that is coherent through the observation profile, we can tell that the waves generated by the ice-quakes are not coherent and show large fluctuations of the waveforms in distance range of about 100 km.

5. X-PHASES (UNKNOWN EVENTS)

Among the recorded events during the SEAL-2002 exploration, there are a few number of undefined events which are not likely both for the teleseismic events and for the local ice-quakes. This third category is named as unknown phases (hereafter referred to as the "X-phases"). A typical example of X-phases is presented in Figure 12 (event#; eq015) in which the SEAL-2002 profile and the three-component broadband waveforms at SYO station are shown. Continuous arrivals of coherent signals are clearly observed for a long duration approximately 100 s. This coherency is the same as observed in the teleseismic waveform records (Figure 6). The slowness of these identical X-phases is almost zero, which implies that these signals come from a direction perpendicular to the observation profile.

These characteristic features of the slowness may indicate that these signals have the same sources as the teleseismic event. By referring to the worldwide earthquake catalogs (such as the ISC bulletin and USGS PDE listings), a candidate for the event might be the Rat Is. of the Aleutians Earthquake which occurred on January 27, 2002, at 13:42 :44.4 (UTC). However, we present more details in the discussion section that the features of the X-phases cannot be explained only by this teleseismic event.

Although the sources of these signals are so far uncertain, we can find another interesting feature. The upper of Figure 13 represents a time-space variation in the envelope amplitudes of the band-pass filtered (1.0–2.0 Hz) seismograms (modified after [23]). A remarkable feature lies in a time-

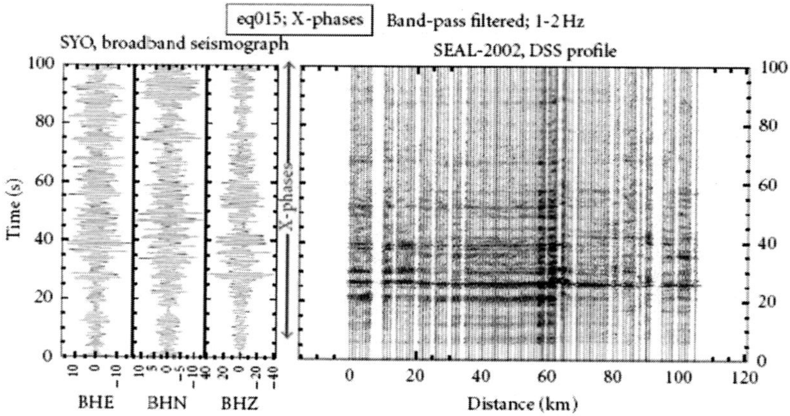

Figure 12: Record section showing the X-phases (event#; eq015) with a band-pass filter (1.0-2.0 Hz). Vertical axis starts from January 27, 2002, 14:02:30 (UTC). (left) Three-component broadband seismograms at SYO (right) SEAL-2002 recordings.

delay of the signal arrivals at around 65 km (at about 27, 31, 41, 45, 53 s, etc.) compared with the corresponding arrivals found at other locations on the SEAL-2002 profile. Moreover, arrivals at about 27 and 31 s show large amplitudes at 65 km with a time delay of about 1 s. This is not identified in other frequency bands and is also not found in the teleseismic waveform records. The distance of 65 km, as mentioned before, is located just above the valley structure of the bedrock topography (Figure 2).

The lower of Figure 13 demonstrates the PWS stacking of the same X-phases (event#; eq015). As in the case of the Kermadec Is. event (Figure 7), we utilize the relative slowness with respect to the first arrival of the X-phases. Consecutive arrivals with high coherency are found for about 100 s after the onset. This long duration is not found in the cases of either the Kermadec Is. event, or the ice-quakes. The slownesses of all the arrivals are almost the same. This implies that these wave energies reach the observation profile with the same incident angle and the back-azimuth.

The dominant frequency of the X-phases (event#; eq015) is found to be 2.0 Hz, similar to the teleseismic events. In addition to the gap of 2.0 Hz at about 65 km, we can identify a strange feature appeared in the spectra, which is not found in the teleseismic event. The clearly recognized gap of 2.0 Hz at 65 km turns into a peak of 1.5 Hz with a delay of only one second (middle and bottom in Figure 14). This peak of 1.5 Hz corresponds to the energy peak at (27 s, 65 km) in the upper of Figure 13.

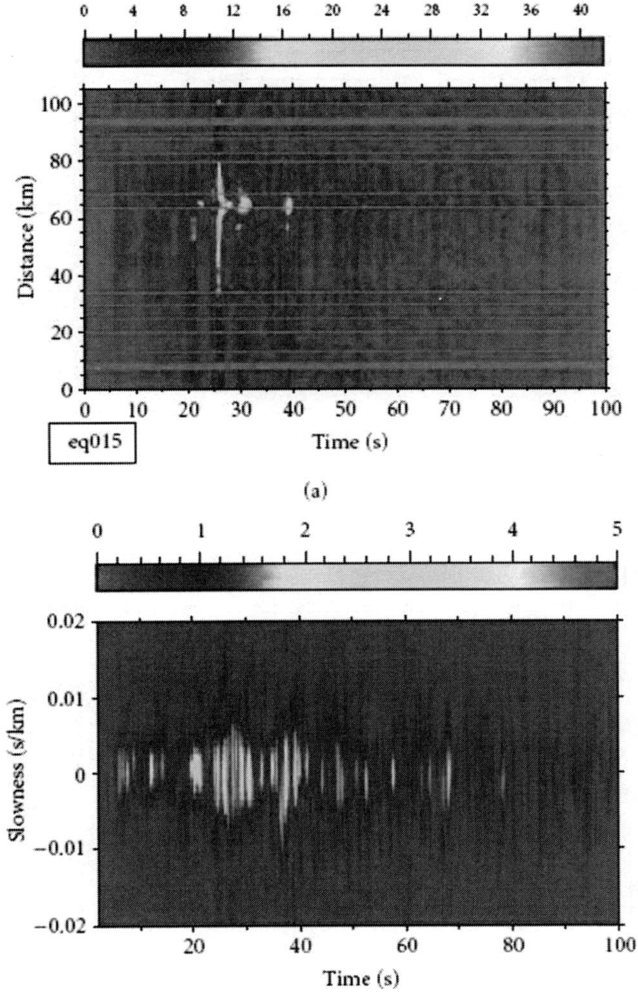

Figure 13: (a) Contour map showing envelope amplitudes of band-pass (1.0–2.0 Hz) filtered traces of the X-phases (event#; eq015). Traces with low signal-to-noise ratios are not used here and are colored in gray. Since the traces here are band-pass filtered, the number of traces with high signal-to-noise ratio is different from that in Figure 12. Color scale is arbitrarily set to see an apparent amplitude variation within the map. (b) Same as in Figure 7 but for the X-phases (event#; eq015).

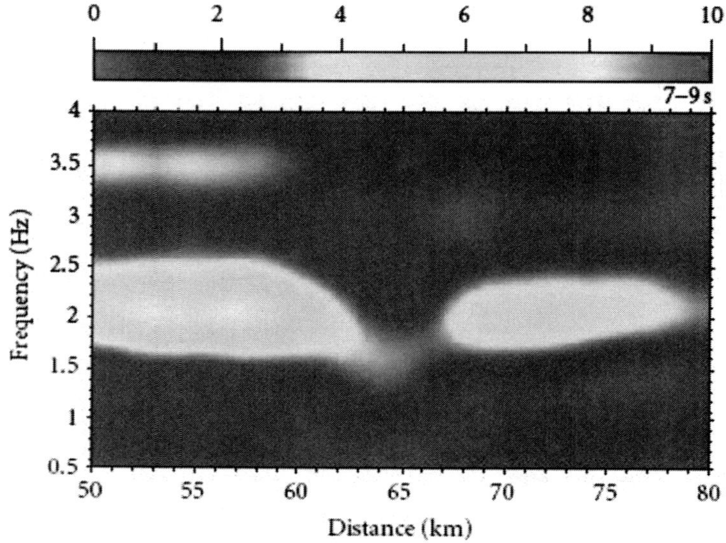
0 2 4 6 8 10
7–9 s
Frequency (Hz)
4 3.5 3 2.5 2 1.5 1 0.5
50 55 60 65 70 75 80
Distance (km)

(a)

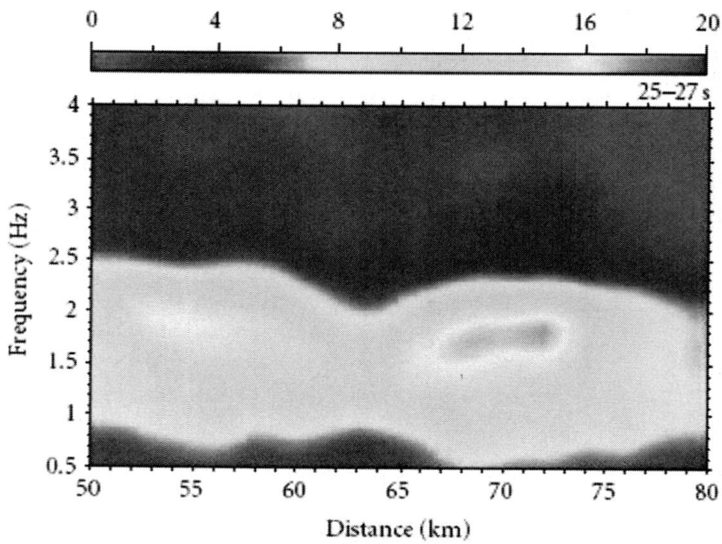
0 4 8 12 16 20
25–27 s
Frequency (Hz)
4 3.5 3 2.5 2 1.5 1 0.5
50 55 60 65 70 75 80
Distance (km)

(b)

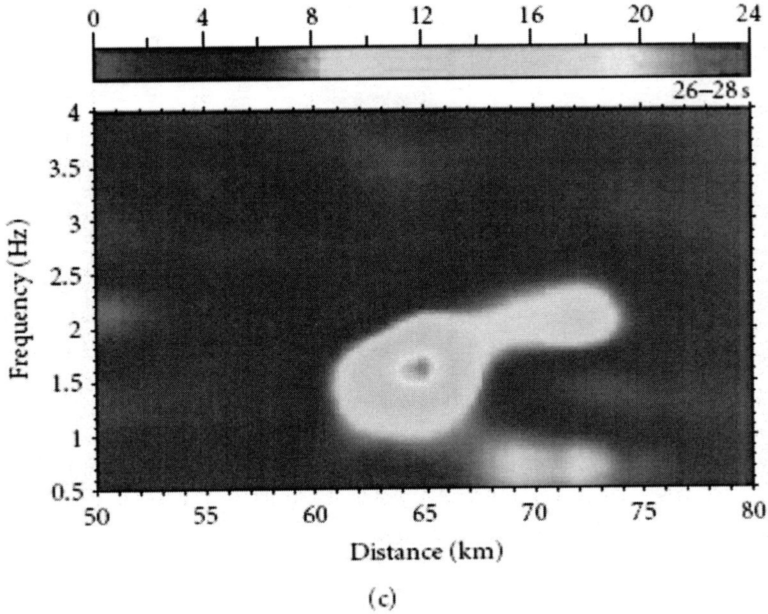

Figure 14: Same as in Figure 8 but for the X-phases (event#; eq015). Time windows are shown at the upper right of each plot (from upper to lower; 24–26 s, 25–27 s, and 26-28 s, resp.). Top and middle show the gap of high energy of 2.0 Hz at around 65 km, and the bottom shows the peak of 1.5 Hz just a second late after the time window showing the gap in the middle.

Another example of the X-phases is demonstrated in Figure 15 (event#; eq006). The band-pass filtered (0.5–4.0 Hz) waveforms of the record section image significantly show the existence of X-phases, together with several local events (ice-quakes). The long-time durations with high amplitudes in the X-phases are evidently identified at the time-window after 300 s. Here, when considering the negative slowness of the X-phases, they may be injected from the northern direction against the seismic profile. In addition, several local (ice-related) events are recorded as the earlier phases in the same seismograms.

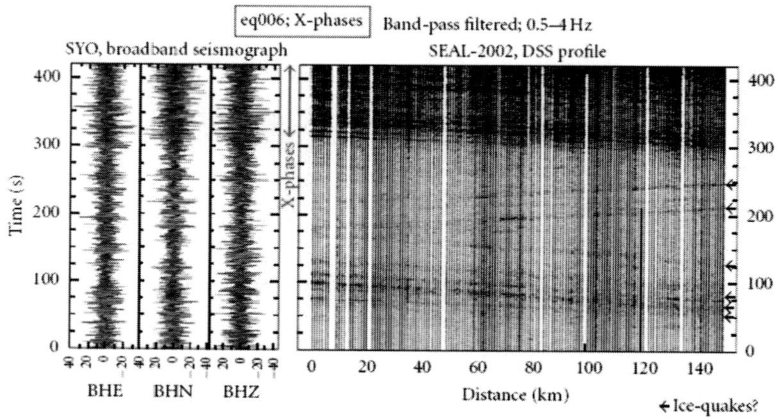

Figure 15: Band-pass filtered (0.5–4.0 Hz) waveforms of the record section for X-phases together with local events (ice-quakes) (event#; eq006). (left) Three-component broadband seismograms at SYO (right) SEAL-2002 recordings. The unknown originated X-phases are evidently identified at the time-window after 300 s. Several local (ice-related?) events are recorded as the former phases, which are raveled as "ice-quakes?".

6. DISCUSSION

In order to identify the location of occurrence of the unknown X-phases, we try to compare the waveforms recorded by the SEAL-2002 exploration with a three-component broadband seismograph at SYO in the Lützow-Holm Bay, near the Mizuho Plateau. Figures 12 and 15 are the comparison of the (event#; eq015) and (event#; eq006), respectively. When looking at a glance, however, we cannot identify the significant difference in the arrival times between SYO and the stations at the most northeastern end of the SEAL-2002 profile. In particular, the signal-to-noise ratio at SYO for (event#; eq006) is poor for resolution, so as to judge the difference in arrival times. In the following few paragraphs, therefore, we focus on the event#; eq015 for more detailed discussion.

A remarkable feature of the X-phases (event#; eq015) is that successive arrivals of waveform energy have almost zero slowness for about 100 s after the onset (Figure 13). A candidate of the teleseismic event might be the Rat Is. (Aleutians) earthquake (included in the event#; eq015; Table 1). However, for three reasons in the following, we cannot state that only the teleseismic event is the source that produced the X-phases.

First reason is that the long duration of 100 s cannot be explained by only teleseismic event. Epicentral distance from the Rat Is. (Aleutians) event to the center of the SEAL-2002 observation profile is about 152°. At the distance, we can observe at most 9 waves of PKP phase, that is, three branches (df, bc, and ab) and those depth phases (pPKP and sPKP). These waves are expected to be arrived in the first half of the duration (about 0–40 s in Figure 14). Actually, some peaks in Figure 14 match a part of expected PKP phases but not for all. However, there is no candidate phase for the latter half (40–100 s in Figure 14). If the long duration is caused by complex structures beneath the observation line, similar phenomenon would appeared in the cases for other recordings of natural teleseismic events observed here. For example, the recordings of the P-wave from the Kermadec Is. Event (Figure 6, Event#; eq012) do not show such long durations. This discrepancy is hard to be explained by a difference in incoming direction between the events. For the Rat Is. event, the back-azimuth from the center of the seismic profile on the Mizuho Plateau is 107.4°, and the incident angle to the boundary between ice-sheet and topmost crust ranges from 4.6° (PKPdf) to 13.4° (PKPab), respectively. They are 143.3° and 18.0° for the direct P-waves of the Kermadec Is. event. If we consider the raypaths of the PKPs of the Rat Is. event and the direct P of the Kermadec Is. event to a station on the SEAL-2002 seismic profile, the piercing points of these waves to the boundary have a difference in horizontally only a few hundred meters. This distance is much smaller than the horizontal length of the complex structures beneath the observation line (e.g., valley structure, which is approximately 10 km (Figure 2)). Therefore, we cannot expect that the different kinds of waveforms are generated by a complex topography of the bedrock boundary for the Rat Is. event and the Kermdec Is. event.

Second reason is successful recordings with long durations similar to eq015. As listed in Table 1, we categorize three recordings (Event#; eq003, 006, and 011) as the X-phase. They show successive arrivals of energies with at least several ten seconds; however, a beginning or ending of wave duration is out of a recording range for each three due to the short duration of seven and half minutes for each recording,. For these three, there is no possible teleseismic earthquake unlike the Rat Is. Event for eq015, leading to an origin other than teleseismic events. Resemblance of the waveforms of eq015 and additional 3 recordings (eq003, 006, 011 and 015) can categorize these 4 into a phenomenon caused by an unidentified origin.

Third reason is a significant difference in the amplitudes of seismic energies in the frequency-distance domain, between the X-phases (Figure 14) and the Kermadec Is. Event (Figure 8). The clear gap of 2.0 Hz in the middle part of the seismic profile is identified in both two events. However, the peak of 1.5 Hz (delayed about a second after the gap) is seen only in the X-phases and the peak of 3.0 Hz (which is not time delayed) is identified only in the Kermadec Is. Event.

These reasons may lead to a conclusion that the X-phases would not be the waves of teleseismic events. Note that, of course, we cannot discard a possibility that PKP phases overlap with the phenomenon for eq015. Since back-azimuth of 107.4° of Rat Is. Event for eq015 is almost perpendicular to the direction of the SEAL-2002 seismic profile, separation and identification of PKP phases by slowness analysis are not possible. Therefore, the waveforms of eq015 might be a composite of the unidentified phenomenon and a remarkable success of recordings of PKP phases from small-magnitude event.

Another striking phenomenon is recognized in Figure 13. In addition to a delay of the arrivals at around 65 km just above the valley structure in the middle of SEAL-2002 profile, we can find, by close inspection, a small arrival without delay just before the late arrival, at about 25 s. Such normal arrivals are also found at 6 s (onset) and 35–40 s, respectively. The X-phases consist of the wave packets with normal and delayed arrivals. It is clear from the records of the Kermadec Is. Event (event#; eq012) that the P-wave arrivals are not delayed at about 65 km. Therefore, the normal arrivals of the X-phases may be the P-waves, but the delayed arrivals might not be the P-waves. A possibility could be the generation of the S-waves. Whether the S-wave arrival is delayed at around 65 km or not can be clarified by checking the dataset of S-wave arrivals of the Kermadec Is. Event. Unfortunately, the S-wave arrivals are out of range of the recording period in the SEAL exploration data. If the above hypothesis is valid, it is plausible that the origin of the X-phases is not a teleseismic event but a regional event around Antarctica.

Next we examine the broadband data distributed in the wide area of Antarctica, as contribution to the Federation of Digital Seismographic Networks (FDSN; [26]). Figure 16 represents waveforms recorded by four existing FDSN stations (PMSA, SPA, SYO, and VNDA) in a time range corresponding to the X-phases (event#; eq015). Instrumental response of each station is removed from the original waveform and a band-pass filter (1.0-2.0 Hz) is applied for comparison with Figure 12. We assume energies in the waveforms as the X-phases though some ambiguity remains for the identification. The arrival times of the X-phases, comparing maximum amplitudes in each waveform, indicate the earliest for the station PMSA (64.8°S, 64.0°W) in Antarctic Peninsula. Then, the ordering is recognized as from SPA (Amundsen-Scott South Pole Station), SYO, and VNDA (77.5°S, 161.9°E) in the Terra Nova Bay of West Antarctica. A determination of a hypocenter by using the arrival times as P-wave indicates a source region of the Weddell Sea (shaded ellipse in the map of Figure 16). Note that the determination is not well constrained and leads to a wider area with large errors due to the difficulty of picking and the ambiguity of the identification as the X-phases. Thus, we propose here, as a preliminary and tentative suggestion, that a location of the X-phases (event#; eq015) is speculated at the Weddell Sea and near the Antarctic Peninsula region (Figure 17, (a)).

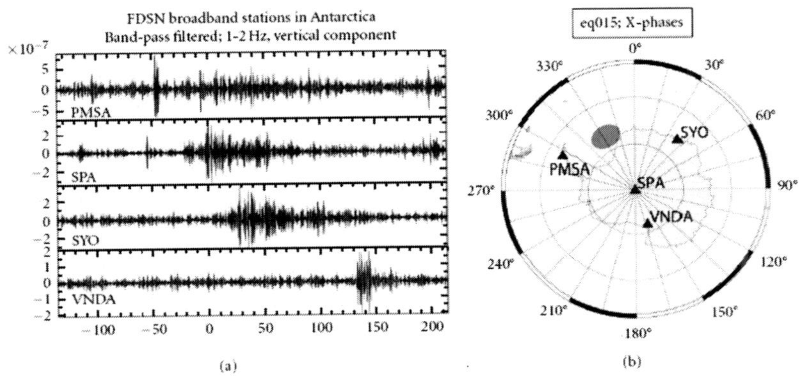

Figure 16: (left) Comparison of the X-phases (event#; eq015) recorded by four existing FDSN stations in Antarctica (from top to bottom; PMSA, SPA, SYO, and VNDA). All waveforms are band-pass filtered (1.0-2.0 Hz) vertical component. (right) Location of the four FDSN stations. Shaded ellipse indicates a possible source for the energies in the waveforms as the X-phases.

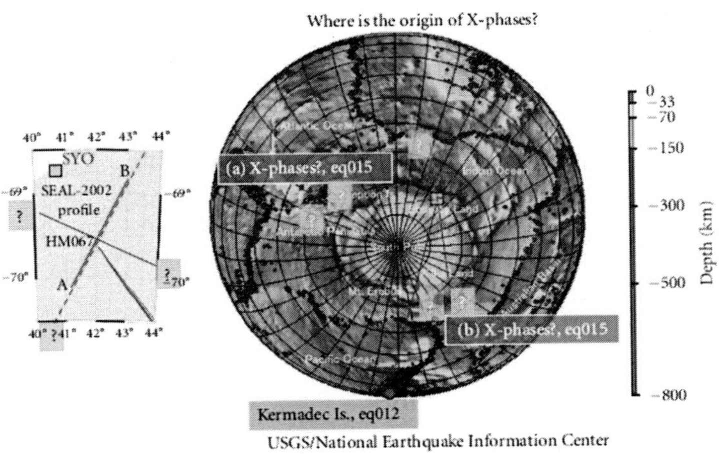

Figure 17: Plausible candidates of the location for X-phases (event#; eq015), overprinting on the background seismicity (by USGS/NEIC) around Antarctic Plate. Hypocenter of the Kermadec Is. earthquake (event#; eq012) is also presented for comparison.

In contrast, on the basis of PWS images of the X-phases (Figure 13), the incident angle to the SEAL-2002 profile is almost perpendicular direction (Figure 17) so as to explain the zero slowness of the incident wavelets. Therefore, when assuming the horizontal injection from the Earth's surface, two directions are considered from both the back-azimuths in eastern side (inside East Antarctic continent) and in western side (inside Indian Ocean, within Antarctic Plate to oceanic ridges) (Figure 17, (b)). As a possibility of the origin of X-phases (event#; eq015) may be the intracontinental earthquakes, such the regional events around Antarctica are compiled in recent one decade [2, 3].

Around continental area in East Antarctica from 90°E to 180°E, seismicity is low in an average by general understanding. The Wilkes Land, in contrast, has been identified as the most tectonically active region among the Antarctic (Figure 18). Poorly located earthquakes in Wilkes Land are lined from north to south along the 140°E longitude, crossing several subglacial highlands such as the Resolution, the Adventure, and the Belgica [1, 3, 27]. It is also noticed that there is a significant number of subglacial lakes (Figure 19) in this area; then it can be considered a relation to produce the occurrence of local seismic activities. The formation, distribution and stability of subglacial lakes could give rise to the fundamental question how do the tectonic processes control the existence of the lakes [28]. The poorly located earthquakes, accordingly, could be a kind of large ice-quakes, because complicated subglacial topography with overlying ice-sheet in this area may efficiently cause ice-related seismic phenomena.

A complexity of the surface bedrock structure of Wilkes Land, moreover, might be influenced by tectonically weakened upper crust along with the present mobile belts, forced by the Tasmania microplate [4, 5]. The aftershock area of the large Balleny Island Earthquake (1998 March,) in the Antarctic Oceanic Plate appears to continue with the continental area of Wilkes Land (Figure 18). Some ideas are supposed to generate the large earthquakes in this area, by assuming thermal stress of the young lithosphere, associated with unusual deformation due to Macquarie triple junction [29, 30].

From more detail, comparison with arrival times of the X-phases recorded at FDSN stations, the maximum amplitudes of seismic phase appear to come to SYO with the delay of several seconds (Figure 12). On the basis, of several discussion above mentioned, the X-phases (event#; eq015) could travel into the SEAL-2002 profile and then to the station SYO at a coast of the Lützow-Holm Bay. Hypocenters in the Antarctic Peninsula (Figure 17, (a)), otherwise in the relatively active intra-plate seismogenic region of Wilkes Land (Figure 17(b), could be the most plausible candidates to generate the X-phases on January 27, 2002.

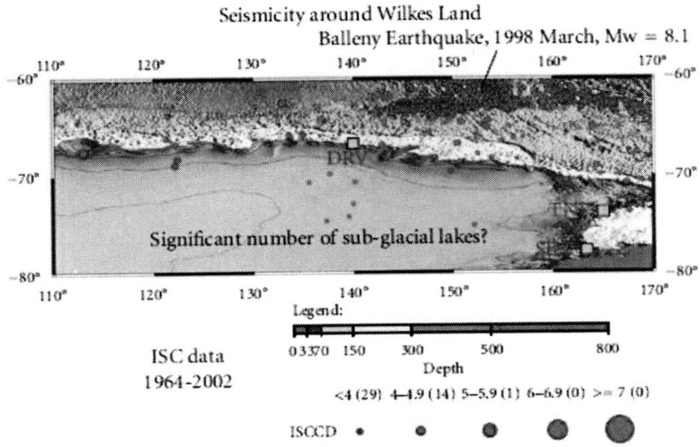

Figure 18: Globally complied seismicity around Wilkes Land, East Antarctica for 1964–2002 by the International Seismological Centre (ISC). The aftershock area of the large Balleny Island Earthquake (1998 March,) is included on the map in the Antarctic Oceanic Plate. A complexity of the surface bedrock structure might be influenced by tectonically weakened upper crust along with the present mobile belts, forced by the Tasmania microplate. It is also noticed that there is significant a number of subglacial lakes and occurrence of local seismic activities including ice-quakes.

Recently, a distribution of the events that are likely to correspond to glacial earthquakes is determined around Antarctica in 1993–2008 ([12]; Figure 20; Table 2). The origin times and epicenters are obtained by the surface-wave detection algorithm. Many of the hypocenters concentrate to locate at the coastal area of the Antarctic Peninsula, the Weddell Sea, and the Wilkes Land, particularly near Vincennes Bay. There are no exactly corresponding events for the X-phases (event#; eq015) in their determined fourteen events, unfortunately, but the events on March 18 and March 30, 2002, in Table 2, show within the two months of time difference between the occurrences of X-phases on January 27. A similar analysis using surface waves by global dataset from 1997 to 2009 is conducted [31]; 17 glacier calving events are detected at the similar regions obtained by [12]. However, there are no corresponding events to the X-phases on January 2002 by our study.

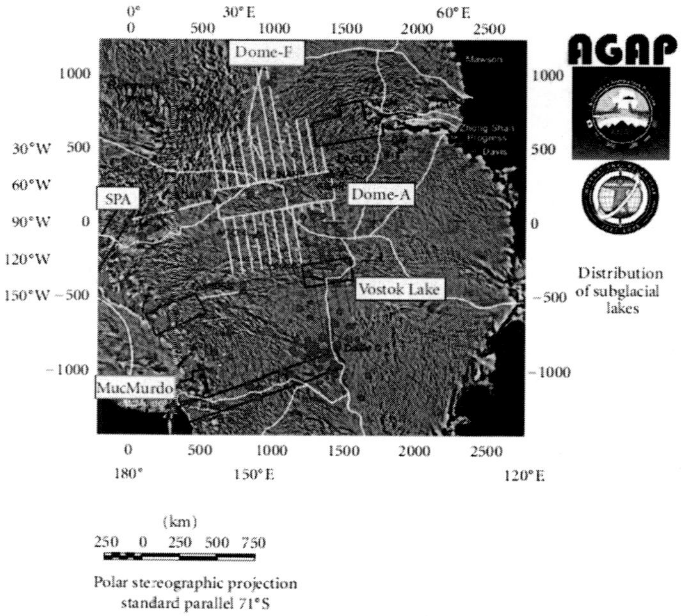

Figure 19: Distribution of subglacial lakes (blue solid circles) in wide region of Wilkes Land, East Antarctica. The target areas for several geophysical surveys including the International Polar Year (IPY 2007–2008) campaign period are demonstrated, by yellow lines (air-borne geophysics by AGAP), red line (Chinese traverse by AGAP), green triangles (seismic stations by AGAP) and several black circulated areas (the other air-borne geophysics projects). (The map is by courtesy of Professors R. Bell and M. Studinger; http://www.ldeo.columbia.edu/~mstuding/AGAP/). The formation, distribution and stability of subglacial lakes tell us the fundamental question how the tectonic processes control them.

It is also notified that the huge area of ice mass (3,275 km^2) disintegrated from "Larsen B Ice Shelf," in the northeast Antarctic Peninsula, between January 31 and March 7 ([32, 33]; Figure 21) A large floating ice mass on the eastern side of the Antarctic Peninsula has shattered and separated from the continent. The break-off process of the floating ice mass could possibly generate the large energy to arrive at the X-phases on observation stations in SEAL-2002 profile, in Eastern Dronning Maud Land. The X-phases (event#; eq015) on January 27, 2002 could presumably be the pre-cursor of the discharge events from the Larsen B Ice Shelf. There is a possibility that the other X-phases (event#; eq006) on January 14 could also be another pre-cursor phases. It is a great pity that we do not record the SEAL exploration data until January 31, 2002, so as to detect directly the signals of the initiation process for break-off at the Larsen B Ice Shelf. Regarding the floating ice shelves in Antarctica, moreover, calving mass loss occurred mostly by large tabular icebergs, which do not capsize and thus produce little low-frequency energy. Calving events of icebergs at the western Greenland are detected by recent available broadband seismic network [34]. The maximum energy at the 2–5 Hz band is very similar frequency range observed by the X-phases in this study of Antarctica.

Table 2: Source parameters for 14 detected plausible glacial earthquakes in Antarctica during 1993–2008 (modified after 2010 [12]). Origin times and epicenters are determined by the surface-wave detection algorithm.

Date	Time	Latitude	Longitude	M
1996/07/14	19:05:52	−62.25	−55.75	4.8^{C}
1998/05/11	23:17:44	−67.00	109.00	4.9^{C}
1999/06/27	15:37:52	−77.75	−47.25	4.8^{C}
2000/04/09	23:25:36	−66.25	110.75	4.9^{B}
2000/06/22	23:10:48	−67.00	109.00	4.9^{C}
2001/11/24	03:20:48	−67.00	109.00	4.9^{E}
2002/03/18	08:22:40	−65.25	−62.75	4.7^{C}
2002/03/30	04:15:28	−67.00	109.00	4.8^{C}
2004/01/24	08:38:24	−68.00	150.00	4.8^{C}
2004/01/31	04:31:52	−67.00	111.00	4.9^{C}
2007/10/14	02:10:08	−66.75	109.75	4.9^{B}
2008/05/18	08:35:04	−65.50	124.50	4.8^{C}
2008/06/18	00:04:00	−68.75	151.25	4.8^{B}
2008/07/21	07:34:48	−69.75	−67.75	4.8^{C}

The letter code following the magnitude M: indicates the quality of the detection and location: B: isthe highest quality and C/E: is the lowest.

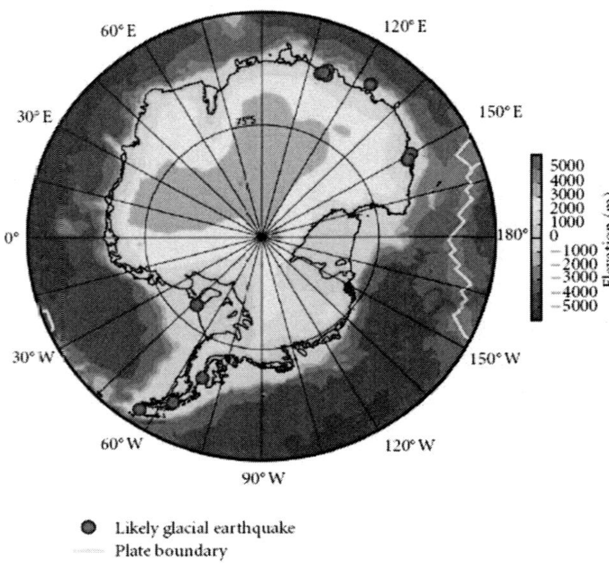

Figure 20: Hypocenters for the detected events that are likely to correspond to glacial earthquakes around Antarctica in 1993–2008 (after [12]).

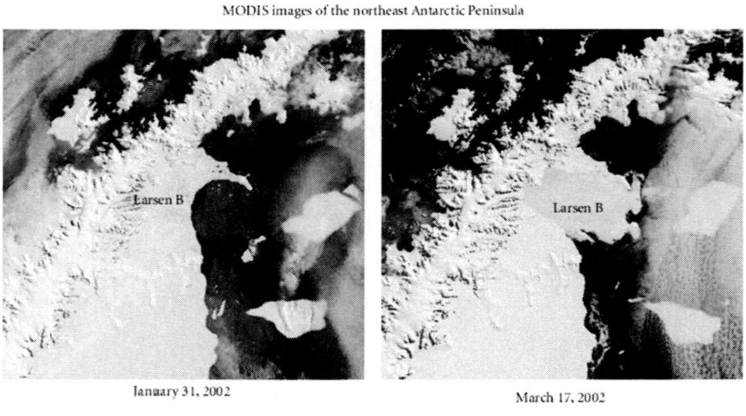

Figure 21: Moderate-resolution Imaging Spectroradiometer (MODIS) satellite imagery at the northern section of the Larsen B ice shelf on 31 January, 2002 (left), and on March 17, 2002 (right) respectively. A large floating ice mass on the eastern side of the Antarctic Peninsula has shattered and separated from the continent. The images are a courtesy of the National Snow and Ice Data Center (NSIDC), University of Colorado.

Identification of the exact sources that produced these characteristic unknown signals has not yet been completed, and theoretical modeling will most likely be required to explain the physical processes. Many kinds of cryoseismic phenomena have been reported around the Ross Sea [35], the marginal sea of the Antarctic Peninsula [36, 37], and the continental margin of Dronning Maud Land [38]. In particular, iceberg-originated harmonic tremor emanating from tabular icebergs is observed by both seismoacoustic and broadband seismics [35]. The tremor signals consisted of extended episodes of stick-slip ice-quakes generated when the ice-cliff edges of two tabular icebergs rubbed together during glancing, strike-slip iceberg collisions. Source mechanisms of the harmonic tremors might provide useful information for ice-dynamic behavior, and remotely monitoring of the activity.

Finally, a significant number of seismic events with small-to-middle magnitude could not actually be obtained by the global network particularly around Antarctica, since they have ambiguous arrivals in the waveforms and triggering algorism by the existing stations. The processes that result in seismic events associated with ice-sheet, sea-ice, ice-shelf, and glaciers are all integral to the overall dynamics of cryosphere, and more condensed observations provide a means of monitoring changes in their behavior over time. Because seismology detects processes internal to the ice and the Earth, the monitoring capability that it provides is complementary to that of other methods, such as satellite remote sensing and GPS geodesy.

7. CONCLUSIONS

The characteristic seismic waveforms from various natural sources (teleseismic, local ice-quakes, and unknown X-phases) are obtained by the SEAL-2002 exploration in Eastern Dronning Maud Land, East Antarctica. Interesting features of the seismic wave propagation around Antarctica are significantly demonstrated. Anomalous behavior of the waves characterized by the focusing/defocusing effects is possibly caused by a valley structure beneath the stations located at the middle of the seismic profile. Several characteristics were identified by detailed spectra analyses. A difference of the response generated from the valley structure might exist for different kinds of incident waves that is, P-wave incidence on the valley results in a "frequency gap" while on the other hand, S-wave incidence produces both the "gap" and the "peak" with a sufficient delay of the arrival time. Although the origin of X-phases is not accurately identified, the most plausible candidates are an intraplate earthquake or a large ice-quake (glacial earthquake) in the Antarctic. Maybe the pre-cursor vibration of the break-off process at the Larsen B Ice Shelf could be the most plausible candidate to cause the X-phases.

ACKNOWLEDGMENTS

The authors would like to express thier appreciation to the participants of the Japanese Expedition members for their great efforts to carry out the SEAL seismic deployments on the ice sheet. Great thankfulness, especially, are to Professors H. Miyamachi, S. Toda, and T. Matsushima for their management of the experiments. The authors would like to express their sincere thankfulness to Professors Meredith Nettles of Lamont-Doherty Earth Observatory, Columbia University, Professors D. Blankenship and T. Diehl of the Institute for Geophysics, University of Texas at Austin, USA, and Professors I. Nakanishi of Kyoto University, for their critical reading and useful comments to improve the paper. The authors would like to express sincere thanks to Professors K. Shiraishi, Director-General of the National Institute of Polar Research, and Professors K. Kaminuma, K. Shibuya, Y. Motoyoshi, K. Doi, Y. Nogi, T. Hokada, and Y. Aoyama of NIPR for their useful discussions about geoscientific issues about Antarctica. The authors would like to express acknowledgement to the volume editor (Dr. M. S. Zhdanov) and two anonymous referees of the special issue on "Seismic Imaging 2012" of International Journal of Geophysics, for their sincere and useful comments. Parts of the figures have been prepared with generic mapping tools (GMT) [39].

REFERENCES

1. K. Kaminuma, "A revaluation of the seismicity in the Antarctic," Polar Geoscience, vol. 13, pp. 145–157, 2000.
2. A. M. Reading, "Antarctic seismicity and neotectonics," in Antarctica at the Close of a Millennium, J. A. Gamble, D. N. B. Skinner, and S. Henrys, Eds., vol. 35, pp. 479–484, The Royal Society of New Zealand Bulletin, Wellington, New Zealand, 2002.
3. A. M. Reading, "On seismic strain-release within the Antarctic Plate," in Antarctica: Contributions to Global Earth Sciences, D. K. Futterer, D. Damaske, G. Kleinschmidt et al., Eds., pp. 351–356, Springer, Berlin, Germany, 2006.
4. M. Kanao, Y. Nogi, and S. Tsuboi, "Spacial distribution and time variation in seismicity around the Antarctic Plate-Indian Ocean region," Polar Geosciences, no. 19, pp. 202–223, 2006.
5. T. Himeno, M. Kanao, and Y. Ogata, "Statistical analysis of seismicity in a wide region around the 1998 Mw 8. 1 Balleny Islands Earthquake in the Antarctic Plate," Polar Geosciences, vol. 5, pp. 421–431, 2011.
6. S. Anandakrishnan and R. B. Alley, "Tidal forcing of basal seismicity of ice stream C, West Antarctica, observed far inland," Journal of Geophysical Research B, vol. 102, no. 7, pp. 15183–15196, 1997.

7. M. Kanao and K. Kaminuma, "Seismic activity associated with surface environmental changes of the Earth system, around Syowa Station, East Antarctica," in Antarctica: Contributions to Global Earth Sciences, D. K. Futterer, D. Damaske, G. Kleinschmidt et al., Eds., pp. 361–368, Springer, Berlin, Germany, 2006.
8. M. Kanao, A. Maggi, Y. Ishihara et al., "Interaction on seismic wves between Atmosphere-Ocean-Cryosphere and Geosphere in Polar Region," in Seismic Waves: Research and Analysis, M. Kanao, H. Takenaka, Y. Murai et al., Eds., InTech, Rijeka, Croatia, 2012.
9. M. Kanao, M. Ishikawa, M. Yamashita, K. Kaminuma, and L. D. Brown, "Structure and evolution of the East Antarctic lithosphere: tectonic implications for the development and dispersal of Gondwana," Gondwana Research, vol. 7, no. 1, pp. 31–41, 2004. ·
10. M. Kanao, A. Fujiwara, H. Miyamachi et al., "Reflection imaging of the crust and the lithospheric mantle in the Lützow-Holm complex, Eastern Dronning Maud Land, Antarctica, derived from the SEAL transects," Tectonophysics, vol. 508, no. 1–4, pp. 73–84, 2011. r · ·
11. H. Miyamachi, S. Toda, T. Matsushima et al., "Seismic refraction and wide-angle reflection exploration by JARE-43 on Mizuho Plateau, East Antarctica," Polar Geoscience, vol. 16, pp. 1–21, 2003. ·
12. M. Nettles and G. Ekström, "Glacial earthquakes in greenland and antarctica," Annual Review of Earth and Planetary Sciences, vol. 38, pp. 467–491, 2010. r · ·
13. M. Takada, S. Toda, D. Kamiya, T. Matsushima, and H. Miyamachi, "Radio echo sounding survey along the profile of the JARE-43 seismic exploration on the Mizuho Plateau, East Antarctica," Antarctic Record, vol. 47, no. 3, pp. 380–394, 2003. ·
14. S. Mae, "The bedrock topography deduced from multiple radar echoes observed in the Mizuho Plateau, East Antarctica," Antarctica Recreation, vol. 61, pp. 23–31, 1978.
15. A. P. Kapitsa, J. K. Ridley, G. D. Q. Robin, M. J. Siegert, and I. A. Zotikov, "A large deep freshwater lake beneath the ice of central East Antarctica," Nature, vol. 381, no. 6584, pp. 684–686, 1996. ·
16. M. J. Siegert, J. C. Ellis-Evans, M. Tranter et al., "Physical, chemical and biological processes in Lake Vostok and other Antarctic subglacial lakes," Nature, vol. 414, no. 6864, pp. 603–609, 2001. r · ·
17. M. Schimmel and H. Paulssen, "Noise reduction and detection of weak, coherent signals through phase-weighted stacks," Geophysical Journal International, vol. 130, no. 2, pp. 497–505, 1997. ·
18. B. L. N. Kennett and E. R. Engdahl, "Traveltimes for global earthquake location and phase identification," Geophysical Journal International, vol. 105, no. 2, pp. 429–465, 1991. ·
19. R. Tibi and D. A. Wiens, "Detailed structure and sharpness of upper mantle discontinuities in the Tonga subduction zone from regional broadband arrays," Journal of Geophysical Research B, vol. 110, no. 6, Article ID B06313, 14 pages, 2005. r · ·
20. S. Kaneshima, T. Kubo, and S. Yoshioka, "Geophysical and mineralogical constraints on the post-spinel transformation for the Tonga slab," Physics of the Earth and Planetary Interiors, vol. 196-197, pp. 23–31, 2012.
21. S. Kaneshima, "Seismic scatterers at the shallowest lower mantle beneath subducted slabs," Earth and Planetary Science Letters, vol. 286, no. 1-2, pp. 304–315, 2009. r · ·
22. S.-X. Zang, Y.-Z. Zhou, J.-Y. Ning, and R. Q. Wei, "Multiple discontinuities near 660 km beneath Tonga area," Geophysical Research Letters, vol. 33, no. 20, Article ID L20312, 2006. r · ·
23. A. Yamada, M. Kanao, and M. Yamashita, "Features of seismic waves recorded by seismic exploration in 2002: responses from valley structure of the bedrock beneath Mizuho Plateau," Polar Geoscience, vol. 17, pp. 139–155, 2004.

24. D. A. Wiens, S. Anandakrishnan, J. P. Winberry, and M. A. King, "Simultaneous teleseismic and geodetic observations of the stick-slip motion of an Antarctic ice stream," Nature, vol. 453, no. 7196, pp. 770–774, 2008. r · ·
25. R. A. Bindschadler, M. A. King, R. B. Alley, S. Anandakrishnan, and L. Padman, "Tidally controlled stick-slip discharge of a West Antarctic ice stream," Science, vol. 301, no. 5636, pp. 1087–1089, 2003. r · ·
26. R. Butler and K. Anderson, "Global Seismographic Network (GSN)," IRIS Annual Report, 2008.
27. P. J. Browne-Cooper, G. R. Small, and R. Whithworth, "Probable local seismicity at Wilkes Land, Antarctica," New Zealand Journal of Geology and Geophysics, vol. 10, pp. 443–445, 1967.
28. T. Wilson and R. Bell, "Earth structure and geodynamics at the Poles," in Understanding Earth'S Polar Challenges: International Polar Year 2007-2008, pp. 273–292, 2011.
29. D. A. Wiens, M. E. Wysession, and L. Lawver, "Recent oceanic intra-plate earthquake in Balleny Sea was largest ever detected," EOS, Transactions American Geophysical Union, vol. 79, pp. 353–354, 1998.
30. M. Nettles, T. C. Wallace, and S. L. Beck, "The March 25, 1998 Antarctic plate earthquake," Geophysical Research Letters, vol. 26, no. 14, pp. 2097–2100, 1999. ·
31. X. Chen, P. M. Shearer, F. Walter, and H. A. Fricker, "Seventeen Antarctic seismic events detected by global surface waves and a possible link to calving events from satellite images," Journal of Geophysical Research B, vol. 116, no. 6, Article ID B06311, 2011. r · ·
32. E. Rignot, G. Casassa, P. Gogineni, W. Krabill, A. Rivera, and R. Thomas, "Accelerated ice discharge from the Antarctic Peninsula following the collapse of Larsen B ice shelf," Geophysical Research Letters, vol. 31, no. 18, pp. L18401–4, 2004. r · ·
33. T. A. Scambos, J. A. Bohlander, C. A. Shuman, and P. Skvarca, "Glacier acceleration and thinning after ice shelf collapse in the Larsen B embayment, Antarctica," Geophysical Research Letters, vol. 31, no. 18, pp. L18402–4, 2004. r · ·
34. J. M. Amundson, J. F. Clinton, M. Fahnestock et al., "Observing calving-generated ocean waves with coastal broadband seismometers, Jakobshavn Isbrae, Grenland," Annals of Glaciology, vol. 53, no. 60, 2012. r ·
35. D. R. MacAyeal, E. A. Okal, R. C. Aster, and J. N. Bassis, "Seismic observations of glaciogenic ocean waves (micro-tsunamis) on icebergs and ice shelves," Journal of Glaciology, vol. 55, no. 190, pp. 193–206, 2009. r · ·
36. D. R. Bohnenstiehl, R. P. Dziak, M. Parlk et al., "Seismicity of the polar seas: the potential for hydroacoustic monitoring of tectonic and volcanic processes," in Proceedings of the 12th Seoul International Symposium on Polar Sciences, pp. 11–14, Ansan, Republic of Korea, 2005.
37. R. P. Dziak, M. Parlk, W. S. Lee et al., "Tectono-magmatic activity and ice dynamics in the Bransfield Strait back-arc basin, Antarctica," in Proceedings of the 16th International Symposium on Polar Sciences, pp. 59–68, Incheon, Republic of Korea, 2009.
38. C. Müller and A. Eckstaller, "Local seismicity detected by the Neumayer seismological network, Dronning Maud Land, Antarctica: tectonic earthquakes and ice-related seismic phenomena," in Proceedings of the 9th International Symposium on Antarctic Earth Sciences Programme and Abstracts, vol. 236, 2003.
39. P. Wessel and W. H. F. Smith, "New version of the generic mapping tools released," EOS, Transactions American Geophysical Union, vol. 76, p. 329, 1995.

CHAPTER 12

Archaeo-Geophysical Survey around Itay El-Baroud Area, Nile Delta, Egypt

Hany Shaaban[1], Gad El-Qady[1,2*], El-Said Al-Sayed[1], Ashraf Khozaym[1], Ahmed Al-Emam[1], Hosni Ghazala[3]

[1]*National Research Institute of Astronomy and Geophysics, Helwan, Egypt*

[2]*Egypt-Japan University of Science and Technology (EJUST), New Borg ElArab, Alexandria, Egypt*

[3]*Geology Department, Faculty of Science, Mansoura University, Mansoura, Egypt*

ABSTRACT

Around the Nile Delta branches, the ancient settlements had been created and left their remains to be good witness for the paleoenvironment during the Holocene time. Therefore, tracing of these remains and their associated environments are of great importance. The area of study is located in the central part of the western Nile Delta (west of Rosetta branch), around Itay El-Baroud city. It is located between Latitudes 30°50'N - 31°N and Longitudes 30°35'E - 30°45'30"E covering an area of about 380 Km^2. It is mainly covered with cultivated lands and is characterized by its numerous archaeological hills such as Kom Geif. It also includes the Greek trading centre of Naukratis. The city was extinct about the beginning of the third century A.D. The archaeological site of Naukratis is now occupied by Kom Geif village. Integrated geophysical studies including magnetic and multi-frequency electromagnetic methods were conducted at Kom Geif village. The study area is divided into two main sectors sites: A (100 m × 20 m in length) and B (40 m × 40 m in length), respectively in 18 grids. Each gird is

10 m × 20 m and subdivided into a number of parallel traverses with spacing of 0.5 meter. Site A was selected to carry out a detailed electromagnetic survey in E-W direction, 20 zigzag profiles with 100 m length and 1 m interval were carried out. The station intervals along each line were 1 m. Based on the interpretation of the acquired geophysical data, seven structures of possible archaeological interest (mud bricks) have been recognized in a relatively small area at the southern end of Kom Geif Lake. Further detail geophysical studies are important to establish the nature, date, depth and state of preservation of these buried archaeological remains.

Keywords: Geophysics, Nile Delta, Kom Geif, Canopic Branch

1. INTRODUCTION

Egypt is known by its deep roots civilization that extends thousands of years through the history of the human being. This civilization has left a huge amount of traces and antiquities. The monuments of Egypt also vary from era to era. They involve Pharaonic, Greek, Roman, Coptic and Islamic traces.

The area of study is located in the central part of the western Nile Delta (west of Rosetta branch), around Itay El-Baroud city (Figure 1). It is located between Latitudes 30°50'N and 31°N and Longitudes 30°35'E and 30°45'30"E covering an area of about 380 Km^2. The area of study is mainly covered with cultivated lands and is characterized by its numerous archaeological hills such as Kom Geif and El-Nukrash. Of many geophysical exploration techniques, the electromagnetic (EM) method provides significant advantages for shallow geophysical exploration (Won, 1980). This is mainly a result of their high acquisition speed. Inductive coupling allows data to be collected without the need to have a physical contact between the electromagnetic equipment and the ground (Fitterman & Stewart, 1986).

The New broadband electromagnetic (EM) sensors have been used to detect landfills, unexploded ordnance, buried drums, trench boundaries, contaminant plumes, underground facilities and for other applications (Keiswetter et al., 1997, Keiswetter & Won, 1997; Murray & Keiswetter, 1998, Sternberg & Birken, 1999, Witten & Calvert, 1999; Won et al., 1996, 1997, 1998). In such applications, interpretations is commonly based on analysis of the measured inphase and quadrature parts per million (ppm) components and mapping the apparent conductivity derived from a single component of the EM responses (Won et al., 1996, 1997).

On the other hand, magnetic method has been first used in the 1950s and has since become the backbone of archaeological prospection. It is a fast and effective technique for mapping the distribution of archaeological remains in the shallow subsurface (Clark, 1990, Scollar et al., 1990;Reynolds, 1997). However, magnetic method can be used mainly to detect the archaeological features that contain magnetic minerals. These minerals have been found in the features, made of mud bricks and fired bricks, which constitute the majority of the archaeological remains in the study area.

It becomes a standard tool for delineation of contaminated waste sites and measures variation in electrical conductivity (EC) levels (Nabighian & Macanae, 1991). The output of an EM survey is generally a map, or 3D surface, showing the variation in EC across a site in units of mS/m.

The archaeological record has two basic properties or mechanisms by which it distorts the Earth's magnetic field. These are called remnant magnetization (a permanent magnetic effect) and the magnetic susceptibility (a bulk magnetic property similar to density). Both magnetisms are dependent on the presence of iron (iron oxides in soils, sherds, hearths etc.) and both magnetisms alter the magnetic field at the surface of the site. They are thus mapped as distortions of the Earth's magnetic field (Herwagner et al., 2000).

The contrast in magnetic susceptibility can be observed in the top-soil and sub-soil by using different types of magnetic instruments. Soils compacted by human occupation or disturbed by a burial will also show a variation from background values of magnetic susceptibility.

2. ARCHAEOLOGICAL CONTEXT

Ancient settlements in the Nile Delta are frequently encountered on the high parts of the alluvial plain, away from the danger of the annual Nile flood and close as possible to the flood basins for cultivation and agriculture after they become dry (Van Wesemael, 1988). Human settlements which are indicated by the occurrence of large amounts of sherds are located on the Pleistocene sand mounds (buried Geziras) and Holocene natural levees. These settlements grew gradually due to the accumulation of debris and the construction of new houses on the top of the remains of collapsed ones. In this way the Tells kept their elevation above the alluvial plain. During the annual Nile floods sediments were deposited and the alluvial plain rose steadily and Geziras were partially buried.

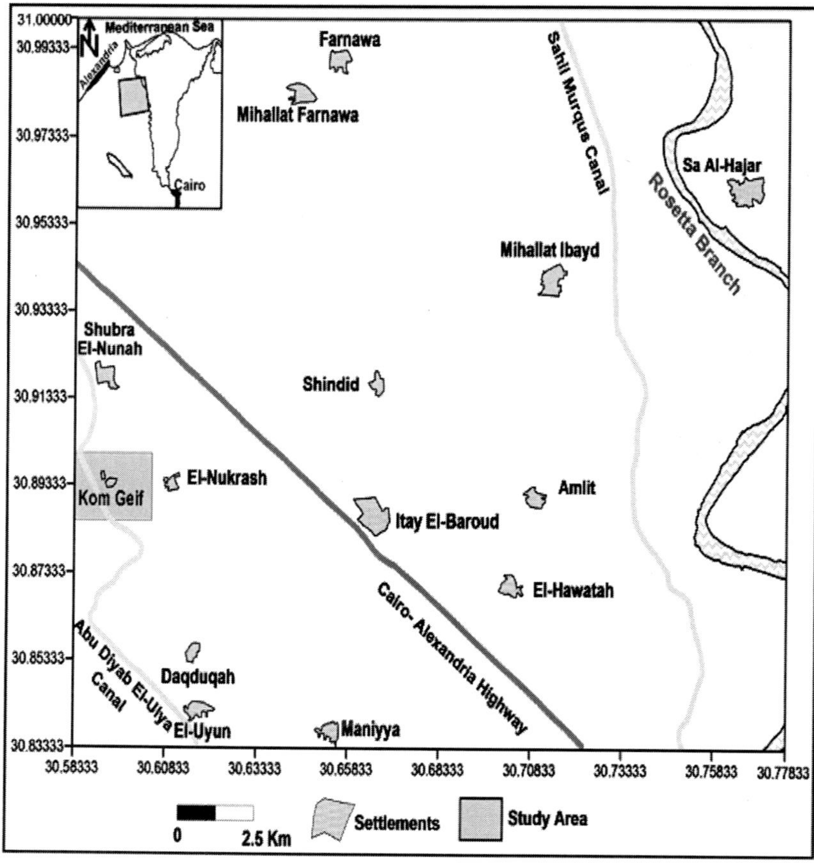

Figure 1. Location map of the study area around Itay El-Baroud area, Nile Delta, Egypt.

The following is a brief account on the archaeological importance of the Greek city at Naukratis; extracted mainly from Coulson and Leonard (1982). About 80km south-east of Alexandria, just off the Cairo-Alexandria highway, are the villages of Kom Geif, El-Nubeira and El-Nukrash, which cover the scattered archaeological site of Naukratis (Figure 1). The settlement, founded at least as early as Dynasty 26 when foreign settlers were granted a monopoly on Aegean trade, later became one of the most important Greek commercial centers, serving the needs of its various communities (Coulson & Leonard, 1982). Naukratis was the principal centre of cultural relations between Greece and Egypt during the late pharaonic era until the founding of the great city of Alexandria. Petrie discovered Naukratis early in his career and excavated parts of this unique site in 1884-1885. When Petrie first visited Naukratis, the area to the south of the site at Kom Geif, consisted of a large mound measuring 400 m by 800 m, then in poor condition but now completely flattened and waterlogged (Figure 2). To the east of this a large temenos wall enclosed the town's main temple, thought to be dedicated to Amun and Thoth. The enclosure also contained a large platform, at least 15 m high, which may have been used for military or administrative purposes. Trade was regulated by Egyptian law and subject to taxes paid to the Saite pharaohs and seems to have consisted of Mediterranean luxury commodities such as silver and olive oil in exchange for Egyptian corn, linen and papyrus (Coulson & Leonard, 1982). To the northern end of the site were three temples dedicated to Dioscuri, Apollo, Hera, constructed in Hellenistic style and each within their own sacred enclosure, thought to be built during the early 6th century BC (Coulson & Leonard, 1982). Here Petrie found remains of Ionic columns from the Apollo Temple and there are still a few extant remains of the Greek monuments, although now in poor condition (Coulson & Leonard, 1982). To the east of these shrines was a "Hellenium" and a tiny shrine of Aphrodite was found further to the south. Several vessels have been recovered from around the sanctuaries, inscribed with dedications to the town's Greek deities. Some of these were donated by well-known characters from history, including a fragment of a cup incised with the name of Herodotus himself. The centre of the site contained the main town, with many houses built during the Ptolemaic period.

It was in this area that Petrie found a faience workshop which had produced both Egyptian and Greek objects, including numerous glazed scarabs dating as far back as the reign of Psamtek I. Many silver and bronze Greek coins have been found during excavation of the site, the only coins known from pharaonic Egypt, although probably not generally circulated at that time as most of them appeared to be in mint condition. The most significant finding from Naukratis is the Dynasty 30 stela of Nectanebo I, one of the last truly Egyptian pharaohs. The perfectly preserved carvings on the Naukratis stela (now in Cairo Museum) announce a decree by the king that a tax of one tenth of all goods from Naukratis be paid into his treasury for the benefit of the Temple of Neith at Sais (Coulson & Leonard, 1982).

3. GEOPHYSICAL DATA ACQUISITION

In the present study, a micro-magnetic survey has been first applied as a rapid and accurate way to detect the archaeological features made of mud bricks and fired bricks which represent the majority of the structures in the

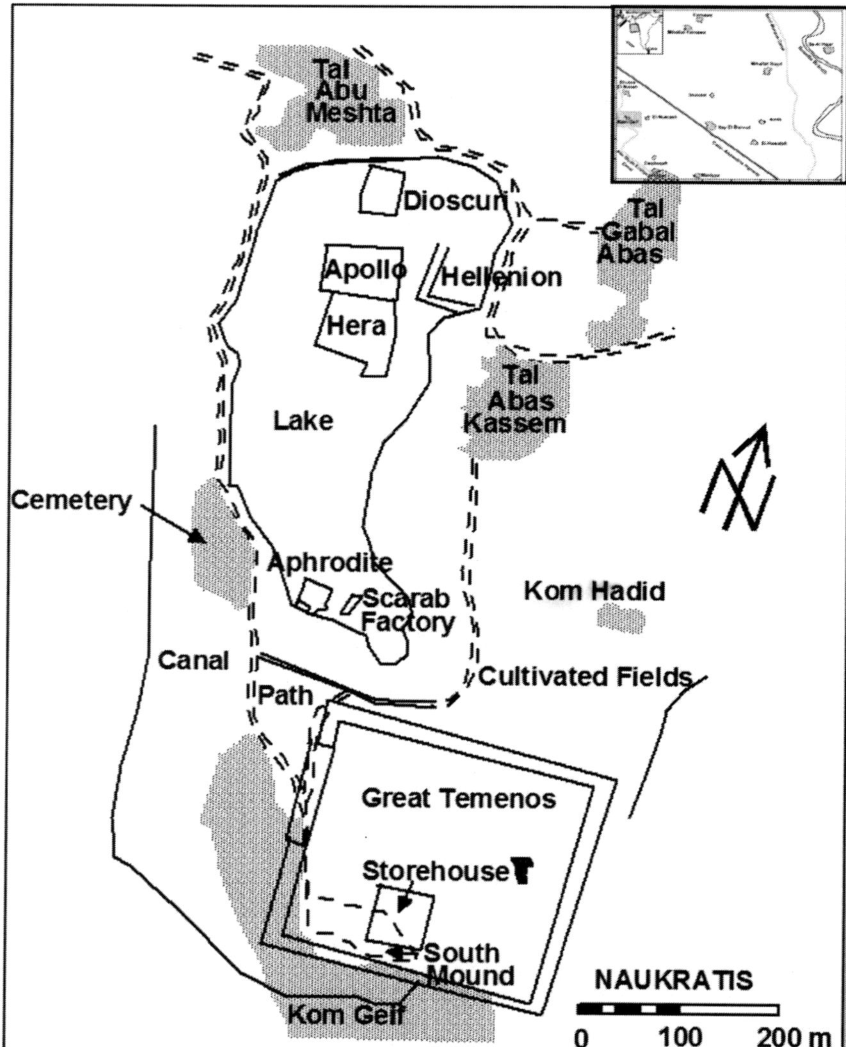

Figure 2. Plan of Naukratis showing the buildings uncovered by the early excavators in relation to the present state of the site (after Coulson & Leonard, 1982).

area of study. The micro magnetic survey was carried out by the fluxgate gradiometer model FM36 through a raster of 0.5 m × 0.5 m. The processing of magnetic data have been achieved by using Geoplot software (Geoscan Research, 1994) and then extracted to Surfer software to be more representative for interpretation. Secondly, multi-frequency electromagnetic method of a selected area had been utilized to support the results of the magnetic survey.

3.1. Magnetic Data Acquisition

Field procedures for the magnetic prospection method have been refined considerably in the last two decades. Some archaeologists now use magnetic gradiometer and microcomputer-controlled automatic data collection systems to make low-noise, high-productivity surveys. It is not unusual for a field crew to acquire as such as a third of a hectare of 1 meter grid data in a single day.

The use of two magnetometers sensors mounted at different heights allows the recording of vertical magnetic gradient in nano-Tesla per meter (nT/m). Such measurements usually reflect the distribution of archaeological remains at relatively shallow depths; can be within few meters. The vertical gradient field is given by:

$$\partial F/\partial Z = \left(F1 - F2\right)/\Delta Z \quad (1)$$

where F1 and F2 are the reading at the upper and lower sensors, and ΔZ is the separation between the sensors. The vertical magnetic gradient has several interesting properties (Scollar et al., 1990; Herwagner et al., 2000). It filters out the effects of the regional magnetic field, is only weakly influenced by magnetic time variations and better divines shallow magnetic features (Bossuet et al., 2001).

The measurements of the vertical magnetic gradient for the purposes of archaeological investigations has been extensively used in Egypt (Kamei et al., 2002; Abdallatif et al., 2003, 2005; Ghazala et al., 2003; Herbich, 2003; Abbas et al., 2005; Odah et al., 2005; Wilson, 2006).

Prior to the measurements, the study area has been cleaned from any visible surface iron materials and the corners of the surveyed areas are marked with wooden sticks. The micro-magnetic survey was carried out using the fluxgate gradiometer through a raster of 0.5 m × 0.5 m.

A pilot area was selected to investigate the buried remains in the archaeological site of Kom Geif village (Naukratis). The selected area has been extended later according to the obtained results. The area under investigation (Figure 3) is divided into two main sectors; Sites A (100 m × 20 m) and Site B (40 m × 40 m), respect tively in 18 grids. Each gird is 10 m × 20 m and subdivided into a number of parallel traverses with spacing of 0.5 meter. Non-magnetic tapes were used to measure the distances and traverses.

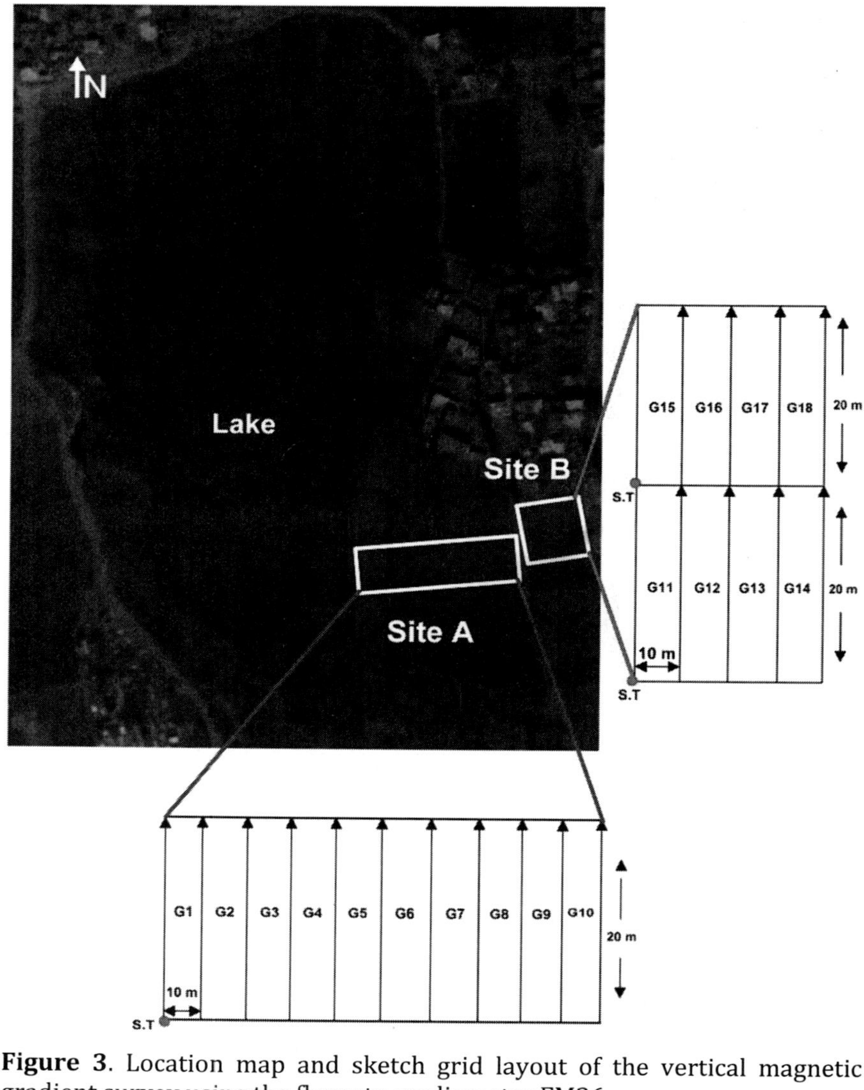

Figure 3. Location map and sketch grid layout of the vertical magnetic gradient survey using the fluxgate gradiometer FM36.

3.2. Multi-Frequency Electromagnetic Technique

Electromagnetic induction survey included conductivity mapping with conductivity meter (GEM-300) (GSSI, 1998). The area (A (100 m × 20 m)) was selected according to the magnetic survey and the availability of places to do the survey can easily be mapped with a conductivity meter. The measurements are done at walking speed and it is suitable to map large areas in a short time. Site A in the magnetic survey (Figure 3) was selected to carry out a detailed electromagnetic survey using GEM-300. The device (GEM-300) is mainly used to record multiple frequency (low, medium and high) electromagnetic induction data. The effective depth of exploration for a given earth medium is determined by the operating frequency of the primary electromagnetic signal. In fact, the lower EM frequencies penetrate deeper than the higher frequencies. Therefore, measuring the areas response to multiple frequencies [low, 750 Hz), medium, (1710 Hz & 3870 Hz) and high, (8790 Hz & 19,950 Hz)] was equivalent to measuring the Earth response from multiple depth levels in terms of the few meters range (deep, intermediate and shallow). This powerful tool allows for more detailed of the subsurface than ever before, and is ideally suited for a wide range of applications.

The EM survey parameters through the surveyed site were designed in a grid pattern which has a dimension of 100 m × 20 m. In E-W direction, 20 zigzag profiles with 100 m length and 1m interval are carried out. The station interval along each line was 1 m.

4. DATA PROCESSING AND INTERPRETATION

4.1. Magnetic Method

The initial presentation of the obtained raw magnetic data for the two sites is presented in Figure 4. The main objective of data processing is to remove the expected noises and to enhance the archaeological remains responses and make it smoother for presentation and expressive for interpretation.

The handled raw data have been initially presented in shade-plots to find out the main field errors and noises. Many processing functions such as Zero Mean Grid, Zero Mean Traverse, Despike, Low Pass Filter and Clip have been used to remove these errors and to enhance the presentation of the obtained magnetic data. Finally, the obvious features are recognized easily in the sense of archaeological interest.

4.1.1. The Applied Processing Functions

The field errors concerning grids and traverses normally occur with the gradiometer survey due to several reasons such as the sensitivity of the instrument to any vertical movement , repeat measuring of the last points at the former and the later grids, irregularities of the ground surface and instability of the operator (Geoscan Research, 1994).

The acquired raw data have been presented as a grayscale image (Figure 4). The gradient of the vertical component of the geomagnetic field ranges from −220 to +140 nT. At a glance, the resultant magnetic image (Figure 4) exhibits foggy and disturbed with many noises that make it difficult to be interpreted clearly to the comparable archaeological meaning. To remove the field errors and enhance the obtained data, the following processing sequences have been achieved:

- The grid edge discontinuities are treated by the application of Zero Mean Grid (ZMG) with a threshold value = 0.25 standard deviation.
- The stripes between traverses are removed by the application of Zero Mean Traverse (ZMT) with least mean square fit.
- The data are clipped initially for Site A (−5/+5 nT) and for Site B (−10/+10 nT) to remove the high frequencies resulting from any expected surface iron spikes. Also, the Despike function (K) is applied for more enhancement and good presentation.
- A Gaussian Low-Pass Filter (L) is applied to remove the high frequencies as well as to smooth and enhance the weak anomalies of the deep archaeological features. The Low-Pass Filter parameters are set to be X and Y radii = 2.

The resultant processed magnetic image is presented in (Figure 5). It is well represented to the expected buried remains at the study area and could be considered as a base for the interpretation process.

4.1.2. Magnetic Data Interpretation

Geophysical data images typically present a two dimensional representation of a complex sequence of archaeological events and natural conditions. Therefore, interpretation of geophysical data must be a cooperative process involving both geophysicists and archaeologists that are familiar with the specific cultural context of the site being studied. The understanding of the geological background and geomorphology are also important and valuable for a complete interpretation of the obtained results.

Interpretation of magnetic anomalies to the comparable monuments is the main target of the previous steps. It is interesting to know the archaeological findings resulting from the application of magnetic method. However, the interpretation process is not an easy job and needs a lot of experience to differentiate between what is archaeological and what is non-archaeological. The interpretation rules in the present study are based mainly on the clarity of the deduced features, their location, polarities, dimensions, shapes etc.

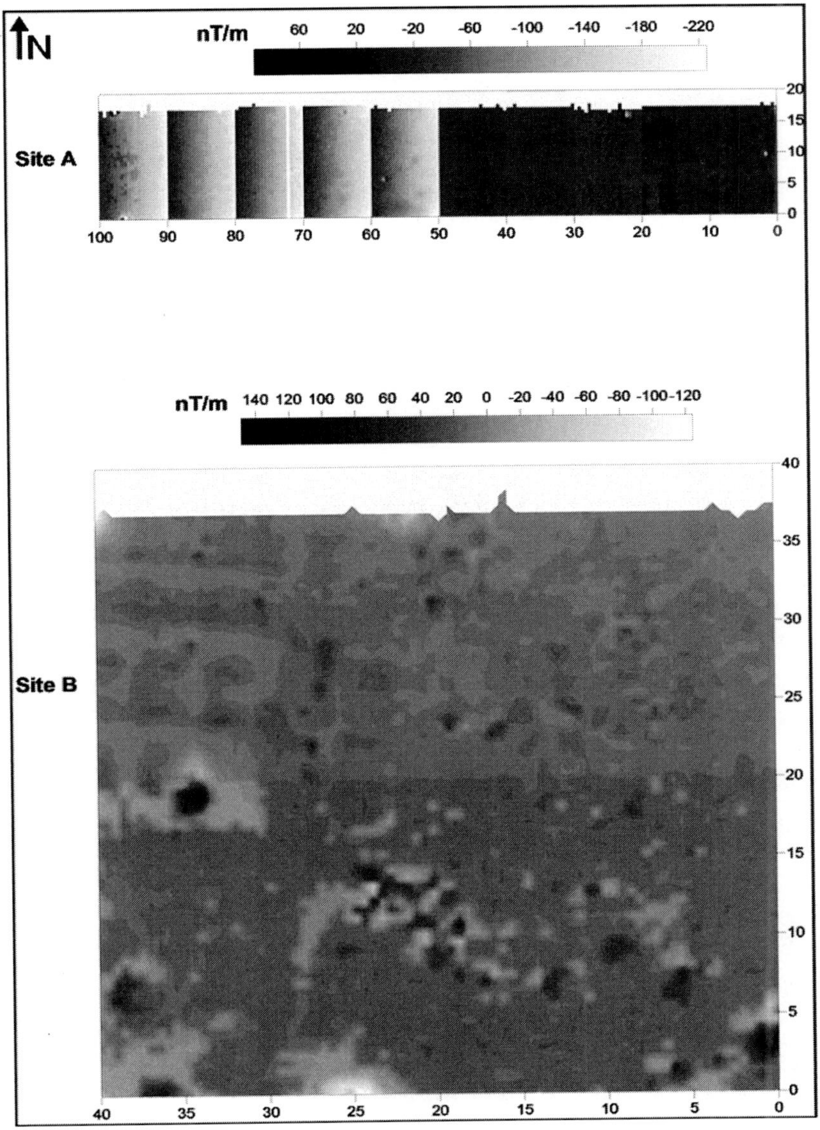

Figure 4. Raw magnetic data of the studied archaeological sites (A and B).

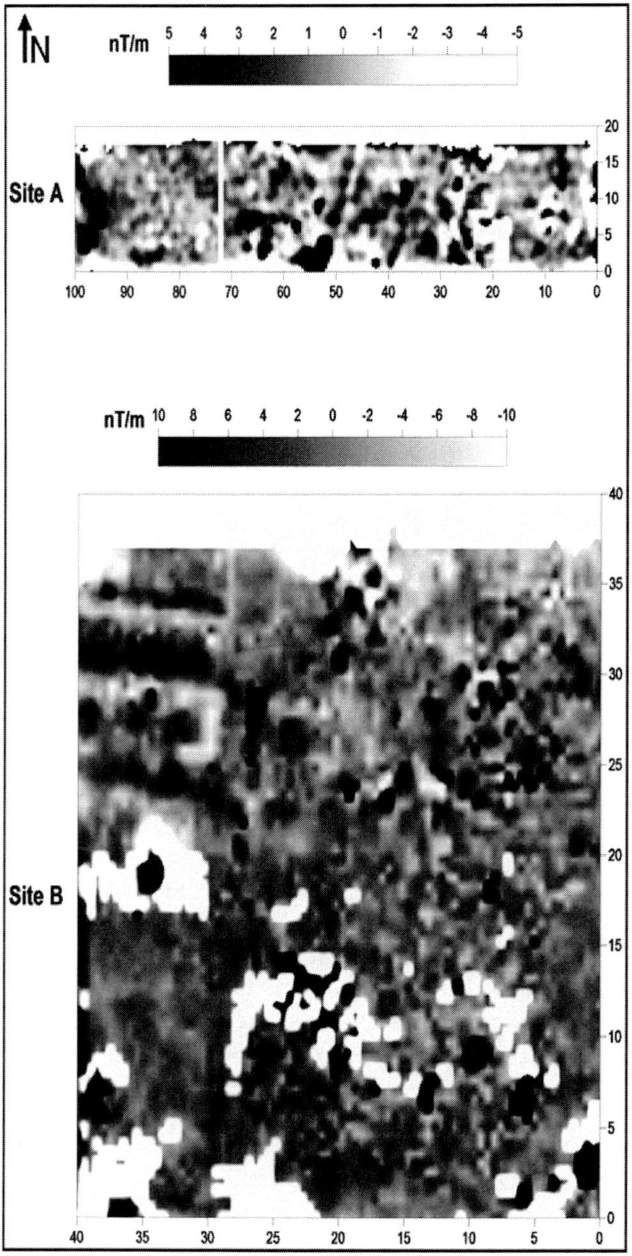

Figure 5. Magnetic image of the studied sites (A and B) after application of the processing functions.

The dark areas in the processed gray scale magnetic image (Figure 5) indicate positive anomalies, while the lighter areas indicate negative anomalies. The magnetic image appears as if it is an aerial photo of surface foundation, so it can be used to guide archaeologists for any future excavations. The clearance of the observed features, their shapes, locations, dimensions and polarities facilitates interpretation of the obtained results to produce a good final archaeological view of the study area.

On the ground, wide field observations have been made in and around the study area indicate that, there was no evidence of any surface mud bricks. Meanwhile, the surface discoveries around the study area indicate the existence of some broken pottery made of fired mud. This is an important note, which could help in the illustration of the obtained results, and could guide to good comprehensive conclusions.

4.1.3. Interpretation of Magnetic Anomalies at Site (A)

The analysis of magnetic image showed scattered and dissected anomalies associated with negative anomalies (Figure 6). The magnetic anomalies within Site (A) shows three different distinctive anomaly features with different colors, traced with colored lines (Figure 6). Their magnetic intensity depends on the amount/density of the magnetic carrier's minerals; for example, the blacker anomaly indicates the presence of more magnetic minerals.

The first anomaly zone (S1) is represented by a connected positive magnetic anomaly and suggests a part of rectangular shape. The rest of this anomaly probably occurs in the adjacent lake (Figure 3). This anomaly may be a part of buried mud brick wall.

The second anomaly (S2) has a triangular shape and may refer to the entrance of tomb. Besides, it has been noticed that the small random negative anomalies (white color) are not representing a magnetic dipole effect. These negative parts may refer to the presence of subsurface archaeological structure that often filled with sediments. The third anomaly (S3) has a rounded shape with big positive magnetic anomaly. This may be a part of a buried wall made of mud bricks.

4.1.4. Interpretation of Magnetic Anomalies at Site (B)

Four interesting anomalies can be distinguished in this site (Figure 7). The first and the second anomalies (S1 & S2) lie at the west of the study area, while the third anomaly (S3) lies at the north and the fourth anomaly (S4) lies at the south right corner of the study area. The scattered and dissected anomalies associated with negative anomalies are found in some grids all over the site.

The first anomaly (S1) is represented by a big connected positive magnetic anomaly elongated from east to west with rectangular shape. The first anomaly may represent a subsurface archaeological structure which made of mud bricks.

The second anomaly (S2) is represented by positive magnetic anomaly. It has a semi-circular shape. This anomaly is surrounded by obvious negative anomaly which is magnetic dipole of the positive one. This anomaly started to appear again after some distance towards the south of the area. This new anomaly is incomplete but it has the same shape and the same behavior of the first one. This anomaly may be supposing the existence of some buried archaeological feature with high magnetic response; it may be kind of metallic body!!

The third and fourth anomalies (S3 & S4) are represented by scattered positive magnetic anomalies. These anomalies may be a result of collapsed buried mud bricks wall.

The magnetic image of Site (B) reveals the presence of many scattered and separated positive magnetic anomalies all over the study area. These anomalies may be a result of some archaeological features or from the existence of highly magnetized minerals.

4.2. Multi-Frequency Electromagnetic Technique

The EM data have been analyzed at different frequencies from low to high. Figure 8 shows the apparent electrical conductivity in the surveyed site. Variations between the five maps are due to the variations of depth of penetration resulted from the different applied frequencies.

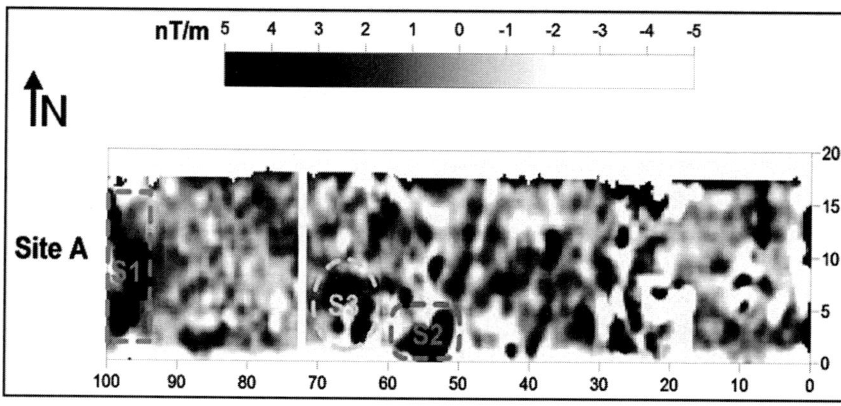

Figure 6. Interested magnetic anomaly zones derived from magnetic image at Site A.

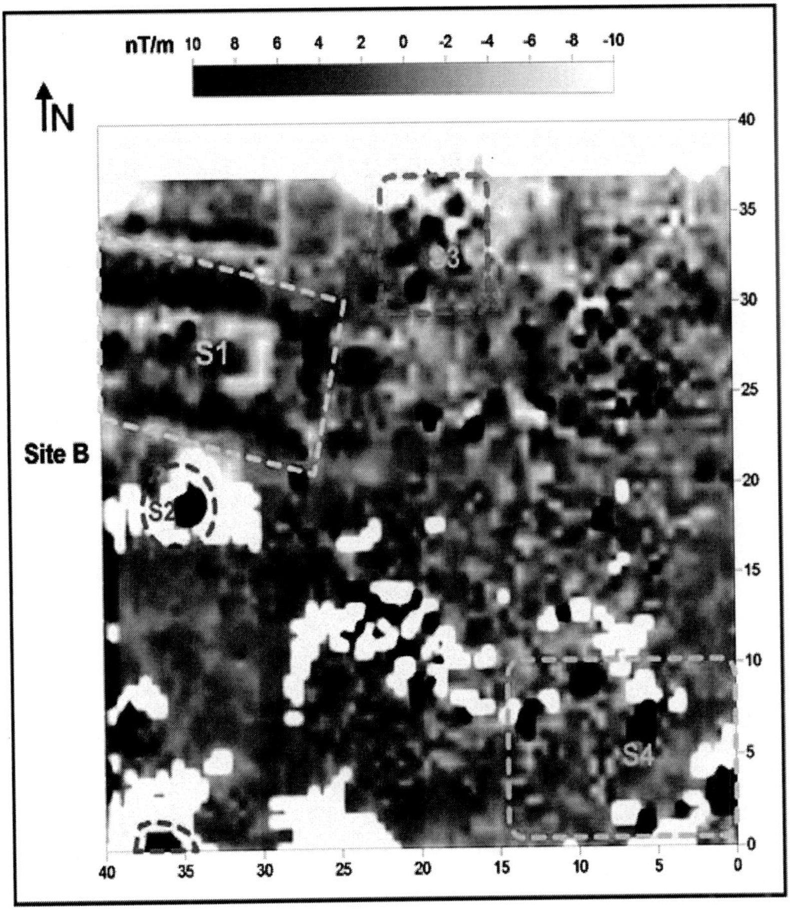

Figure 7. Interested magnetic anomaly zones derived from magnetic image at Site B.

In practice, the apparent conductivities are more representative of the conductivity structure than EM data (Fitterman & Stewart, 1986). For example, quadrature is proportional to conductivity in low to mid induction numbers but is inversely proportional at mid to high induction numbers (Huang & Fraser, 1998), which is true for the case of multi frequency EM data we used in this work.

Figure 9(a) & Figure 9(b) shows the magnetic results of Site (A) and the conductivity maps at five different frequencies from high to low (i.e. from shallow to deep). Initial comparison between the magnetic gradient and electromagnetic results, shows matching between each other.

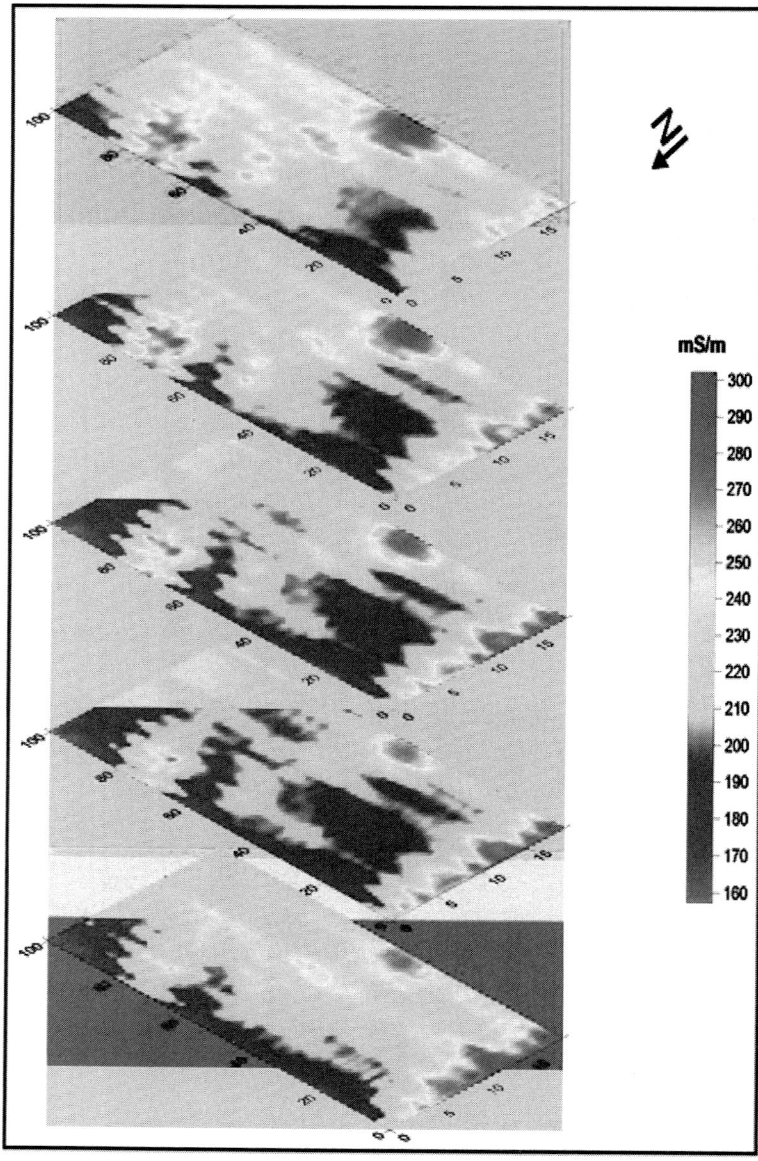

Figure 8. Conductivity maps for Site A at frequencies from bottom to top 750, 1710, 3870, 8970 and 19,950 Hz, respectively.

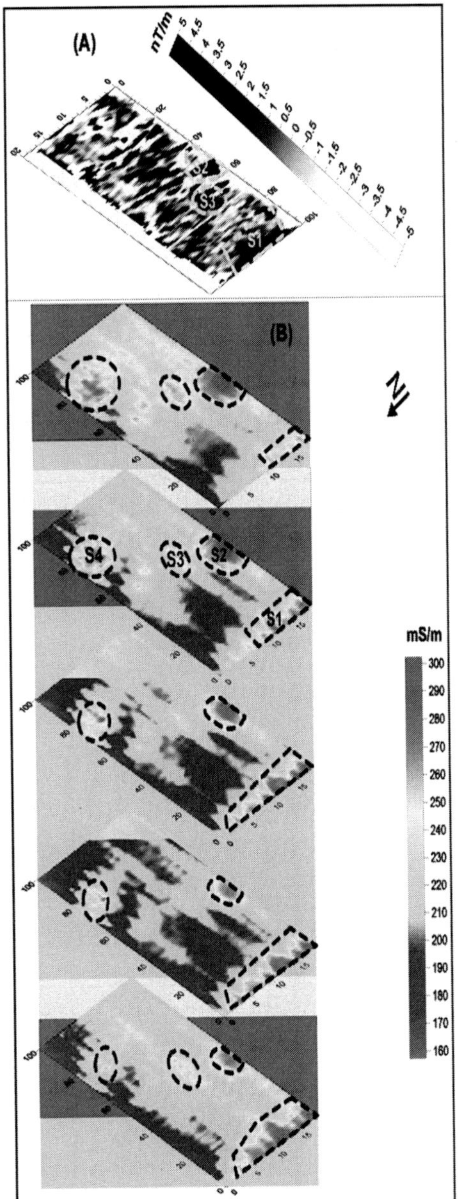

Figure 9. (A) Magnetic image of Site A., (B) Conductivity maps for frequencies from bottom to top 750, 1710, 3870, 8970 and 19950 Hz, respectively.

The correlation between the magnetic results and the conductivity maps (Figure 9(a) & Figure 9(b)) reveals the existence of a big elongated anomaly with rectangular shape (S1). This anomaly started to appear at the high frequency and then become more obvious and clear in intermediate and low frequency. This anomaly suggests the existence of archaeological feature at relatively shallow and intermediate depths of the explored sections which is within few meters depth. The high conductivity with high magnetic anomaly suggests that this archaeological feature is made of mud bricks. The conductivity map (Figure 9(b)) shows a big anomaly of high conductivity (S2) which has a rounded shape. The high frequency of this big anomaly supposes to be originated from a near-surface archaeological feature. This high conductivity anomaly with high magnetic anomaly suggests the existence of archaeological feature made of mud bricks, which agrees well with the magnetic results of (Figure 9(a)). There is also another big anomaly (S3) appears at high frequency and then disappeared at intermediate frequency and then appear again at low frequency. This big anomaly has a semi-circular shape with high conductivity. The high frequency of this anomaly may suggest a near-surface archaeological feature.

There is also another big anomaly (S4) appears at high frequency and then become less obvious at intermediate frequency. This big anomaly has a semi-circular shape with high conductivity. The high frequency of this anomaly may suggest a near-surface archaeological feature.

5. SUMMARY AND CONCLUSIONS

Two sites (A & B) around Kom Geif village (Naukratis) were selected to carry out the detailed archaeo-prospection survey. The micro-magnetic survey was carried out by the fluxgate gradiometer model FM36 through a raster of 0.5 m × 0.5 m. The measurements of the vertical magnetic gradient have been taken through successive parallel traverses separated by 0.5 m interval.

The acquired data were then represented through the Geoplot program as a grayscale image that represented the raw data. The processed magnetic image is very clear and appears as if it was an aerial photo of surface foundation, so it can be used to guide archaeologists during future excavation. Seven archaeological structures of mud bricks have been recognized in Sites A & B.

Site A was selected to carry out a detailed electromagnetic survey using GEM-300. There is very clear similarity between the magnetic data and the five conductivity maps and the anomaly associated between them which indicate that these anomalies are made of mud brick.

ACKNOWLEDGEMENTS

Hany Shaaban, Gad El-Qady, El-Said Al-Sayed, Ashraf Khozaym, Ahmed Al-Emam, Hosni Ghazala Sincere thanks for all of NRIAG stuff, who helped in acquiring the geophysical data using NRIAG facilities. We also would like to thank the anonymous reviewers for their valuable comments. The authors are indebted to the anonymous referees and their valuable comments that enrich the manuscript to its present form.

REFERENCES

1. Abbas, M. A., Abdallatif, T. F., Shaaban, F., Salem, A., & Suh, M. (2005). Archaeological Investigation of the Eastern Extensions of the Karnak Temple Using Ground-Penetrating Radar and Magnetic Tools. Geoarchaeology, 20, 537-554.
2. Abdallatif, T. F, Abd El-All, E. M., Suh, M., Mohamad, R. M., & El-Hemaly, I. A. (2005). Magnetic Tracing at Abu Sir (Land of Forgotten Pyramids), Northern Egypt. Geoarchaeology, 20, 438-503.
3. Abdallatif, T. F., Mousa, S. E., & El-Bassiony, A. (2003). Geophysical Investigation for Mapping the Archaeological Features at Qantir, Sharqyia, Egypt. Archaeological Prospection, 10, 27-42.
4. Bossuet, G., Camerlynck, C., Brehonnet, C., & Petit, C. (2001). Magnetic Prospecting of Diachronic Structures (Antiquity to First World War) on the Site of Sanctuary of Ribemont-sur-Ancre, Somme, France. Archaeological Prospection, 8, 67-77.
5. Clark, A. J. (1990). Seeing Beneath the Soil Prospecting Methods in Archaeology. London: B.T. Batsford Ltd., 176p.
6. Coulson, W. D. E., & Leonard Jr., A. (1982). Investigations at Naukratis and Environs, 1980 and 1981. American Journal of Archaeology, 86, 361-380.
7. Fitterman, D. V., & Stewart, M. T. (1986). Transient Electromagnetic Sounding for Groundwater, Geophysics, 51, 995-1005.
8. Geoscan Research (1994). Instruction Manual Version 1.01. (Geoplot 2.01). Bradford, West Yorkshire, England, Geoscan Research.
9. Ghazala, H. H., El-Mahmoudi, A. S., & Abdallatif, T. F. (2003). Archaeogeophysical Study on the Site of Tell Toukh EL-Qaramous, Sharkia Governorate, East Nile Delta, Egypt. Archaeological Prospection, 10, 43-55.
10. GSSI (1998). Operating System Version 1.10 (GEM-300). North Salem, NH 03073-0097 USA, Geophysical Survey System Inc. 79p.
11. Herbich, T. (2003). Archaeological Geophysics in Egypt. The Polish contribution. Archaeologia Polona, 41, 13-55.
12. Herwagner, J., Muarer, H., & Green, A. G. (2000). 3-D Inversion of Magnetic Gradiometer Data in Archaeological Prospecting. Possibilities and Limitations. Geophysics, 65, 849-860.
13. Huang, H., & Fraser, D. C. (1998). Magnetic Permeability and Electrical Resistivity Mapping with a Multi-Frequency Airborne EM system. Exploration Geophysics, 29, 249-253.

14. Kamei, H., Atya, M. A, Abdallatif, T. F., Mori, M., & Hemthavy, (2002). Ground-Penetrating Radar and Magnetic Survey to the West of Al-Zayyan Temple, Kharga Oasis, Al-Wadi Al-Jadeed (New Valley), Egypt. Archaeological Prospection, 9, 93-104.
15. Keiswetter, D. A, Novikova, E., Won, I. J., Hall, T., & Hanson, D. (1997). Development of a Monostatic, Multi-Frequency Electromagnetic Mine Detector. Society of Optical Engineering, 3079, 831-839.
16. Keiswetter, D. A., & Won, I. J. (1997). Multi-Frequency Electromagnetic Signature of the Cloud Chamber, Nevada Test Site. Journal of Environmental and Engineering Geophysics, 2, 99-104.
17. Murray, C., & Keiswetter, D. A. (1998). Application of Magnetic and Multi-Frequency EM Techniques for Landfill Investigations, Case Histories. Symposium on the Application of Geophysics to Environmental and Engineering Problems, Proceedings, 445-452.
18. Nabighian, M. N., & Macanae, J. C. (1991). Time-Domain Electromagnetic Prospecting Methods. In M. N., Nabighian, (Ed.), Electromagnetic Methods in Applied Geophysics, Vol. 2 (pp. 427-514), Society of Exploration Geophysics.
19. Odah, H., Abdallatif, T. F., El-Hemaly, I. A., & Abd El-All, E. M. (2005). Gradiometer Survey to Locate the Ancient Remains Distributed to the Northeast of the Zoser Pyramid, Saqqara, Giza, Egypt. Archaeological Prospection, 12, 61-68.
20. Reynolds, J. M. (1997). An Introduction to Applied and Environmental Geophysics. Chichester: John Wiley and Sons Ltd. 796p.
21. Scollar, I., Tabbagh, T., Hesse, A., & Herzog, I. (1990). Archaeological Prospecting and Remote Sensing. Cambridge: Cambridge University Press, 692p.
22. Sternberg, B. K., & Birken, R. A. (1999). A New Method of Subsurface Imaging, The LASI High Frequency Ellipticity System. Part 3, System Tests and Field Surveys. Journal of Environmental and Engineering Geophysics, 4, 227-240.
23. Van Wesemael, B. (1988). The Relation between Natural Landscape and Distribution of Archaeological Remains in the Northeastern Nile Delta. In E. C. M. Van Den Brink (Ed.), The Archaeology of the Nile Delta: Problems and Priorities (pp. 125-134). Amsterdam: Netherlands Foundation for Archaeological Research in Egypt.
24. Wilson, P. (2006). The Survey of Sais (Sa El-Hagar) 1997-2002. London: Egyptian Exploration Society, 336p.
25. Witten, A. J., & Calvert, G. (1999). Characterizing the Distribution of Near-Surface Solution Channels Using Electromagnetic Induction and Ground Penetrating Radar. Journal of Environmental and Engineering Geophysics, 4, 35-43.
26. Won, I. J, Keiswetter, D. A., Hanson, D. R., Novikova, E., & Hall, T. M. (1997). GEM-3: A Monostatic Broadband Electromagnetic Induction Sensor. Journal of Environmental and Engineering Geophysics, 2, 53-64.
27. Won, I. J. (1980). A Wide Band Electromagnetic Exploration Method, Some Theoretical and Experimental Results. Geophysics, 45, 928-940.
28. Won, I. J., Keiswetter, D. A., & Novikova, E. (1998). Electromagnetic Induction Spectroscopy. Journal of Environmental Engineering Geophysics 3, 27-40.
29. Won, I. J., Keiswetter, D. A., Fields, G. R. A., & Sutton, L. C. (1996). GEM-2: A New Multi-Frequency Sensor. Journal of Environmental and Engineering Geophysics, 1, 129-138.

CHAPTER 13

Mineral Chemistry of Two-Mica Granite Rare Metals: Impact of Geophysics on the Distribution of Uranium Mineralization at El Sela Shear Zone, Egypt

Ibrahim Gaafar[1], Michel Cuney[2], Ahmed Abdel Gawad[1]

[1]*Nuclear Materials Authority, Cairo, Egypt*

[2]*G2R, Nancy-Université, CNRS, CREGU, Lorraine, France*

ABSTRACT

The present work aims at identifying Nb-Ta-, Zr-Hf-, REE-, Th-U-bearing two-mica granite from geological, geophysical cross-sections and mineral chemistry studies from three boreholes at G. El Sela shear zone. Microscopically, the three boreholes are composed mainly of two-mica granite. They are composed of K-feldspar, plagioclase, quartz, biotite and muscovite. Accessories are pyrite, zircon, fluorite, rutile, monazite with Th-U-mineralization identified by scanning electron microscope (SEM) and electron probe-microanalyses (EPMA). Chlorite, muscovite, sericite, kaolinite are secondary minerals. Geochemically, two-mica granite boreholes are A-type granites and peraluminous characteristics. They are enriched in large ion lithophile elements (LILE; Ba, Rb and Sr), high field strength elements (Y, Zr and Nb), and LREE but depleted in HREE with negative Eu anomaly. U-enrichment associated with chloritization, muscovitization, albitization, sericitization, ka- olinization and argillization results from convective hydrothermal circulation of fluids through brittle structures along the ENE-WSW main shear zone. The ratios Nb/Ta (7.7 - 17.7) and Zr/Hf (16.9 - 26.4) are relatively enriched in the lighter isovalents Ta and Hf. The accessory minerals ob- served in the two-mica granites are represented by metallic sulfides (pyrite, arsenopyrite, chalco- pyrite, galena and sphalerite), Nb-rutile, Hf-zircon, fluorite, monazite, columbite, betafite, thorite, phosphothorite, uranothorite, brannerite, uraninite, coffinite and pitchblende at G. El Sela shear zone. Uraninite with a low Th content indicates a hydrothermal origin of U-mineralization, Thorite, uranothorite, monazite and zircon is the main uranium bearing minerals of magmatic origin with- in the enclosing granite. The primary U-mineralization has been observed in two boreholes. In order to illustrate the geophysical signature of El Sela U-mineralization, the radiometric, magnetic, and VLF-EM data as well as radon concentration are included. The magnetic, electrical conductivity and radiometric profiles were produced from detailed ground surveys. The shear zone is characterized by relatively weak levels for both K and eTh, but very high eU anomalies (<3500 ppm), Therefore, the Sela shear zone acts as a good trap for U-mineralization. The Sela Shear zone coincides with positive conductivity anomalies, which are the most prominent features on the respective profiles. The magnetic field over the Sela shear zone is also conspicuous by the sharp contrast which makes with the strong negative signatures of the altered microgranite. The radon distribution map showed the presence of seven high anomalies that are mostly controlled by the structures due to the easy movement of radon through them.

Keywords: Boreholes, Shear Zone, Radon, Spectrometric, Magnetic, Conductivity

1. INTRODUCTION

Uranium ore is the indispensable raw material for nuclear fuel preparation and is currently heavily prospected. Unlike the majority of metals, the metallogeny of uranium is specifically characterized by an extreme diversity of deposits that are directly related to the various conditions under which U deposits formed in geological environments [1] . More than 1500 uranium deposits are identified in the UDEPO data base of the AIEA (http://www.iaea.org/OurWork/ST/NE/NEFW/Technical_Areas/NFC/infcis-UDEPO.html) and 16 major deposit types are distinguished by the International Atomic Energy Agency [2] .

Generally, granitic rocks are one of the most favourable host rocks for U-mineralization in many parts of the world. Genetically, the economic uranium deposits associated with granites are mostly located in analectic melts or in strongly peraluminous two-mica granites [3] - [5] . In general, the granites are rich in incompatible elements (Rb, Nb, Sn, Th, and U) [6] [7] . The genetic relationship between granite intrusions and associated U-minerali- zation has been discussed for a long time. Background uranium concentrations at the scale of the granite plutons crystallize at the magmatic to late magmatic stage as accessory minerals (zircon, monazite, allanite, uranothorite, uraninite···). They represent the primary source of uranium for many deposits and their enrichment rarely reach- es economic grades except in some highly fractionated peralkaline granites or anatectic pegmatoids [8] . The for- mation of an economic uranium concentration generally requires the remobilization of the uranium from these primary sources followed by deposition in tectonic structures with higher grades.

Most Egyptian uranium occurrences in granite rocks of G. Um Ara [9] [10] , G. Gattar [11] G. Missikat [12] and G. Erediya [13] [14] are associated with metaluminous to slightly peraluminous granite. Uranium mineralization in these granites belongs to the vein-type except G. Gattar and G. Um Ara, where U-mineralization belongs to the fracture-filling type. The marginal parts of all these granitic intrusions are generally more radioactive than their central parts in relation with the development of different types of alteration, which include silicification, episyenitization, albitization, fluoritization, ferruginization, sericitization, chloritization, kaolinitization and illitization.

El Sela two-mica uraniferous peraluminous granite [15] - [18] is located in the south-eastern part of Egypt between latitudes 22°16'26" - 22°18'38"N and longitudes 36°13'33" - 36°16'26"E.

2. GEOLOGIC SETTING

El Sela granitic pluton is composed of two-mica (muscovite-biotite) granite intrusions trending ENE-WSW (Figure 1). It is jointed in different directions, and defines low to moderate reliefs. This granite is injected by microgranite, dolerite and bostonite dikes, and quartz and jasperoid veins. They are mostly injected along ENE- WSW and/or NNW-SSE to N-S directions which represent the most important tectonic trends for U anomalies in the study area. Pegmatite pockets are found as irregular bodies in two-mica granite. They are very coarse grain- ed, buff to reddish pink colors, highly sheared and composed of K-feldspar, quartz, plagioclase, biotite with mu- scovite. They show slightly higher total radioactivity than two-mica granite.

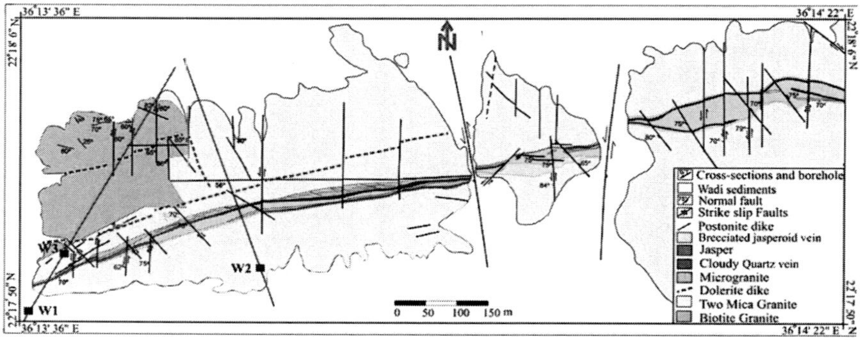

Figure 1. Detailed geological map of the main shear zone at G. El Sela area, south E D, Egypt (after [16]).

A microgranite dike hosts the most radioactive anomalies in the study area. It is injected into the two-mica granite along an ENE-WSW shear zone with dip of about 75 to south. It is whitish pink, pale pink, reddish pink to pale grey colors and is 3 to 20 m thick.

It is also essentially composed of quartz, K-feldspar, plagioclase, biotite and muscovite. It is enriched with mineralized with uranophane and pyrite cubes. Ferrugination, silicification, albitization, episyenitization, fluoritization and illitization are different alterations mainly occurring with the ENE-WSW shear zone. Some of these alterations are possibly related to U remobilization, but probably not all. To be discussed in detail later.

Dolerite dikes are greyish green to dark grey colors, fine- to medium-grained and range in thickness from 0.5 to 6 m . They are injected into the two-mica granite along ENE-WSW, NNW-SSE and N-S directions with dip of about 75 to south. They are mainly composed of plagioclase, olivine and pyroxene. Accessories are apatite and opaque minerals while chlorite, sericite and calcite are alteration minerals. The dolerites have higher U-contens which may reach up to 3500 ppm eU and 75 ppm eTh. They are highly altered to clay minerals and cavities are filled with pyrite cubes and/or secondary uranium minerals (uranophane and autunite). Many open cuts have been made to recognize the importance of the U mineralization within the microgranite and dolerite dikes. Many semi-industrial technological procedures for the leaching and recovery of the U incorporated within these mineralized dikes have been performed.

Along the main ENE-WSW shear zone three generations of quartz veins with different colors have been identified. The milky quartz vein occurs at the shear margins, is the oldest one, highly brecciated, barren, and range in thickness from 1 to 4 m . The milky quartz vein is crosscut by two younger generations of quartz with obvious displacement. Red and grey to black jasper veins are commonly parallel to the shear zone and the older milky quartz vein and are mineralized. They are strongly jointed, fragmented, brecciated and are 0.5 to 1 m thick, containmacroscopically visible pyrite and uranophane. Within the ENE-WSW shear zone, the quartz veins brecciate the microgranite and dolerite dikes. Some other quartz veins are trending NNW-SSE and N-S. Bostonite dikes invade two-mica granite along N-S and NNE-SSW structures. They are usually fine-grained, pale brown to deep red colors, fractured, jointed, sheeted, and range in thickness from 0.5 to 2 m . They are essentially composed of K-feldspar, plagioclase, little quartz and iron oxides set in a finely crystalline groundmass of K-feldspar microlites. They have U-contents from 4.1 to 25.4 ppm eU.

The Sela shear zone is clearly evidenced from its rugged topography by a sharp increase in the eU values as well as from a marked decrease of the K and eTh values. These characteristics result from multiple magmatic, tectonic and hydrothermal activities, confined to a wide elongated domain of average 20 m in width and 1.5 km [19] . The relatively narrow and elongated shape of the Sela shear zone magnetic anomaly is consistent with the steep dip of the magnetized shear zone. This should be related to the nature of the intruding dykes rather than from the mineralization. The prominence of this anomaly can be attributed to the strong contrast between the magnetic susceptibilities of the Sela shear zone, and the surrounding granitic rocks. The northern contact between the granitic body and the wadi sediments is characterized by an elongated low magnetic gradient, trending in the ENE-WSW direction, parallel to the main shear zone indicate the presence of a major basement fault [20] . The main zone of strong conductivity has been separated into parts of high amplitudes, resulting in an obvious exploration target. This zone is found on a broader discrete area, extending over a distance of about 1500 m and subjected to northeast displacements, due to a parallel system of NNW left-lateral strike-slip faults [21] .

3. METHODOLOGY

Solid state nuclear track detectors (SSNTDs) have been recognized by IAEA as a standard method for estimation of radon, thoron and their daughter products in the environment. The detectors that are commonly used in environmental monitoring are generally made from cellulose nitrate (LR-115) and polycarbonates (CR-39). These Track Etch cups are placed in shallow holes (30 - 50 cm) and left undisturbed for one month to accumulate the tracks produced by the radon soil emanations. Track Etch detectors are responsive to the alpha radiation from radon (Rn-222) which originates from radium (Ra-226) in the uranium (U-238) decay chain as well as to thoron (Rn-220) which originates from radium (Ra-224) in the thorium (Th-232) decay chain. The track density is converted to radon concentration in the environment. The radon concentration was calculated through the following equation [22] : $C = D/k \cdot t$. Where; C is the radon concentration (Bq/m^3), D is the track density ($track/cm^2$), k is the calibration factor and t is the exposure time (d).

Track Etch detectors are responsive to the alpha radiation from radon (Rn-222) which originates from radium (Ra-226) in the uranium (U-238) decay chain as well as to Thoron (Rn-220) which originates from radium (Ra-224) in the thorium (Th-232) decay chain. Thus, if significant amounts of thorium are present in the area that is being explored, it can produce an unwanted thorium related signal when looking for uranium. In the present work the cups which have been used have been produced by ALGADE. They use a solid-state nuclear track detector type Kodak LR115R films. They have a water tight polyethylene membrane to avoid water entrances and which is used also as a thoron filter. This membrane decrease de rate of passage of throne into the cup (because of its short half-life) bu allows radon to pass into the cup. The radon concentration in the cup is not reduced significantly by this kind of filter.

The ground geophysical studies get information about the subsurface configuration of the shear structures controlling uranium mineralized zones. In studied area, Track Etch survey (60 cups) for radon gas were placed in shallow bore holes (30 - 65 cm deep) and undisturbed left for 1 - 1.5 month to accumulate the reading produced by the varying radon soil activities.

The ground gamma-ray spectrometric measurements were conducted using a Geophysica Brno GS-256 spectrometer having a 0.35 L sodium iodide (NaI) thallium activated detector. All the ground magnetic and VLF measurements were collected using the Scintrex proton magnetometer and ENVI-VLF. The data were generally recorded of 20 m spacing interval along each of the two studied profiles.

Twenty samples were selected from the host rocks two-mica granite of three drill cores from the El Sela shear zone. Minerals were investigated in thin polished and thick double-polished sections using optical microscope and scanning electron microscope (Hitachi S-4800, SCMEM, Henri Poincaré University, France). Minerals were analysed by electron microprobe (CAMECA SX 100, SCMEM, Henri Poincaré University, France) at 10 nA and at 20 kV, counting time 10 s for peak and 5 s for background. Natural albite, orthoclase, wollastonite, hematite, olivine and monazite were used as standards for Al (Kα), Si (Kα), Ca (Kα), Fe (Kα), Mg (Kα) and Yb (Lα), respectively; synthetic V, ZrO_2, $PbCrO_4$, UO_2, $MnTiO_3$, ThO_2, $MnTiO_3$, YPO_4,

$CePO_4$, $NdPO_4$, $SmPO_4$, $DyRu_2Ge_2$, $EuRu_2Ge_2$, GdTiGe and $ErNi_2Si_2$ were used as standards for V (Kα), Zr (Lα), Pb (Mα), U (Mβ), Ti (Kα), Th (Mα), Mn (Kα), Y (Lα), Ce (Lα), Nd (Lα), Sm (Lα), Dy (Lα), Eu (Lα), Gd (Lα), Er (Lα), respectively. After agate mortar crushing, the rocks were analysed by ICP-AES and ICP-MS (SARM, CRPG, Nancy, France) for 11 major and 44 trace elements.

4. PETROGRAPHY OF BOREHOLES

The three boreholes are composed mainly of two-mica granite. It is whitish to pale pink, buff, reddish brown, medium- to coarse-grained, highly fractured and jointed. Microscopically, it is composed essentially of K-feldspar, quartz, plagioclase, biotite and muscovite. Monazite, pyrite, Zircon, galena, sphlerite, Nb-rutile, columbite, titanite, apatite, fluorite and uraninite are accessories, while sericite, chlorite, muscovite, epidote and kaolinite are alteration minerals.

K-feldspar is represented by orthoclase microperthite. It occurs as subhedral to anhedral prismatic crystals showing simple twinning and compressed patchy, string and flame perthitic intergrowths (Figure 2(a)). It encloses fine plagioclase, quartz, biotite inclusions and can be highly altered into kaolinite (Figure 2(b)). Quartz occurs as interstitial subhedral to anhedral crystals. It shows wavy and undulose extinction. Some small anhedral

crystals poikilitically occur in perthite, plagioclase and biotite. Plagioclase of albite and oligoclase composition (An_{9-15}) occurs as subhedral to euhedral tabular crystals. Biotite occurs as subhedral flakes, which are commonly altered to ferrichlorite (pennite) and/or muscovite along the crystal margins and cleavage planes (Figure 2(c)). It is corroded by K-feldspar and quartz its margins. Muscovite occurs as minor irregular fine or medium grained flakes of seldom primary and generally secondary origin (Figure 2(d)). The overgrowths on the primary muscovite grains tend to replace adjacent K-feldspars and quartz grains. The secondary muscovite is formed after biotite and contained elongated iron-titanium inclusions along its cleavage.

Zircons occur as metamict, zoned, mostly euhedral crystals with prismatic a shape and bipyramidal terminations, associated with quartz, perthite, plagioclase and opaques (Figure 2(e)). Nb-rutile occurs as crystal of pale brown color enclosed in feldspars (Figure 2(f)). Opaque minerals occur as cubic crystals, small irregular aggregates or scattered grains as well as tiny inclusions in biotite and quartz. Uraninite filled micro-fractures in kaolintized and sericitized feldspars and generally surrounded by pyrite cubes (Figure 2(g), Figure 2(h)).

Figure 2. Photomicrographs for El Sela shear zone granites: (a) String and patchy perthites. (b) Feldspars altered to kaoliniteand sericite with opaques enclosed zircon. (c) Biotite crystals highly altered to ferrichlorite (pennite) and muscovite. (d) Muscovite flake corroding adjacent plagioclase, and poikilitically encloses opaques and quartz. (e) Zircon crystals show zonation enclosed in pyrite (opaque). (f) Nb-rutile has elongated shape. (g), (h) Uraninite filled microfractures with pyrite in kaolinized and sericitized feldspars.

5. GEOCHEMISTRY

The geochemical data have been used to determine the characteristics of granites enclosing the Sela shear zone and their uranium fertility. The uranium fertility of the granites depends essentially on the level of their U background concentration and the leache ability of the uranium bearing accessory minerals. Peraluminous, high-K and low Na and Ca contents and presence of two-mica are also good indicators of the U-fertility of granite rocks. Under certain conditions of alteration, the high-calc-alkaline (HKCA) granite can form uranium ore but generally with lesser tonnage than the peraluminous leucogranites [8] .

5.1. Major Elements

El Sela two-mica granites have high SiO_2, Al_2O_3, Na_2O and K_2O contents with the following average contents 76.5 wt.%, 12.95 wt.%, 3.9 wt.% and 4.3 wt.%, respectively. They are also poor in Fe_2O_3, CaO, MnO and MgO with average concentrations of 0.73 wt.%, 0.53 wt.%, 0.04 wt.% and 0.17 wt.%, respectively Table 1.

deficiency of Al with respect to the quantity needed to make the feldspars. The granites have compositions varying from slightly metaluminous (A = −7) toperaluminous (A = 62). The peraluminous composi- tions correspond mineralogically to the presence of Al-rich biotite and small amounts of muscovite. The four most peraluminous samples (W_2-5, 6, 8 and 9) present the lowest Na contents (3.2 to 2.6 wt% Na_2O, instead of more than 4 wt% for the other samples) and the highest ignition losses (1.9 to 3.1 wt% instead of about 1 wt% for the other samples).

These features indicate an argilic alteration of the plagioclase in these samples. The metaluminous samples tend to have the highest Ca contents (0.7 to 1.1 wt% CaO), without a corresponding increase of the ignition loss indicating the presence of Ca silicates and the absence of a significant carbonate contribution (Figure 3). Quartz and feldspars are located at the origin in this diagram and represent the eutectic composition of the granite system. The differentiation parameter B = Fe + Mg + Ti, which also represents a coloration index of the rock. The granite are leucocratic with low B-parameter values (B < 15).

The granites of the three drill holes tend to define three distinct magmas batches in the A-B and Q-P diagrams (Figure 3 and Figure 4). The samples of drill hole W3 have all very low peraluminosity for very leucoratic compositions in the A-B diagram and are among the richest in plagioclase for low quartz contents in the Q-P diagram. Samples from W1 form a trend parallel to the one defined by the samples from the W3 drilling with higher average peraluminosity. Two samples have higher A parameter due to clay alteration especially for W1-2 (lower Na_2O content and higher ignition loss). In the Q-P diagram the samples define a fractionation trend with simultaneous increase of quartz and K-feldspar content. Sample W1-3 is slightly silicified. The W2 drill hole hosts the 4 samples with the highest degree of clay alteration as discussed above and well evidenced in the A-B and Q-P diagrams (Figure 3 and Figure 4). The other samples define a different than the two other drill holes with a decrease of the A parameter (20 to 10) with a decrease of the B parameter (13 to 4). These samples are also the richest in K-feldspar for similar quartz contents.

Table 1. Chemical analyses of the samples from the bore holes of the El Sela two-mica granite.

S. No.	W1-1	W1-2	W1-3	W1-4	W1-5	W1-6	W1-7	W1-8	W2-1	W2-2	W2-3	W2-4	W2-5	W2-6	W2-7	W2-8	W2-9	W3-1	W3-2	W3-3
depth m.	20	24	35	53	55	59	62	63	1	4	10	17	26	35	37	42	44	12	30	43
Major oxides (wt.%)																				
SiO_2	76.2	76.3	77.7	74.6	76.9	76.2	76.9	76	77.4	76.8	76.7	76.8	78	77.5	76.8	74.8	76.8	75.5	77	74.7
TiO_2	0.13	0.13	0.13	0.10	0.12	0.10	0.09	0.13	0.10	0.10	0.10	0.14	0.10	0.14	0.12	0.16	0.14	0.13	0.13	0.09
Al_2O_3	13.3	13.3	12.5	12.6	12.8	13.7	12.7	13	12.9	13.3	13.2	13	12.6	12.3	13.1	13.1	12.8	12.7	12.5	13.7
Fe_2O_3	0.7	0.9	0.9	1.2	0.5	0.6	0.9	0.9	0.6	0.5	0.6	0.9	0.7	0.8	0.7	0.6	0.6	0.8	0.7	0.5
MnO	0.05	0.06	0.04	0.10	0.02	0.03	0.08	0.06	0.01	0.01	0.01	0.01	0.01	0.01	0.03	0.02	0.03	0.09	0.08	0.06
MgO	0.12	0.22	0.15	0.12	0.07	0.11	0.12	0.16	0.10	0.09	0.10	0.19	0.23	0.25	0.17	0.35	0.38	0.20	0.16	0.12
CaO	0.5	0.3	0.2	0.9	0.2	0.4	0.7	0.8	0.4	0.4	0.4	0.5	0.3	0.3	0.4	1	0.4	1.1	0.7	0.7
Na_2O	4.4	3.8	4	4.1	4.1	4.5	4.2	4.3	4	4.4	4.1	4	3.2	3	4.1	2.8	2.6	4.1	4.3	4.3
K_2O	4.1	4.2	3.7	4.2	4.2	4.5	4.3	4.3	4.6	4.5	4.8	4.3	4	4.2	4.4	4.2	4.4	4	3.8	5
P_2O_5	0.1	0.1	0.1	0	0	0.1	0.1	0.1	0.1	0.1	0.1	0.2	0.1	0.1	0.1	0.2	0.2	0.1	0.1	0.1
IL	0.8	1.5	0.9	1.6	1	1	1.1	1.1	0.7	0.7	0.7	1	1.9	2	1.1	3.1	2.8	1.4	1.2	0.9
Total	100	101	100	100	100	101	101	101	101	101	101	101	101	101	101	100	101	100	101	100
Trace elements (ppm)																				
Be	5.4	4.4	4.9	8.4	3.2	5.4	7.4	7.6	3.1	3.5	2.8	4.2	2.9	4.1	5.2	3.4	3.5	7.8	8.5	8.7
Sc	4.9	5.1	5.2	6.8	3.9	5.1	6.7	7.7	3.4	2.6	2.6	4.1	2.3	3.8	4.1	3.8	3.6	8.4	7.7	6.2
Ti	1.6	1.6	1.6	1.2	1.5	1.3	1.2	1.6	1.3	1.2	1.3	1.8	1.2	1.7	1.5	2	1.7	1.6	1.6	1.2
V	9.7	8.7	9.1	5.8	6.3	5.6	4.4	5.8	5.2	5.1	3.5	8.2	3.6	7.9	6.6	8.2	6.8	8.2	6.7	5.5
Cr	62.2	132	95	100	62.7	89	70.3	78.9	80.2	50.3	62.7	93	93	88	92	78.2	83	77.8	95	69
Mn	0.8	0.8	0.6	1.4	0.2	0.4	1.1	0.9	0.1	0.1	0.1	0.2	0.2	0.1	0.4	0.3	0.4	1.3	1.1	0.8
Co	0.8	1.5	1.4	1.8	0.7	0.9	1.1	1.1	0.6	0.8	0.7	1.1	0.9	0.8	1	0.9	0.9	1	0.9	0.7
Zn	35.2	39.5	30	57.9	24.7	33.9	45.9	78	39.7	33	29.4	59	38.9	27.3	48.8	62.1	36.7	51.2	57.3	43.2
Ga	24	24	23	26	21	25	25	27	21	21	22	23	20	21	25	23	23	25	26	30
Ge	1.5	1.4	1.7	2.2	1.3	1.8	2.2	2.2	1.3	1.4	1.4	1.5	1.3	1.5	1.9	1.4	1.3	2	2.2	3
Rb	248.8	275.5	244.6	427	211	280	412	394	195	184	206	188	174	204	267	196	198	364	357	464

Sr	151	121	107	79	101	72	72	100	55	103	47	99	62	110	89	116	96	91	110	87
Y	7.2	6.7	6.9	4.8	5.1	4.3	4.9	5.3	11.3	4.2	3	6.4	3.4	5.9	5.8	19.9	5.4	5.4	7.2	5.7
Zr	90.4	88.2	85.6	66.9	80.9	69.5	66.2	92.5	62.8	62.2	53.7	79.3	68.4	82	73	89	80	88	97	61
Nb	13.7	13.3	14.8	16.4	9.5	13.2	16	22.9	7.2	5.7	5.3	6.9	5.3	8.9	14.5	9.3	7.1	22.3	21.1	31.4
Mo	0.8	2.7	1.3	2.6	0.9	2	2.9	3.4	1.2	0.8	1	1.6	1.3	1.2	1.3	1.1	1.1	3.7	5.3	1.3
Sn	2.1	2.6	2.2	5.8	1.5	1.6	2.3	2.5	1.1	0.9	1.1	1.5	0.7	1.4	1.5	1.7	1.7	2.8	2.7	2.1
Cs	5.9	6.3	4.6	11.5	4	6.9	10.2	10.9	4.5	5.2	4.4	7.5	2.9	3.8	6.6	4.8	5	10.2	10.1	11.7
Ba	351	296	222	137	263	155	132	173	147	261	83	198	94	179	167	204	213	154	168	167
Hf	4.2	3.9	4.4	3.5	3.3	3.3	3.6	4.7	2.7	2.8	2.2	3	2.8	3.5	3.3	3.5	3.4	4.4	4.7	3.6
Ta	1.7	1.6	1.8	1.8	1.1	1.5	1.8	2.2	0.8	0.5	0.5	0.9	0.3	0.9	1.7	1	0.7	2.1	2.2	3.9
W	3.7	12.9	10.6	6.4	4.3	5.7	4.5	5.1	5.5	3.3	4.6	6.8	6.2	6.2	6	5.9	5.7	5.4	6.8	4.5
Pb	27.4	23.9	30.1	36.3	18.3	21.8	43	32.4	17.9	16.6	14.9	19	20.5	13.8	25.7	22	21.8	34.8	67	50
Th	18.3	18.1	16.3	22.5	17.3	19.9	23.2	29	21	13	23	17	24	18	18	20	16	28	29	19.6
U	6	7.8	8.1	15	4.1	9.5	18	33	5.8	3.2	3.4	4.7	3.9	4.4	9.1	7.5	3.9	15	23	25
REE (ppm)																				
La	14.7	15.1	12.9	13	12.5	13.4	14.8	21	17	13.4	17.4	16.3	12.4	19.8	15.7	22.9	17.8	18.5	20.1	14
Ce	29.2	29.3	27.8	18.6	20.9	18.3	22	29.5	27.7	21.8	25.2	25.5	20.3	26.6	25.1	39.5	30.4	26.6	27.9	20
Pr	2.9	2.8	2.4	1.5	2.1	2	1.8	2.2	3	2.2	2.3	2.5	1.9	3.2	2.4	3.8	2.9	2.1	2.1	1.6
Nd	10.6	9.5	8.1	4.6	7.4	6.5	5.8	6.7	10.4	7.8	7.5	8.8	6.3	11.5	8.5	13.2	10.2	6.5	6.5	4.9
Sm	1.7	1.5	1.3	0.7	1.2	0.9	0.9	1	1.7	1.3	0.9	1.4	0.9	1.8	1.4	2.5	1.7	0.9	1	0.8
Eu	0.3	0.2	0.2	0.1	0.2	0.2	0.1	0.2	0.3	0.2	0.2	0.3	0.2	0.3	0.2	0.3	0.3	0.2	0.2	0.2
Gd	1.3	1.1	1	0.6	0.9	0.6	0.7	0.7	1.2	0.9	0.7	1	0.7	1.2	1.1	2	1.2	0.7	0.8	0.6
Tb	0.2	0.2	0.2	0.1	0.1	0.1	0.1	0.1	0.2	0.1	0.1	0.2	0.1	0.2	0.2	0.4	0.2	0.1	0.1	0.1
Dy	0.9	0.9	0.9	0.5	0.7	0.6	0.5	0.6	1.8	0.6	0.5	1	0.5	0.9	0.8	3.2	0.9	0.6	0.9	0.7
Ho	0.2	0.2	0.2	0.1	0.1	0.1	0.1	0.1	0.4	0.1	0.1	0.2	0.1	0.2	0.2	0.7	0.2	0.1	0.2	0.2
Er	0.6	0.6	0.6	0.4	0.4	0.4	0.4	0.4	1.6	0.4	0.3	0.7	0.3	0.5	0.6	2.9	0.5	0.4	0.7	0.6
Tm	0.1	0.1	0.1	0.1	0.1	0.1	0.1	0.1	0.3	0.1	0.1	0.2	0.1	0.1	0.1	0.6	0.1	0.1	0.2	0.1
Yb	0.9	0.8	0.9	0.7	0.6	0.7	0.6	0.8	2.7	0.5	0.4	1.2	0.5	0.7	0.9	5	0.6	0.8	1.3	1.3
Lu	0.1	0.1	0.2	0.2	0.1	0.1	0.1	0.2	0.4	0.1	0.1	0.2	0.1	0.1	0.2	0.8	0.1	0.2	0.2	0.3
Parameters and ratios																				
P	−64	−39	−54	−59	−46	−57	−56	−62	−38	−53	−37	−46	−23	−13	−46	−19	3	−67	−70	−45
Q	188	208	222	182	203	178	192	182	198	184	187	200	241	241	196	224	244	189	200	162
B	9.0	12.8	11.0	11.8	6.4	7.8	9.8	11.3	7.5	6.6	7.5	12.1	11.4	13.0	10.1	14.5	15.0	11.6	10.0	7.3
A	13.8	38.1	30.2	−6.7	22.2	13.4	−3.0	−3.9	11.8	8.8	10.2	16.5	48.0	44.3	16.7	41.5	59.2	−7.6	0.5	−1.5
Th/U	3.1	2.3	2.0	1.5	4.2	2.1	1.3	0.9	3.6	4.1	6.8	3.6	6.2	4.1	2.0	2.7	4.1	1.9	1.3	0.8
Rb/Sr	1.6	2.3	2.3	5.4	2.1	3.9	5.7	3.9	3.5	1.8	4.4	1.9	2.8	1.9	3.0	1.7	2.1	4.0	3.2	5.3
Nb/Ta	8.1	8.3	8.2	9.1	8.6	8.8	8.9	10.4	9	11.4	10.6	7.7	17.7	9.9	8.5	9.3	10.1	10.6	9.6	8.1
Zr/Hf	21.5	22.6	19.5	19.1	24.5	21.1	18.4	19.7	23.3	22.2	24.4	26.4	24.4	23.4	22.1	25.4	23.5	20.0	20.6	16.9

The peraluminosity is calculated as the difference between the total amount of Al present in the rock and the amount of Al bound to the feldspars: A = Al – (Na + K + 2Ca), calculated in cations. This index indicates the excess or the

The compositions and the global trends defined in the A-B and Q-P diagram (Figure 3 and Figure 4) correspond to highly fractionated high-K calc-alkaline granites. Initially derived from metaluminous sources they become slightly peraluminous by extreme fractionation of Ca-rich, Al-poor or absent minerals (amphiboles and or pyroxenes), and possible assimilation of peraluminous crustal material during their ascent, and/or incipient alteration of the feldspars in K-micas or clay minerals.

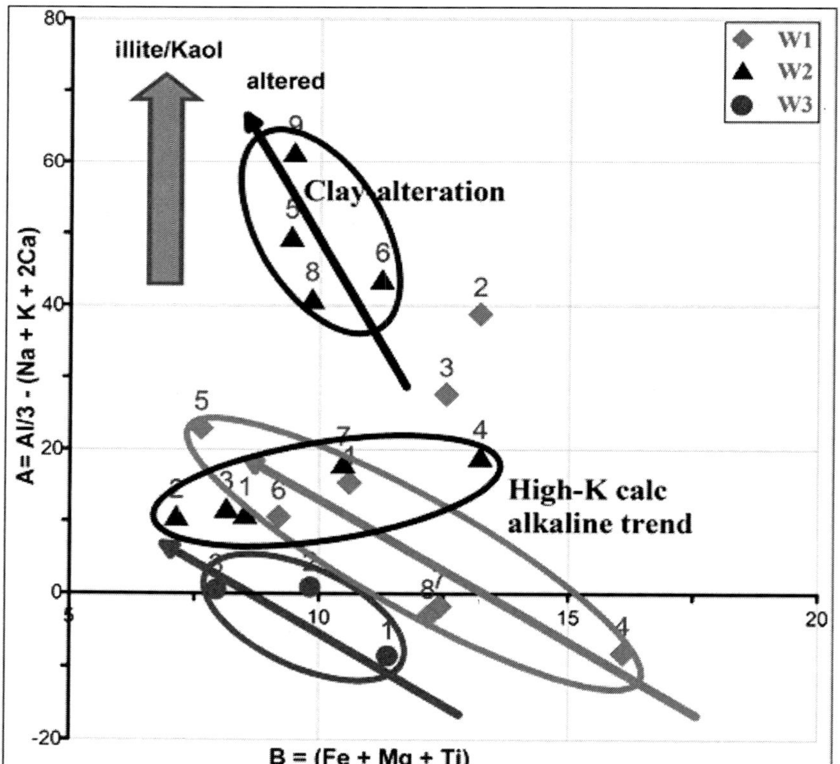

Figure 3. Variation of the peraluminous index (A) with fractionation index (B), for the two-mica granite at G. El Sela shear zone. A-B mine-ralogical-chemical diagram, from [23] parameters in thousands of cations.

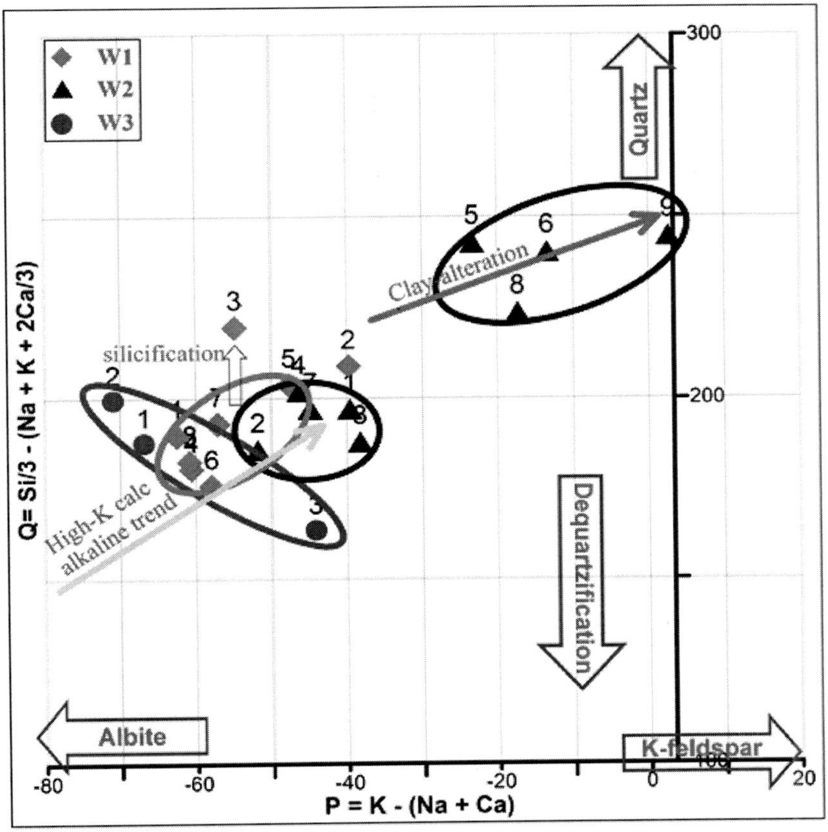

Figure 4. Variation of the K-feldspar/albite (P parameter) versus quartz contents (Q parameter) for the two-mica granite at El Sela shear zone. Q-P mineralogical-chemical diagram from [23] parameters in thousands of cations.

5.2. Trace Elements

The granitesfrom the El Sela shear zone areatend to been riched in most large ion lithophile elements (LILE; Ba, Rb, Sr) and high field strength elements (HFSE; Y, Zr and Nb) Table 1. The biotite ± muscovite granite is moderately rich in Th (16 to 29 ppm) and U (3 to 33 ppm). The highest U values are observed in borehole W1 and W3. Sample W1-8 at a depth of 63 m has the highest Th and U contents (29 and 33 ppm respectively) (Figure 5) Most samples of the W2 drilling tend to follow a simultaneous enrichment in Th and U, with a Th/U ration of 4. However the Th richest samples (W2-3 and 5) are slightly depleted in uranium (Th/U > 4). Sample

5 show a clay alteration which may explain the U loss but not samples 3. Sample 7 tend to be enriched in U (Th/U = 2). Th samples from drill hole W1 plot along a well-defined trend with increasing Th and U and decreasing Th/U, sample W1-8 being the richest in U and Th with a Th/U below1 indicates the possible presence of uraninite. This trend seems to reflect magmatic fractionation. However, the three Th and U richest samples (W1-4, 7 and 8) are the less fractionated in the A-B diagram (Figure 3). The three samples of drill hole W3 define hydrothermal U enrichment trend with low Th/U (<2).

In the Nb versus Sn diagram (Figure 6) most samples from each drill hole form well defined groups with increasing Sn an Nb contents from W2 to W1 and W3, although W3 is the less peraluminous group of samples (Figure 3). Sample W1-4 is anomalously enriched in Sn relatively to Nb, whereas samples W1-8 and W3-3 are anomalously enriched in Nb. All the enrichments remain however limited (up to 32 ppm Nb and 6 ppm Sn).

In the diagram Zr versus Hf two groups of granites (one for most samples from W2 and another with the samples from W1 and W3) are well defined with distinct Zr/Hf ratios (Figure 7). Granites from the drill hole W2 show the lowest Hf contents.

Nb and Ta are very well correlated (Figure 8). They also define three granite groups corresponding to the three drill holes with increasing Nb and Ta concentrations from W2 to W1 and W3, as for the Sn-Nb correlation, with a nearly constant Nb/Ta ratio of 10. However, the highest Ta contents do not exceed 5 ppm.

The concentrations of trace elements are normalized to the primitive mantle [24] are represented as spider diagrams (Figure 9). The El Sela granites show enrichment of large ion lithophile elements (LILE; Rb, Ba, Sr) and high field strength elements (HFSE; Y, Zr, Nb) and a marked moderate level of Ba, Ce and Ti in the three boreholes which are typical of subduction related magmas [25] [26] (Figure 9).

5.3. Rare Earth Elements

The El Sela granite has variable total REE contents: 44 to 98 ppm in samples W_1-6 and W_2-8 respectively (Table 1). The chondrite-normalized REE patterns enrichment strong fractionation of light rare earth elements (LREE) ($5.34 < (La/Sm)_N < 12.9$) and flat to reversely fractionated heavy rare earth elements (HREE) ($4.6 < (Ga/Yb)N < 58.3$) (Figure 9 and Figure 10). The negative Eu anomaly observed in the granites from the three W_1, W_2 and W_3 drill hole. Two samples from W_2present an enrichment in the heaviest REE and Y with a strongly reverse HREE fractionation, indicating the abundance of a HREE-rich mineral, which may correspond to xenotime. The negative Eu suggests that plagioclase was an early fractionating phase.

6. MINERAL CHEMISTRY

Scanning electron microscope (SEM) and electron microprobe analyses (EMPA) were used to identify the nature of the sulfides and Nb-Ta-, Zr-Hf-, REE-, Th-, U-bearing minerals in the e the El Sela granite (Figure 11) and Table 2 to Table 4.

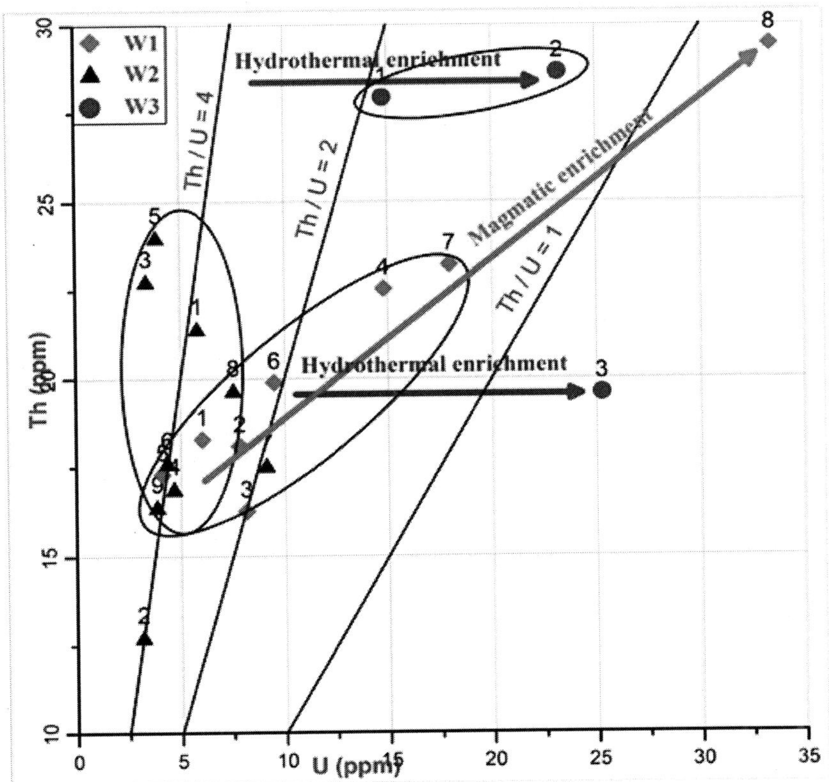

Figure 5. U-Th relationships of the studied boreholes at G. El Sela shear zone.

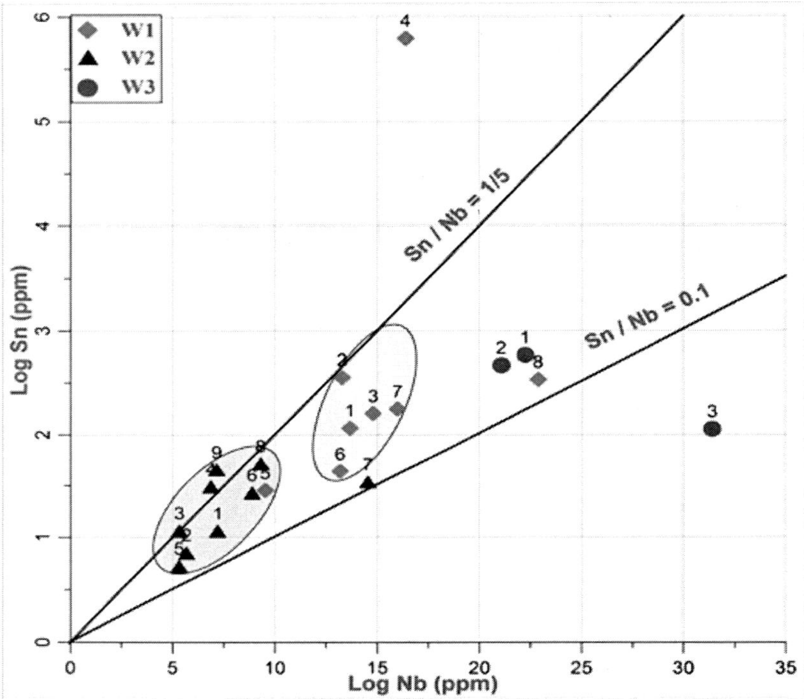

Figure 6. Nb-Sn relationships of the studied boreholes at G. El Sela shear zone.

6.1. Sulfides

Pyrite, chalcopyrite, arsenopyrite, galena, and sphalerite, occur as disseminations and fracture-fillings in mineralized smoky and red jasperoid veins, dolerite and microgranite dikes, in the El Sela graniteas illustrated by Fig- ure 11(a). Magnetite, hematite, illmenite, rutile and apatite are also spatially associated with the sulfide minerals. Pyrite (FeS_2) is the most common sulfide mineral which occurs as well developed octahedral crystals.

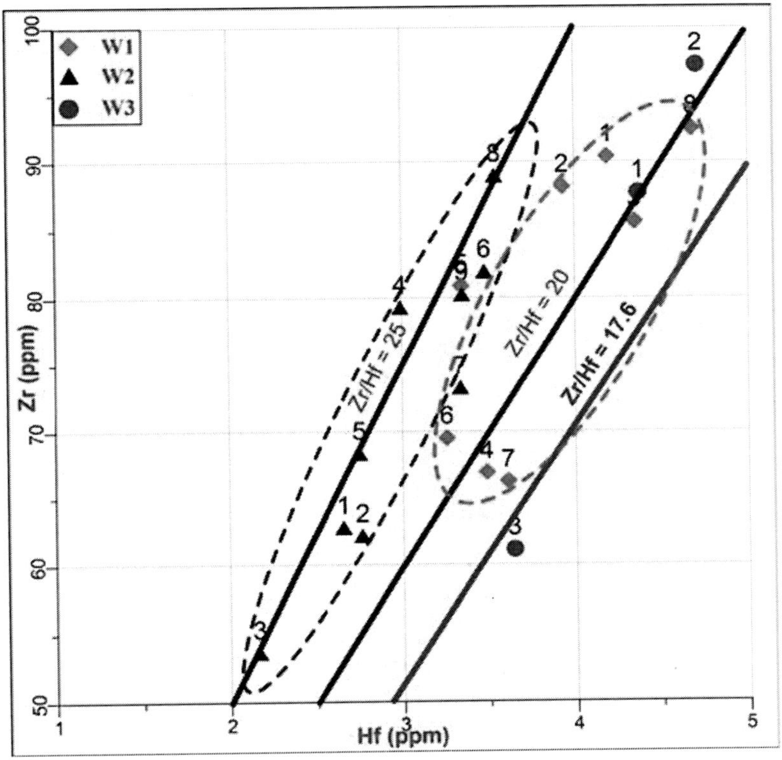

Figure 7. Hf (ppm)-Zr (ppm) relationship for the granites of the G. El Sela shear zone, and show the reference Zr/Hf isovalent ratio of 17.6.

Pyrite is preserved in the samples from the drillings and seems to control the deposition of uraninite, branne rite, pitchblende and coffinite at the margin of the crystals. In the El Sela shear zonethe cavities left after pyrite dissolution are by uranophane and autunite because of the alteration of the pyrite and primaru U minerals by meteoric fluids [28] .

Sulfides are more effective reductant because sulfur can act as reducing agent where one mole of pyrite can reduce six more uranium than Fe^{+2} [29]

6.2. Ti-, Nb-Ta-, Zr-Hf-, REE-Mineralization

Rutile, fluorite, zircon, monazite, LREE-carbonates, columbite and betafite have been also identified by SEM and EMPA analyses.

Rutile is common mineral in most samples from the El Sela granite. It occurs as angular to subangular, elon gated, cylinder and -like grains of reddish brown to blackish brown colors associated with zircon, fluorite, monazite, uranothorite, and uraninite (Figures 11(b)-(d)). The EMPA analyses indicate that the rutile has 0.39 - 8.48 wt.% Fe_2O_3, 0.51 to 22.78 Nb_2O_3 and up to 2.7 wt.% Ta, 5.8 wt.% Th, and 1.57 wt.% U (Table 3). It would be interesting in which samples the highest concentration of each element has been found to interpret the whole rock geochemistry.

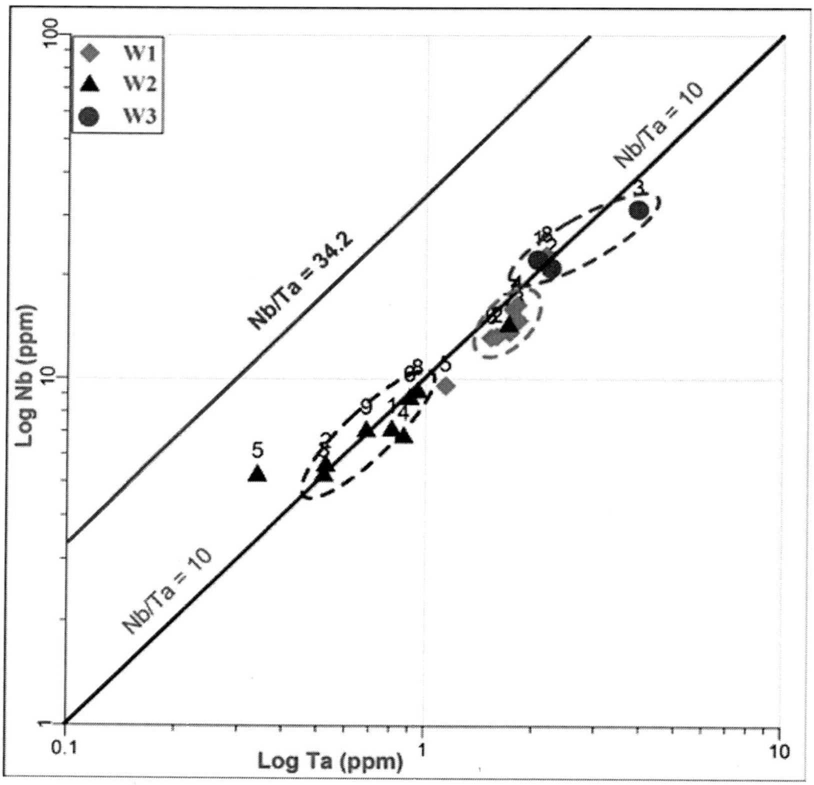

Figure 8. Nb (ppm) - Ta (ppm) for the granites of the G. El Sela shear zone, and show the reference Nb/Ta isovalent ratio of 34.2.

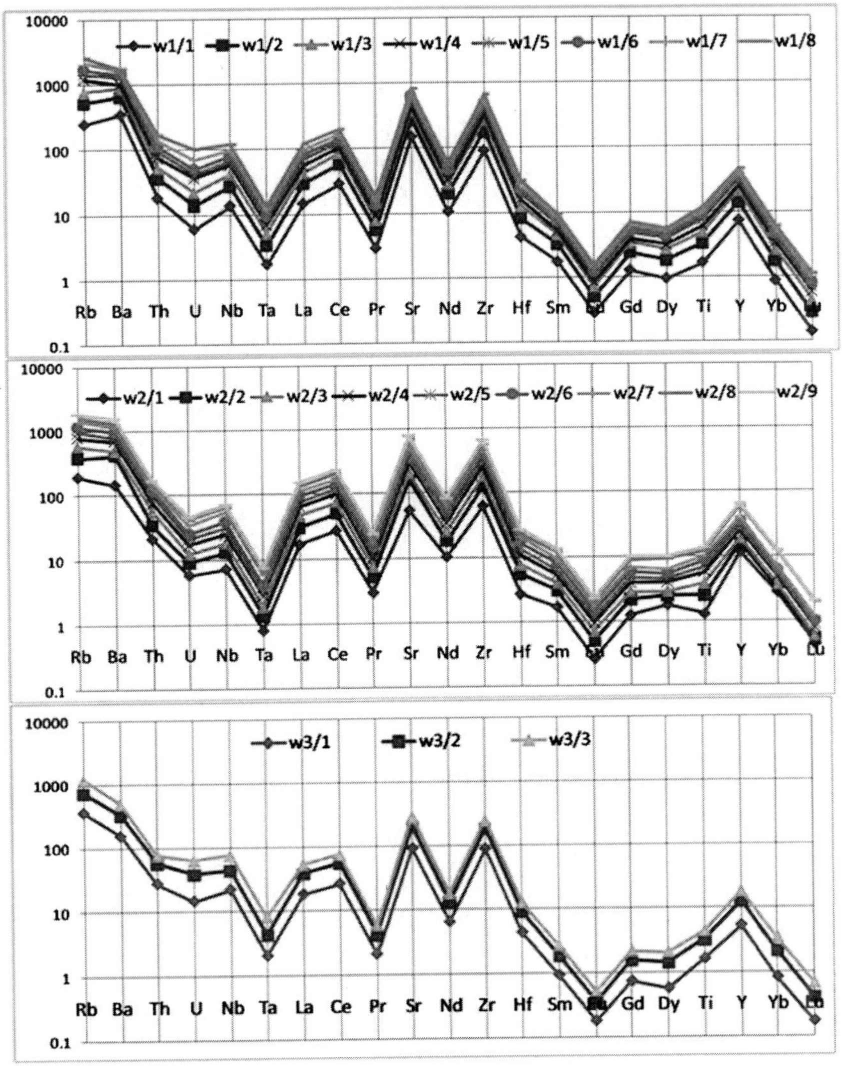

Figure 9. Primitive-mantle-normalized multi-elements spider diagrams (normalization values are from [24] for the El Sela granite.

Table 2. Electron microprobe analyses of Ti-, Nb-Ta-, Zr-Hf-, REE-minerals from the El Sela granite.

S. No.	W3-3 -C3	W3-2 -C2	W1-7 C10-4	W1-4 C1-2	W2-5 -C3	W2-9 -C2	W_{1-4} C_{4-1}	W_{1-4} C_{4-2}	W_{1-7} C_{10-3}	W_{2-5} -C_5	W_{2-9} -C_{10}	W_{3-3} -C_4	W_{3-3} -C_4	W_{3-3} -C_4	W1-7 -C6-2	W1-7 -C6-3	W3-3 -C3
depth m.	43	30	62	53	26	44	53	53	62	26	44	43	43	43	62	62	43
oxides	Beta-fite			rutile						Zircon					REE-rich uranothorite		columbite
SiO_2	1.56	0	0.5	0.56	0.73	4.6	30.4	26.6	32.4	29.1	32.3	29.5	27.7	28.8	19.0	11.4	0.09
P_2O_5	0.05	0	0	0	0	0	0.43	0.85	0.42	0.56	0.45	0.3	0.71	0.44	1.04	1.1	0
SO_2	0	0	0	0	0	0	0	0	0	0	0.01	0	0	0	0.16	0	0
CaO	2	0.34	0.18	0.29	0.04	0.2	0.85	1.48	0.1	0.44	0.02	0.6	1.21	0.68	7.81	7.7	0.14
TiO_2	46.3	85.8	64.1	90.1	92.7	82.3	0.1	0.1	0	0.13	0	0	0.08	0	0.15	0.02	11
MnO	0.24	0.05	1.05	0.08	0.05	0.02	0.63	0.38	0.07	0.1	0	0.49	0.36	0.46	0.08	0.03	9.76
FeO	8.48	4.4	21.6	1.47	0.39	5.78	1.2	0.87	0.22	2.5	0.03	1.41	1.26	1.53	1.89	0.13	7.6
Y_2O_3	0.1	0.04	0	0	0	0.03	0	0.89	0	0	0	0	0.44	0	1.11	1.58	0.84
ZrO_2	0.64	0.04	0.07	0.15	0	0.03	59.9	54.6	63.9	59.9	66.0	59.5	49.3	60.4	2.84	0	0.55
Nb_2O_5	22.8	7.56	1.61	1.18	0.51	0.94	0	0	0	0	0	0	0	0	0.28	0.04	62.3
SnO	0.08	0.3	0	0	0	0	0	0.01	0	0	0.01	0	0	0	0	0	0.15
La_2O_3	0	0	0	0	0	0	0	0	0.02	0	0	0	0	0.02	4.51	4.02	0
Ce_2O_3	0.15	0.26	0.2	0.28	0.27	0.27	0	0.04	0	0	0.01	0	0	0.08	7.74	7.3	0
Nd_2O_3	0.08	0	0.03	0	0	0	0.01	0.13	0.02	0	0.12	0	0.15	0.07	2.24	2.16	0.03
Yb_2O_3	0.08	0	0	0	0	0.02	0	0.4	0.12	0.01	0	0.16	0.43	0	0.02	0	0.28
HfO_2	0.55	1.22	0.9	1.23	1.33	1.27	2.24	3.09	1.68	1.49	1.26	2.2	10.9	1.7	0	0	0.1
Ta_2O_5	2.7	0.33	0.03	0.03	0	0.05	0.28	0.13	0.34	0.22	0.33	0.37	0.56	0.29	0	0	2.91
PbO	0	0	0	0	0	0	0	0	0	0	0	0	0	0	0.11	0.02	0
ZnO	n.d	n.d	0.02	0.06	0	0	0	0	0	0	0	n.d	n.d	n.d	0	0	n.d
Al_2O_3	n.d	n.d	0.26	0.43	0.24	1.3	0.21	0.56	0.03	0.1	0	n.d	n.d	n.d	7.52	0.66	n.d
ThO_2	5.8	0	0.06	0.03	0	0.04	0.04	0.54	0.03	0.6	0	0.27	0.68	1.22	25.9	42.2	0.2
UO_2	1.57	0	0.27	1.21	0.01	0	1	1.57	0	0.79	0.17	1.33	2.24	1.61	1.3	1.19	0.75
Total	93.**	100.**	91	97	96	97	97	92	99	96	101	96	96	97	84	80	97

Zircon is generally included as idiomorphic crystals (Figure 11(c)) variably enriched in hafnium (Table 2)

(1.26 < HfO_2 < 10.87 wt.%), UO_2 (up to 2.23 wt.%) and ThO_2 up to (up to 1.22 wt.%). Similar Hf enrichment has been observed in Rare Metal Granites such as at Cinovec with 3 to 4 wt.% HfO_2[30] .

Uranothorite crystals occur as brownish red and orange crystals, isotropic because of their metamictisation. SiO_2 and ThO_2 (25.93 - 42.2 wt.%) concentrations are quite variable and analytical total are low due to the strong metamictization of the crystals (Table 2). Uranium concentrations are low and frequently below the EPMA detection limit (Table 3 & Figure 11(e)).

Table 3. Electron microprobe analyses (EMPA) of Th-, U-minerals from the El Sela granite.

S. No.	$W_{2\text{-}9}$ -C_{10}	$W_{1\text{-}7}$ -$C_{10\text{-}2}$	$W_{2\text{-}5}$ -C_3	$W_{2\text{-}5}$ -C_3	$W_{2\text{-}5}$ -C_3	$W_{2\text{-}9}$ -C_{10}	$W_{2\text{-}9}$ -C_{10}	$W_{1\text{-}7}$ -$C_{6\text{-}1}$	$W_{1\text{-}7}$ -$C_{10\text{-}1}$	$W_{3\text{-}3}$ -C_4	$W_{3\text{-}2}$ -C_1	$W_{3\text{-}2}$ -C_1	$W_{3\text{-}2}$ -C_2	$W_{3\text{-}2}$ -C_3	$W_{3\text{-}2}$ -C_3	$W_{3\text{-}2}$ -C_3	$W_{3\text{-}2}$ -C_3
depth m.	44	62	26	26	26	44	44	62	62	43	30	30	30	30	30	30	30
oxides	thorite			phosphothorite					uranothorite					brannerite			
SiO_2	33.73	11.5	14.2	16.3	8.77	21.1	28.7	17.9	16.8	20.8	6.8	6.08	13.35	3.02	11.4	5.48	2.85
P_2O_5	1.48	1.51	15.11	4.84	10.28	8.68	10.61	0.47	0.28	1.85	0.14	0.03	0.29	0.18	0.61	0.22	0.07
SO_3	0.07	0.08	0.34	0.05	0.21	0.34	0.08	0.07	0.08	0.05	7.76	0	0	0	0	0	0
CaO	1.96	5.29	7.85	2.61	4.84	8.82	12.95	1.52	1.73	2.63	2.42	3	3.42	3.19	1.51	4.69	1.75
TiO_2	0.04	0.1	0	0	0.01	0	0	0	0.04	0.26	22.9	31.3	17.2	31.4	41.6	22.7	62.8
MnO	0.07	0.02	0	0.05	0.01	0.03	0.01	0.01	0	0.09	0.19	0.72	0.13	0.35	0.07	0.69	0.34
FeO	3.12	0.67	0.57	0.59	0.6	0.55	0.49	0.11	0.06	1.18	7.25	2.07	0.78	2.48	2.17	0.9	0.99
Y_2O_3	0.2	1.22	0.88	1.97	3	1.53	1.24	0.77	0.06	2.23	0	1.23	1.56	0	1.44	0	0
ZrO_2	0.6	1.45	0.07	0	0	1.69	2	0	0.26	11.6	2.72	1.37	9.32	1.63	3.96	1.66	0.64
Nb_2O_5	0.19	0.09	0	0.15	0	0.01	0	0.09	0	0.3	2.21	3.54	0.3	1.59	0.74	1.67	0.96
La_2O_3	0.01	0.28	0.28	0.04	0.06	0.01	0.03	0	0	0	0	0	0	0	0	0	0
Ce_2O_3	0	0.56	0.03	0.22	0.28	0.09	0.11	0.11	0.04	0.15	0.17	0.53	0.21	0.14	0.23	0.15	0.25
Nd_2O_3	0.38	0.47	0.52	0.82	0.27	0.66	0.29	0.24	0.13	0.31	0.01	0.43	0.08	0.07	0.08	0	0.07
Yb_2O_3	0.03	0	0.23	0	0.09	0.19	0	0	0	0.36	0.09	0.37	0.21	0	0.21	0	0
HfO_2	0	0	0	0	0	0	0	0	0	0	0	0	0	0	0	0	0.03
Ta_2O_5	0.06	0	0.03	0	0	0	0.07	0	0	0	0	0	0	0	0	0	0
PbO	0.12	0.17	0.09	0.09	0.13	0.09	0	0.31	0.75	0.12	0	0	0	0.63	0	0	0
ZnO	0	0	0	0	0	0	0.02	0	0.02	n.d	n.d	n.d	n.d	0	0	0	0
Al_2O_3	3.23	0.16	1.62	0.25	0.49	2.85	6.34	0.11	0.03	n.d	n.d	n.d	n.d	n.d	0.44	0.22	0.46
ThO_2	48.1	63.5	54.4	62.7	61.9	45.1	27.5	52.0	53.5	33.0	1.4	2.19	1.28	0.66	0.57	0.59	0.18
UO_2	0	0	0.22	0.84	2.1	0.19	0.45	18.4	14.5	11.4	36.4	39.0	39.8	36.8	26.3	49.1	15.7
Total	93.4	87.1	96.4	91.5	93.0	91.9	90.9	92.1	88.4	86.5	90.5	91.9	88.5	82.1	91.4	88.1	87.1

S. No.	W3-2 -C10	W3-2 -C10	W1-7 -C8-3	W1-4 -C1-3	W3-2 -C5	W3-2 -C5	W3-3 -C2	W3-3 -C2	W1-4 -C2-1	W1-4 -C2-4	W3-3 -C4	W3-3 -C4	W3-3 -C6	W3-3 -C6	W3-3 -C6	W3-3 -C6	W3-3 -C6
depth m.	30	30	62	53	30	30	53	53	53	53	43	43	43	43	43	43	43
oxides		brannerite				coffinite				uraninite					pitchblende		
SiO_2	6.86	5.23	1.8	25.1	18.6	12.0	16.0	22.2	6.89	3.06	9.45	13.55	4.08	4.69	3.56	4.31	1.05
P_2O_5	0.03	0.11	0.08	0.06	1.16	2.28	0.66	0.72	0.25	0.17	1.06	1.56	0.15	0.1	0.11	0.43	0.13
SO_3	0	0	0	0	1.28	0.41	0.25	3.41	0.09	1.09	1.88	0.02	0	1.3	0.04	0.06	6.68
CaO	1.96	2	2.14	2.17	2.15	1.84	7.66	2.63	3.95	7.56	6.59	3.78	4.44	4.18	5.28	3.36	3.36
TiO_2	28.2	41.4	45.1	3	0.26	0.28	0.2	0.17	1.47	1.1	0.63	2.4	0	0	0	0	0
MnO	0.09	0.19	0.49	0.36	0.01	0	1.62	0.14	0.81	0.69	0.44	0.44	0.12	0.04	0.17	0.08	0.13
FeO	1.6	1.14	2.1	3.04	2.28	1.06	4.59	3.36	11.65	4.69	3.66	1.02	1.92	3.98	1.58	0.47	6.51

Y_2O_3	0.04	0	0	0	3.34	3.77	1.53	1.7	0	0	2.02	2.96	0.06	0.08	0.14	0.96	0.22
ZrO_2	1.7	0.87	1.65	2.92	3	3.13	6.34	5.91	8.29	4.34	7.14	8.65	0.01	0	0.16	0.25	0.05
Nb_2O_5	3.99	4.41	1	0.83	0.13	0.1	1.82	1.51	1.99	2.08	2.29	2.12	0	0.03	0.34	0.48	0.08
SnO	0	0	0	0	0	0	0	0	0	0	0	0	0	0	0	0	0
La_2O_3	0	0	0	0	0.07	0.02	0.03	0	0	0.02	0.06	0	0.1	0.13	0.13	0.09	0.07
Ce_2O_3	0.3	0.28	0.45	0.13	0.28	0.26	0.17	0.09	0.29	0.21	0.3	0.26	0.5	0.5	0.45	0.54	0.52
Nd_2O_3	0.03	0.19	0.21	0	0.26	0.24	0.2	0	0.02	0	0.21	0.36	0.21	0.04	0.23	0.34	0.16
Yb_2O_3	0.01	0	0.1	0	0.62	0.47	0.44	0.36	0.03	0	0.27	0.44	0	0	0.07	0.24	0
HfO_2	0	0	0	0	0	0	0	0	0	0	0	0	0	0	0	0	0
Ta_2O_5	0	0	0	0	0	0	0	0	0	0	0	0	0	0	0	0	0
PbO	0	0	0	0.92	0	0	0.34	0.36	2.13	1.16	2.35	0	0.73	0.69	0.38	0.15	0.47
ZnO	0.05	0	0	0	0	0	n.d	n.d	0	0	n.d	n.d	n.d	n.d	n.d	n.d	n.d
Al_2O_3	0.24	0.21	0.64	12.13	1.99	0.49	n.d	n.d	1.19	0.38	n.d	n.d	n.d	n.d	n.d	n.d	n.d
ThO_2	1.9	2.17	0.2	1.1	1.31	2.33	1.29	0.71	2.97	1.65	1.1	2.37	0	0	0	0.03	0
UO_2	43.7	33.4	36.7	34.1	56.8	61.7	32.6	27.8	47.2	63.0	51.2	49.2	76.8	77.8	82.2	78.8	80.4
Total	90.7	91.7	92.6	85.8	93.5	90.4	75.8	71.1	89.2	91.2	90.5	89.1	89.2	93.6	94.8	90.7	99.9

Table 4. The main characteristics El Sela granite in the vicinity of the main shear zone.

Borehole No.	W1	W2	W3
Rock types	• 0 - 12 m: wadi deposits. • 12 - 63 m: biotite ± muscovite granite.	• 0 - 44 m: biotite ± muscovite granite.	• 0 - 43 m: biotite ± muscovitegranite.
Hydrothermal alteration	biotite highly altered into ferrichlorite pennite), muscovitization, albitization, sericitization, kaolinization and argillization.		
Sulfides	Pyrite, arsenopyrite, chalcopyrite, galena and sphalerite		
Nb-Ti-, Zr-Hf-, F-minerals	Columbite, betafite, rutile, zircon	Columbite, rutile, zircon	Columbite, betafite, rutile, zircon, fluorite
REE-minerals	Allanite, LREE carbonate	Allanite	Allanite
Th-minerals	Thorite	Thorite, phosphothorite with apatite	Thorite
U-Th-minerals		uranothorite	
U-minerals	• Brannerite, coffinite, uraninite and pitchblende • W1-4 and W1-7 samples at depth 53 and 62 m, respectively.		• Brannerite, coffinite, uraninite and pitchblende • W3-2 and W3-3 samples at depth 30 and 43 m, respectively.

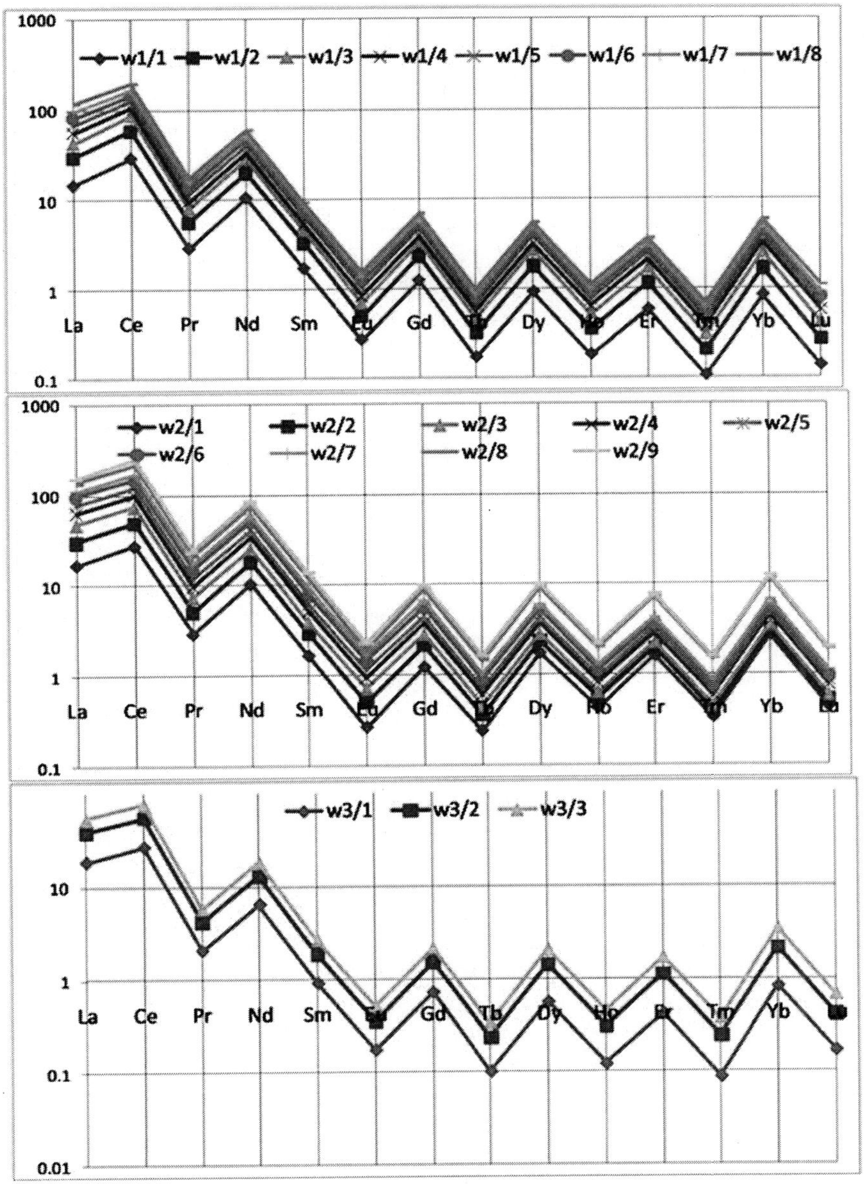

Figure 10. Chondrite-normalized REE patterns (normalization values are from [27] for the El Sela granite.

Columbite [(Fe, Mn)(Nb, Ta, Ti)O_4] occurs as dark gray to black color, flattened, prismatic, massive subhedral to euhedral crystals (Figure 11(f)). The SEM-BSE image and EPMA analyses shows that it is aferro-co- lumbite with 7.59 wt.% Fe_2O_3, 11 wt.% TiO_2, 9.76 wt.% MnO with minor amount of U and Th 0.75 and 0.2 wt.% respectively, (Table 2).

Betafite $(Ca, U)_2(Ti, Nb, Ta)_2O_6(OH)$ is detected under the SEM-BSE image (Figure 11(f)) as Nb-Ti-bearing mineral occurring as subhedral mineral in the boreholes W1 and W3 of two-mica granite. The obtained results confirm from a chemical analyses of the studied betafite minerals by SEM analyses, the major elements are included Nb_2O_5 vary from (12.4% to 21.4%), TiO_2 vary from (13.2% to 45.9%), FeO vary from (1.9 to 8.3 wt.%), Ta_2O_5 reach 4.6 wt.% and MnO reach 1.24 wt.%. It obvious that betafite has higher U concentration reaches

22.5, 26.95 and 31.9 wt.% with lower content in W3-3-C3 1.14 wt.%.

Thorite, phosphothorite, uranothorite, brannerite, coffinite, uraninite and pitchblende have been identified by SEM and EPMA analyses (Figure 11 & Table 3). Thorite and uranothorite is rare in the El Sela granites but phosphothorite was observed in the granites from the W1 and W2 drill hole. Thorite-group minerals form euhedral to subhedral crystals, black to brown colors and enclosed in biotite, quartz, zircon and monazite. Based on SEM and EMPA analyses four types of thorite-group minerals can be distinguished; thorite, Zr-rich thorite, pho- sphothorite and uranothorite (Figure 11(b), Figure 11(c), Figure 11(e); Table 3). The total of the analyses of all these minerals is low (close to 90 wt% or lower) indicating their strong metamictisation and hydration. ThO_2 contents in thorite vary from 20.1 to 48.8 wt.% and SiO_2 from 5.1 to 33.7 wt.% Table 3. SEM analyses reveal locally Zr-rich thorite the W_1-1 and W_2-3 drill holes with 5 to 8 wt.% ZrO_2. The occurrence of thorite within zircon crystal cores indicates a thorite analysis may be contaminated by the enclosing zircon crystal. Phosphothorite composition exhibitsa wide range of U, Zr, Y, LREE, Ca and P contents (Table 4). ThO_2 varies from 27.5 to 63.5 wt.%, P_2O_5 from 1.5 to 15.1 wt.%, CaO vary from 2.6 to 12.95 wt.%, Y_2O_3 vary from 0.8 to 3 wt.% and SiO_2 vary from 8.7 to 28.6 wt.% Table 3.

Uranothorite $(U, Th)SiO_4$ is interstitially associated with iron oxides and/or found as numerous subhedral to anhedral inclusions in zircon. The EPMA analyses indicate that UO_2 contents may reach 18.4 wt% and Y_2O_3 2.23 wt% composition of zircon up to 11.6 wt.% Table 3.

Ti-U oxides are the most U minerals in the El Sela granites (Figure 11(g) & Table 3). They generally form aggregates along the periphery of pyriteor occur as pseudomorphs of former Ti-minerals such as ilmenite and titanite El Sela granite. The major elements composition is variable: TiO_2 to 62.8 wt.% and UO_2 range from 15.7 to 49.1 wt.%. Up to 9.32 wt.% ZrO_2 may also be present. Brannerite is associated with uraninite along pyrite crystals at depths 53 m and 43 m in boreholes W1 and W_3, respectively.

Uranium oxides occur in granites samples (W3-1, -2, -3 and W1-4) as minute crystals filling micro-fractures and/or associated with pyrite, rutile and fluorite or with other U-minerals such as brannerite (Figure 11(d), Figure 11(g)). ThO_2 content of uraninite vary from (1.1 to 3 wt.%), for PbO contents from 1.16 to 2.35 wt.%, P_2O_5 contents from 0.17 to 1.56 wt.% and CaO contents from 3.8 to 7.6 wt.%. Also, minor amounts of LREE and Yare present (Table 3). Under reducing conditions, uraninite may be altered to coffinitization, with loss of radiogenic Pb and Y with HREE, [31] . The variable amounts of SiO_2 indicate Si diffusion in the structure along fractures at the beginning of coffinitization.

EMPA analyses show the major elements are SiO_2 vary from (11.98 to 25.13 wt.%) and UO_2 vary from (27.48 to 61.73 wt.%) correspond to the composition of coffinite Table 3 and Figure 11(h).

EPMA analyses show a wide variation of the pitchblende composition which exhibit very weak zoning and occurs as botryoidally form or vein surrounded metallic sulfides especially pyrite (Figure 11(i), Figure 11(j)). Most samples have no Th or traces (up to 0.03 wt.%) Table 3 but significant contents of SiO_2 (up to 4.69 wt.%), CaO (5.28 wt.%), Y_2O_5 (0.96 wt.%), ZrO_2 (0.25 wt.%), Nb_2O_5(0.48 wt.%), PbO (0.73%) and REE (0.30 wt.%). Moreover the trace elements in pitchblende may occur in the form of minute inclusions (galena, pyrite, sphalerite etc.) or as amorphous mixtures such as Ca^{2+}, Pb^{2+}, Th^{4+} and REE^{3+} which have similar ionic size to that of U^{4+}. The high FeO contents in some samples (3.98 or 6.51 wt.%) is not attributed to substitution for U because of the large difference in atomic radii (Fe^{3+} = 0.64 A; U^{4+} = 0.97 A). It may correspond to some inclusions such as hematite and pyrite. The U contents fluctuate between 76.84 wt.% and 82.15 wt.% in relation with the effect of the substitutions of other elements. Silica contents range between 1.05 to 4.69 wt.%.

The granite from the El Sela shear zone area are enriched in the lighter isovalents Nb and Zr relatively to the heavier isovalents Ta, Hf elements. The lower Zr/Hf ratio is in accordance with the presence of Hf-rich zircons (3 to 4 wt.% HfO_2) (Table 4).

7. RADON MAPPING

Uranium exploration has been aided by the identification of radon anomalies near the surface [32] . These anomalies can detect hidden uranium deposits that are too deep to be seen with gamma rays from the uranium daughter products or with normal surface geochemical

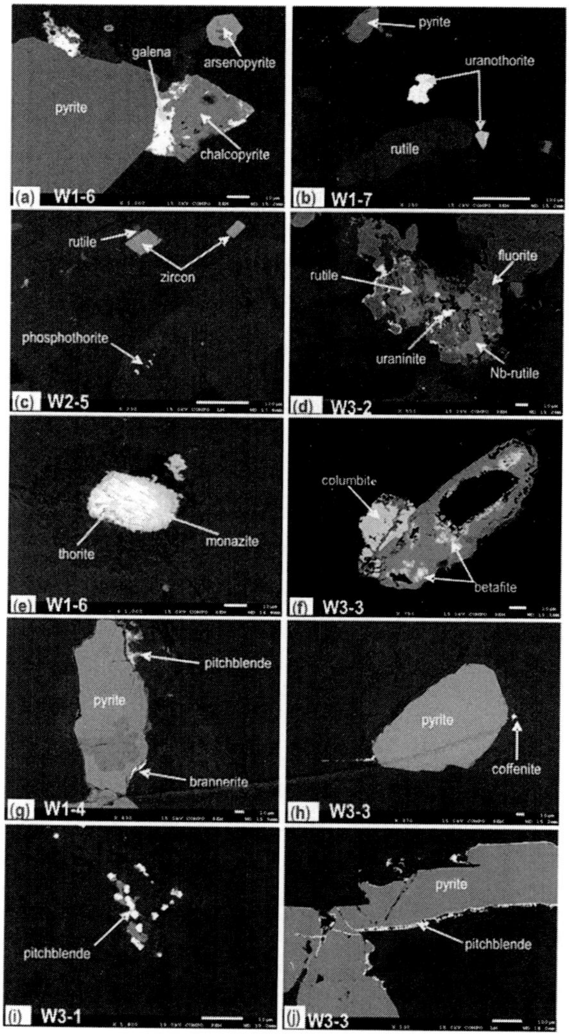

Figure 11. Representative back-scattered electron microscopy (BSEM) images of different minerals from El Sela granites. (a) Pyrite, chalcopyrite, arsenopyrite and galena. (b) Association of uranothorite, rutile and pyrite. (c) Rutile along periphery of zircon with phosphothorite. (d) Uraninite enclosed between fluorite and rutile-(Nb). (e) Monazite crystal enclosed thorite. (f) Nb-minerals (columbite and betafite). (g) Uraninite and brannerite along the periphery of pyrite. (h) Pyrite surrounded by coffinite. (i) Botryoidal pitchblende. (j) Veinlet of pitchblende at the margin of pyrite crystals.

techniques, which analyze uranium or radium. Radon, a relatively mobile gas, can originate from deeply buried deposits and gaseous or aqueous movement to the surface can transport it where it can be readily detected [33] . However a great care has to be taken in the interpretation as radon emanations because it may migrate relatively far from its source.

Shear zones, faults and joints are the main conducts for radon emanations to the surface especially for structure controlled deposits. Radon anomalies can be detected over deeply buried uranium mineralization. At the present time it is known that ore bodies at depths of 100 m to 200 m can generally be detected and that depth of up to 300 meters are possible in very favorable circumstances [33] .

Obviously, the cluster distribution map of radon shows that the radon activity varies along shear zone over a wide range that lies between 960 Bq/m^3 to 300,300 Bq/m^3 (Figure 12). The distribution of radon activity 3D-image map (Figure 13) shows clearly that the radon emission is heterogeneous along shear zone with about seven strong anomalies (Figure 13). The largest of these anomalies, which is located on the eastern side of the study area occupies large area of about 200 m^2 and reaches a radon activity act of 296,000 Bq/m^3. In the west direction another anomaly with a smaller extension presents the highest radon activity from the study area with 300,300 Bq/m^3. This anomaly is characterized by the existence of visible surficial uranium mineralization that is represented by uranophane and autunite.

The western part of the study area is occupied with an anomaly covering an area of medium intensity between the two previous two anomalies but the radon activity is up to half their values which is approximately 150,000 Bq/m^3. In the middle of the study area there is a small radon anomaly reaches about 200,000 Bq/m^3.

8. GEOPHYSICAL SIGNATURES OF SELA GEOLOGIC CROSS-SECTIONS

The variations of K, eU and eTh concentrations and their ratios, beside the VLF-EM and magnetic data are illustrated along two cross sections nearly perpendicular to the Sela shear zone as shown in Figure 14 & Figure 15. Potassium reached its highest values (4% ± 1%) in two mica granites and decrease abruptly in the shear main zone with less than 0.5%. The eTh variations reflect contrasts between the biotite granite and the two-mica granite. The biotite granite produces some of the highest eTh values (up to 50 ppm) and exhibits strong internal magmatic zonation. The two-mica granite tends to have lower Th concentrations. The lowest values (<3 ppm) occur in the shear zone. eU profile is mainly characterized by one anomaly, related to the shear zone and reaching >600 ppm. The remaining rock units in the study area show low uranium concentrations (1 to 20 ppm), but with three small anomalies which coincide with thorium anomalies. The shear zone is also logically characterized by high eU/eTh and eU/K ratios (up to 7 and 90). Field investigations have shown that uranium has been strongly concentrated along fractures. Uraninite has been located along the margins and discrete radioactive sources. In case of the two-mica granites, there is no appreciable variation in eU versus eU/eTh ratio. This indicates that uranium in initially as uraninite in the two mica granite; U may have been leached by weathering in the surface samples which are analyzed by gamma spectrometry. The coincident Th and U anomalies within the biotite granite, may indicate U-Th minerals refractory to weathering.

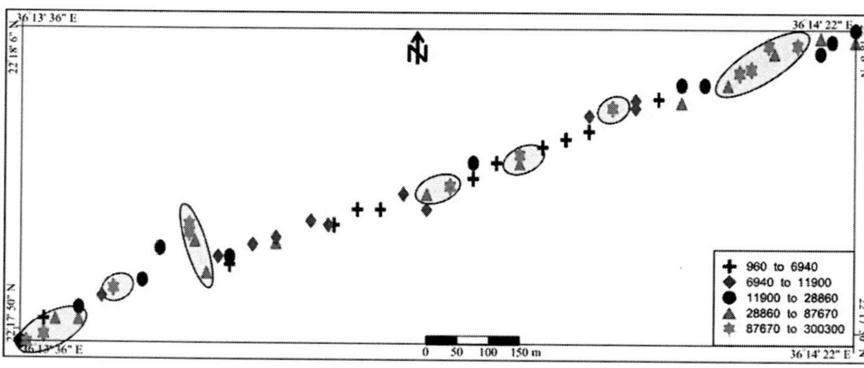

Figure 12. Radon concentration classed map (Bq/m³) of Sela area, S E D, Egypt.

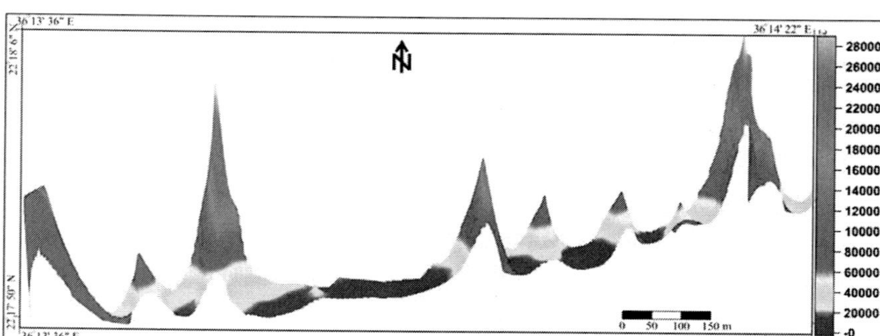

Figure 13. Radon activities 3D-image map (Bq/m³) of Sela area, S E D, Egypt.

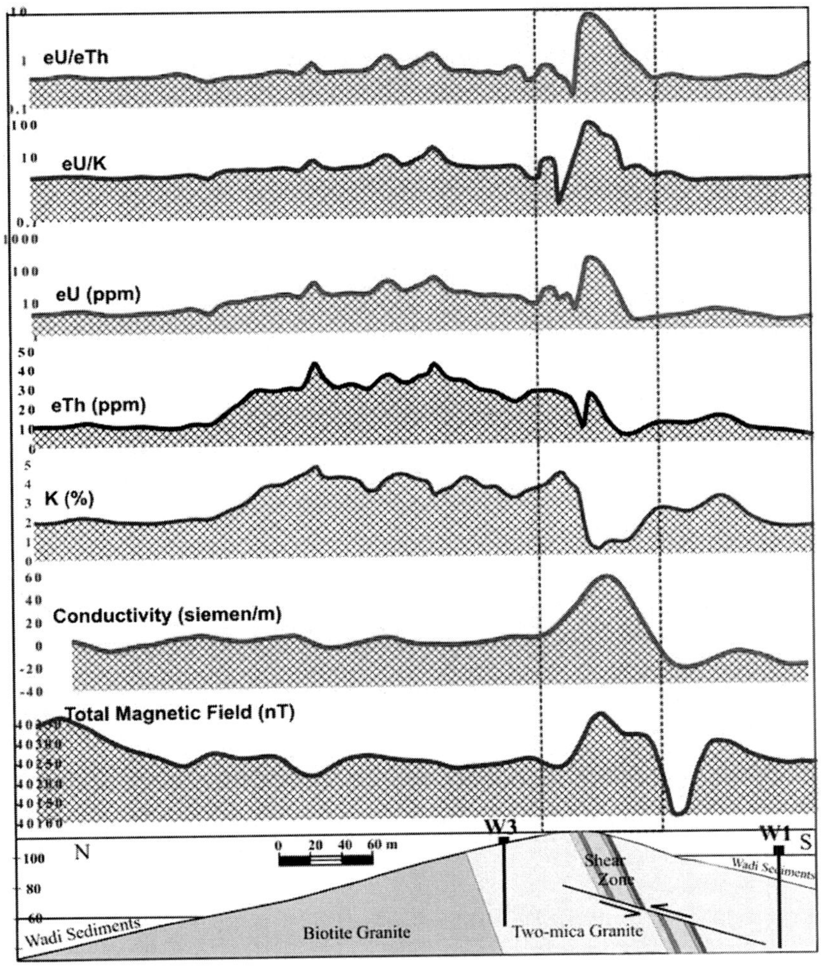

Figure 14. Stacked profiles for radio elements, magnetic field and conductivity across the two wells W1 and W3 of Sela area, Southern Eastern Desert, Egypt

The VLF-EM profiles (Figure 14 and Figure 15) show a marked increase of conductivity directly over the shear zone indicating vertical or steeply dipping sources. The VLF survey does not allow differentiating the bio- tite granite from the two-mica granite. The location of the VLF anomalies also correspond to a strongly positive magnetic anomaly for the first profile passing through the W1 and W3 drill holes, probably caused by the lamprophyre which may have high magnetite content. Immediately to the

East of the magnetic high, there is a sharp magnetite low which may correspond to a strongly oxidized zone at the southern margin of the Sela Shear Zone.

.

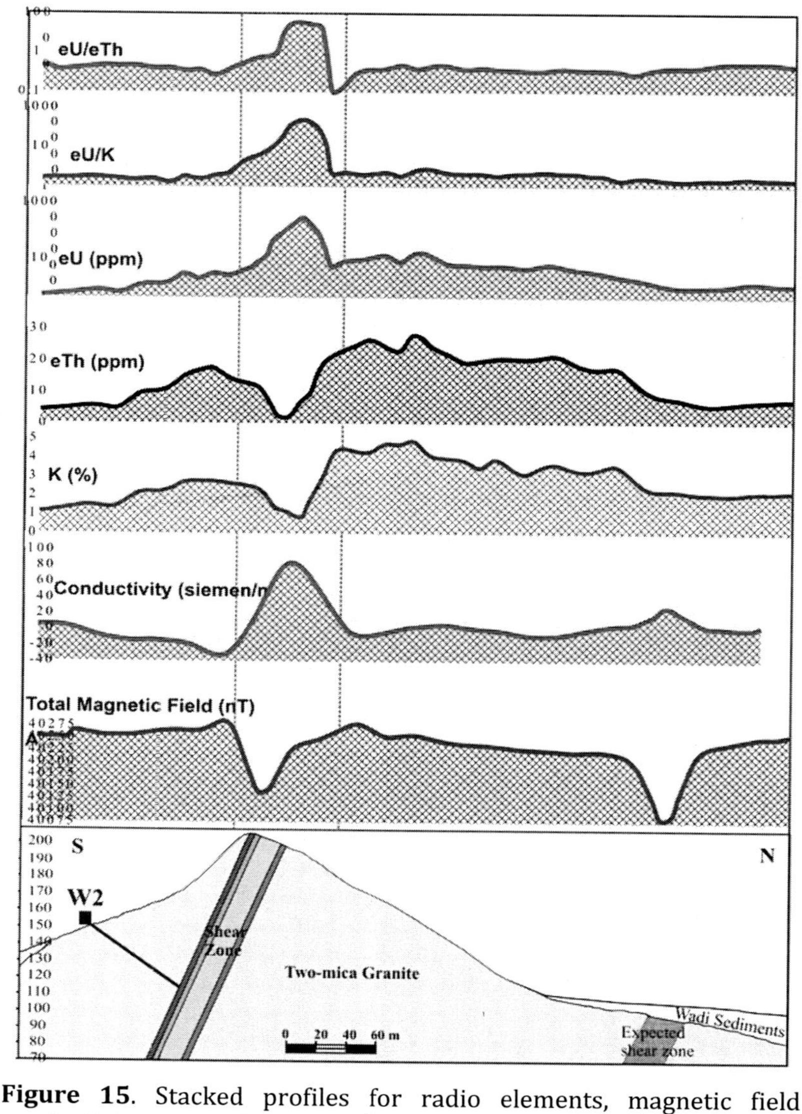

Figure 15. Stacked profiles for radio elements, magnetic field and conductivity across the two well W2 of Sela area, Southern Eastern Desert, Egypt.

In the second profile passing through the drill hole W2, above the Sela shear zone, the conductivity high correspond to the opposite of the preceding profile to a low in the total magnetic field, narrower than the conductivity high. Such a low magnetic field should correspond to a zone were the Fe-Ti bearing minerals have been oxidized or pyritized. A similar total magnetic field low is observed to the south of the Sela shear zone covered by sand, but is not associated with a conductivity high this indicate the causative body is located at depth more than 30m where the sense detection of VLF technique. The magnetic field over the granitic rocks and the wadi sediments is characterized by low magnetic signatures.

9. DISCUSSION AND CONCLUSION

9.1. Uranium Fertility of the El Sela Granite

The El Sela granite complex is composed of biotite and biotite ± muscovite granites. The three samples of drill holes have only intersected the biotite ± muscovite granites. Geochemically, the El Sela granites are very leucocratic, meta- to slightly per-aluminous and enriched in large ion lithophile elements (LILE: Ba, Rb and Sr), high field strength elements (HFSE: Y, Zr and Nb), light rare earth elements (LREE) mostly hosted in allanite, Th mostly hosted in uranothorite and allanite and uranium partly hosted in magmatic Th-bearing uraninite, but relatively poor in Ta, Sn, Be and heavy rare earth elements (HREE). Most of them have a negative Eu anomaly. Their characteristics correspond to those of A2-type granites (high-K cac-alkaline granites).

During most geological processes the chondritic ratios of the isovalent elements Nb/Ta (34.2 ± 0.3) and Zr/Hf (17.6 ± 1) are constant [34] . However Nb and Ta can fractionate in peraluminous magmas because of the higher solubility of Ta (tantalite) relatively to Nb (columbite) in such melts [35] . Nb-Ta, Nb-Ti rich minerals are re- presented by columbite and betafite.

The Th/U ratios of the El Sela granites vary from about 6, indicating uranium leaching, to 0.8 indicating late magmatic and/or hydrothermal enrichment. 6 samples have Th/U ratios below 2 indicating that magmatic or hydrothermal U-oxides may be present as the main U-bearing minerals. Th-bearing uraninite has been identified in two samples (W1-53 with 15 ppm U and 23 ppm Th and W3-43 with 25 ppm U and 20 ppm). Uraninite may have been present in other shallower samples but may have been leached by hydrothermal and/or meteoric fluids. The presence of uraninite in the El Sela granite indicates that it can represent a good uranium source, even if uranothorite was not sufficiently metamict to liberate uranium at the time of the circulation of the hydrothermal fluids.

9.2. The Hydrothermal Ore System

Two sets of structures are observed in the El Sela granite: a major E-W to ENE-WSW shear zone and N-S to NNE-SSW set of fault associated with different types of contemporaneous magmatic injections. These structures are reactivated several times in association with succession fluid circulations leading to the deposition of a series of quartz generations and U mineralization.

The main shear zone at G. El Sela is trending ENE-WSW and steeply dipping from 70° to 85° to the south. This trend has been reactivated many times by NNW-SSE extensional stress field, however to know that a detailed structural study would be necessary. The stress field may be transtensional with opening of tension gashes filled by the microgranite in the eastern part of the shear zone and filled by microgranite and dolerite dikes with white quartz and reddish to black jasperoid veins.

The main shear zone is affected by hydrothermal alteration as ferrugination, silicification, albitiation, epsyenitization, illitization and fluoritization with U-mineralization as hexavalent U minerals visible at the surface. The N-S strike slip sinistral fault cross-cut the ENE-WSW shear zone main trend. The width of the mineralized part of the shear zone varies from 3 to 40 m and is developed over a length of 1.5 km but reached 6 km in the east of G. El Sela area.

One of the aims of the present study is to detect U-mineralization in the drilled core samples and also to identify the depth of the reduced zone where the primary hydrothermal U-mineralization may be present.

The continuation of the uranium mineralization at depth is evidenced by the Track Etch Survey along the El Sela main shear zone with radon emission from 960 to 300,300 Bq/m^3.

The γ-ray spectrometric profiles across the Sela shear zone indicate the strong increase of eU and, eU/eTh and eU/K ratios and a decrease of eTh and eK anomaly in relation with the enrichment of the shear zone in quartz. The strong direct increase of eU/eTh ratio is typical of hydrothermal uranium enrichment with no mobility of Th.

The VLF-EM profiles show marked high positive conductivity correlated with the shear zone within the granitic body, indicating higher fracturation and clay mineral content of the shear zone. The magnetic profile passing through W1 and W3 shows clearly an increase of total magnetic field above the El Sela shear zone associated with a magnetic low at its southern margin. To the opposite the magnetic profile passing through W_2 showed only a strong magnetic low above the El Sela shear zone. Therefore, the association of spectrometric, radon, magnetic and electrical conductivity anomalies for the Sela shear zone clearly identifies the uraniumanomalous zones which can be selected for more drilling.

The occurrence of U-oxides along the margin of the pyrite crystals may indicate that they have played a role on the reduction of U^{6+} from the uranyl ion (UO_2^{2+}) dissolved in the hydrothermal fluids into U^{4+} which precipitates as, brannerite, coffinite and pitchblende. Pb tends to migrate outside the uranium mineral structure [36] . In the present study Pb released from U-oxides commonly forms galena.

Ibrahim Gaafar, Michel Cuney, Ahmed Abdel Gawad (31,38)">31] . Pitchblende occurs as botryoidal form or vein deposited onpyrite.

REFERENCES

1. Cuney, M. (2009) The Extreme Diversity of Uranium Deposits. Mineralium Deposita, 44, 3-9.
2. IAEA (2009) World Distribution of Uranium Deposits (UDEPO) with Uranium Deposit Classification. IAEA, Vienna, IAEA-TECDOC-1629, 117 p.
3. Cuney, M., Le fort, P. and Wang, Z.X. (1984) Uranium and Thorium Geochemistry and Mineralogy in the Manaslu leucogranite (Nepal, Himalaya). Geology of Granites and Their Metallogenetic Relations (Proceedings of a Sympo- sium), Nanjing, 1982, University Sceinces Editions, 853-873.
4. Poty, B., Cuney, M. and Friedrich, M. (1986) Uranium Deposits Spatially Related to Granites in the French Part of the Hercynian Orogeny, Vein Type Uranium Deposits. IAEA-TECDOC-361, IAEA, Vienna, 215-246.
5. Friedrich, M.H., Cuney, M. and Cregu, G. (1989) Uranium Enrichment Processes in Peraluminous Magmatism. Inter- national Atomic Energy Agency (IAEA)-TC-571/2.11-35.
6. Rameshbabu, P.V. (1999) Rare Metal and Rare Earth Pegmatites of Central India. Special Issue on Rare Metal and Rare Rarth Pegmatites of India. Exploration and Resarch, 12, 7-52.
7. Pal, D.C., Mishra, B. and Bernhardt, H.J. (2007) Mineralogy and Geochemistry of Pegmatite-Hosted Sn-, Ta-Nb- and Zr-Hf-Bearing Minerals from the Southereastern Part of the Bastar-Malkangiri Pegmatite Belt, Central India. Ore Geology Reviews, 30, 30-55.
8. Jiashu, R. and Zehong, H. (1982) Forms of Uranium Occurrence and Its Distribution in Uraniferous Granites. In: Geology of Granites and Their Metallogenetic Relations, Sciences Press, Beijing, 621-635.
9. Ibrahim, M.E. (1986) Geologic and Radiometric Studies on Um Ara Granite Pluton, South East Aswan, Egypt. Unpub- lished M.Sc. Thesis, Mansoura University, Mansoura.
10. Abdalla, H.M., Matsueda, H., Ishihara, S. and Miura, H. (1994) Mineral Chemistry of Albite-Enriched Granitoids at Um Ara, Southeastern Desert, Egypt. International Geology Review, 36, 1067-1077.
11. Roz, M.E. (1994) Geology and Uranium Mineralization of Gabal Gattar Area, North Eastern Desert, Egypt. Unpub- lished M.Sc. Thesis, Al Azhar University, Cairo, 175 p.

12. Abu Dief, A. (1985) Geology of Uranium Mineralization in Missikat, Qena-Safaga Road, Eastern Desert, Egypt. Un- published M.Sc. Thesis, Assiut University, Assiut, 242 p.
13. Osmond, J.K., Dabous, A.A. and Dawood, Y.H. (1999) U Series Age and Origin of Two Secondary Uranium Deposits, Central Eastern Desert, Egypt. Economic Geology, 94, 273-280.
14. Abdel Naby, H.H. (2008) Genesis of Secondary Uranium Minerals Associated with Jasperoid Veins, El Erediya Area, Eastern Desert, Egypt. Mineralium Deposita, 43, 933-944.
15. NMA (Nuclear Material Authority) of Egypt (2003-2013) El Sela Development Project, Internal Reports.
16. Gaafar, I.M. (2005) Applications of Geological and Geophysical Survey for Defining the Uranium Potentiality of Some Younger Granites in the Eastern Desert of Egypt. Unpublished Ph.D., Faculty of Science, Mansoura University, Man- soura, 186 p.
17. Ali, K.G. (2011) Structural Control of El Sela Granites and Associated Uranium Deposits, South Eastern Desert, Egypt. Arabian Journal of Geosciences.
18. Abdel Gawad, A.E., Orabi, A.H. and Bayoumi, M.B. (2014) Uranium Evaluation and Its Recovery from Microgranite Dike at G. El Sela area, South Eastern Desert, Egypt. In Press.
19. Gaafar, I.M., Ghazala, H.H., Ibrahim, T.M. and Ammar, S.E. (2006) Gamma-Ray Spectrometry Studies for A Promising Vein Type Uranium Mineralization, South Eastern Desert, Egypt. Proceedings of 4th International Symposium on Geo- physics, Tanta, 2006, 445-456.
20. Gaafar, I.M., Ghazala, H.H., Ibrahim, T.M. and Ammar, S.E. (2006) Application of Ground Magnetic Survey for Defining the Subsurface Structural Pattern of West Abu Ramad Shear Zone, Southern Eastern Desert, Egypt. The 8th SEGJ In- ternational Symposium-Imaging and Interpretation, Kyoto, 2006, 487-492.
21. Gaafar, I.M., Ghazala, H.H., Ibrahim, T.M. and Ammar, S.E. (2006) Conductivity Mapping Using VLF-EM Ground Survey For Promising Vein Type Uranium Mineralization, South Eastern Desert, Egypt. 8th International Conference of the Geology of the Arab World, Cairo University, Giza.
22. Mayya, Y.S., Eappen, K.P. and Nambi, K.S. (1998) Methodology for Mixed Field Inhalation in Monazite Areas Using a Twin-Cup Dosimeter with Three-Track Detector. Radiation Protection Dosimetry, 77, 177-184.
23. Debon, F. and Lefort, P. (1988) A Cationic Classification of Common Plutonic Rocks and Their Magmatic Associations: Principles, Method, Applications. Bulletin de Mineralogie, 111, 493-510.
24. Sun, S.S. and McDonough, W.F. (1989) Chemical and Isotopic Systematics of Oceanic Basalts: Implications for Mantle Composition and Processes. In: Saunders, A.D. and Norry, M.J., Eds., Magmatism in the Ocean Basins, Vol. 42, Geologi- cal Society, Special Publications, London, 313-345.
25. Pearce, J.A., Harris, N.B.W. and Tindle, G. (1984) Trace Elements Discrimination Diagram for the Tectonic Interpreta- tion of Granitic Rocks. Journal of Petrology, 25, 956-983.
26. Saunders, A.D., Nory, M.J. and Tarney, J. (1991) Fluid Influence on the Trace Element Compositions of Subduction Zone Magmas. In: Tarney, J., Pickering, K., Knipe, R. and Dewey, J., Eds., The Behaviour and Influence of Fluids in Subduction Zones, Philosophical Transactions of the Royal Society of London, 335, 377-392.
27. Nakamura, N. (1974) Determination of REE, Ba, Fe, Mg, Na and K in Carbonaceous and Ordinary Chondrites. Geo- chimica et Cosmochimica Acta, 38, 757-775.
28. Abd Elaal, H.A. (2006) Geochemistry and Processing of Some Valuable Metals from Mineralized Two Mica Granites, South Eastern Desert, Egypt. Unpublished M.Sc. Thesis, Cairo University, Giza, 110 p.
29. Speer, J.A., Solberg, T.N. and Becker, S.W. (1981) Petrography of the Uranium-Bearing Minerals of the Liberty Hill Pluton. South Carolina: Phase Assemblage and Migration of Uranium in Granitoid Rocks. Economic Geology, 76, 2162- 2175.

30. Wang, R., Cheng, F.F. and Monchoux, P. (1992) Minéraux disséminés comme indicateurs du caractère pegmatitique du granite de Beauvoir, massif d'Échassières, Allier, France. Canadian Mineralogist, 30, 763-770.
31. Janeczek, J. and Ewing, R.C. (1992) Structural Formula of Uraninite. Journal of Nuclear Materials, 190, 128-132.
32. Miller, J.M. and Ostle, D. (1973) Radon Measurement in Uranium Prospecting. Uranium Exploration Methods, Proceedings of a Panel, Vienna, 10-14 April 1972.
33. Fleischer, R.L. and Mogro-Campero, A. (1978) Mapping of Integrated Radon Emanation for Detection of Long-Distance Migration of Gases within the Earth: Techniques and Principles. Journal of Geophysical Research, 83, 3539-3549.
34. Weyer, S., Münker, C., Rehkomper, M. and Mezger, K. (2002) Determination of Ultra-Low Nb, Ta, Zr and Hf Concen- trations and the Chondritic Zr/Hf and Nb/Ta Ratios by Isotope Dilution Analyses with Multiple Collector ICP-MS. Chemical Geology, 187, 295-313.
35. Linnen, R.L. and Cuney, M. (2008) Granite-Related Rare-Element Deposits and Experimental Constraints on Ta-Nb- W-Sn-Zr-Hf Mineralization. In: Linnen, R.L. and Samson, I.M., Eds., Rare-Element Geochemistry and Mineral Deposits: Geological Association of Canada, GAC, Short Course Notes, 17, 45-67.
36. Mann, A.W. and Deutscher, R.L. (1980) Solution Geochemistry of Lead and Zinc in Water Containing Carbonate, Sul- fate and Chloride Ions. Chemical Geology, 29, 293-311.

CHAPTER 14

Equilibrium Elastic Stress Field of the Earth's Solid Shell

Alexander Ivanchin

Institute of Monitoring of Climatic and Ecological Systems, Tomsk, Russia

ABSTRACT

In modern geophysics, hydrostatic dependence of pressure on the depth in the lithosphere is postulated. It is considered evident and requiring no proof. As shown in the present work, the above postulate is erroneous. Proceeding from one of the fundamental laws of physics related to the minimum of potential energy in the equilibrium state, one can derive a nonhydrostatic solution of the elasticity equation with minimum elastic energy referred to as a Gravitational Equilibrium Field with an energy by an order of magnitude less than the hydrostatic field energy. The Earth's solid shell like a bearing structure carries its own weight, which reduces the pressure on the surface of the liquid nucleus down to zero. The influence of solidity in the subsurface region of the Earth is characteristic. As the calculation shows, although the rock density in the crust is thrice as much as that of the water, the pressure in the ocean at the same depth is higher than the pressure in the solid crust, which is an account for the existence of land. If there was a hydrostatic stress distribution, the pressure under the continents would be thrice as much as that in the ocean and the continents would descend below sea level.

Keywords: Stress and Pressure in Lithosphere

1. INTRODUCTION

At the present-day level of geophysics, the problem on the stress field in the lithosphere has been poorly investigated in terms of theory. It is postulated that the stress field in the lithosphere is hydrostatic in nature and coincides with the pressure field in the liquid of the same density [1] . It means that the stress tensor for the lithosphere has stress components other than zero on the main diagonal and they are equal. The hydrostatic stress distribution in the lithosphere is considered "evident" and requires no explanation or proof. The models of the Earth's structure, which have been suggested, are based on the above postulate.

The lithosphere stress field is fundamental in geophysics. It determines the parameters of the plastic flow, the tectonic plate movement, the formation of earthquake and volcano chamber. The knowledge of these stresses is essential in designing engineering underground structures, as well as for geological exploration and other purposes.

In this work, the main principle determining the formation of the lithosphere stress field is one of the fundamental laws in physics and, namely, a system is in equilibrium at minimum of its potential energy. If its potential energy exceeds its minimum, then there appear some forces that tend to bring a system to equilibrium. During the geological period of time, the processes occurring in the lithosphere must bring it to the equilibrium state with minimum elastic energy, which considerably differs from the hydrostatic field.

2. STRESS FIELD

The lithosphere is a spherical solid crust with the external radius R and the internal radius r_0. Here R is the Earth radius and r_0 is the nucleus radius. Let us consider the following problem. In the absence of gravitation there are no elastic stresses in the crust. They arise after gravitation appears and can be derived from the equation of elastic equilibrium [2] :

$$\frac{\partial \sigma_{ij}}{\partial x_j} = f_i \tag{2.1}$$

Here σ is the elastic stress tensor (summation is performed by the repeated index from 1 to 3), f_i is the volume force density, in this case, the gravitation force per unit of mass is

$$f_r = 4\pi\gamma \frac{\rho(r)}{r^2}\int_0^r \rho(\xi)\xi^2 d\xi \tag{2.2}$$

where γ is the gravitation constant, $\rho(r)$ is the mass density, r is the distance to the Earth center. A spherical system of coordinates with the origin in the Earth center is chosen. The problem is centrally symmetric, therefore, only the radial component of the mass force vector is different from zero which is expressed by the relation (2.2). Further we shall consider the density $\rho(r)$ in the first approximation to be constant in the lithosphere and independent of the radius r. In this case, (2.2) is written as

$$f_r = \frac{4\pi}{3}\gamma\rho^2 r$$

Equation (2.1) for the displacement vector U has the form

$$\nabla^2 U + \frac{1}{2(1-\nu)}\nabla\times\nabla\times U = Ar \tag{2.3}$$

Here

$$A = \frac{4}{3}\frac{\pi\gamma\rho^2(1+\nu)}{K(1-\nu)} \tag{2.4}$$

ν is the Poisson coefficient, K is the modulus of cubic compressibility, ∇ is the nabla operator, ∇^2 is the Laplace operator. Let us also consider the elastic moduli constant. Due to the spherical symmetry of the elastic displacement vector U, only the radial component U_r is different from zero and (2.3) is written as

$$\frac{d}{dr}\left[\frac{1}{r^2}\frac{d}{dr}\left(r^2U_r\right)\right]=Ar \tag{2.5}$$

The deformation tensor in the spherical coordinates looks like

$$\varepsilon_{rr}=\frac{\partial U_r}{\partial r},\qquad \varepsilon_{\vartheta\vartheta}=\varepsilon_{\varphi\varphi}=\frac{U_r}{r} \tag{2.6}$$

The nondiagonal components are zero. Equation (2.5) integrates to the following expression

$$U_r=A\frac{r^3}{10}+\frac{C}{3}r \tag{2.7}$$

Here C is an arbitrary constant. According to the theory, it is necessary to add a general solution of the homogeneous equation to (2.7).

$$\frac{d}{dr}\left[\frac{1}{r^2}\frac{d}{dr}\left(r^2U_r\right)\right]=0$$

Its solution is

$$U_r=\frac{C_1}{r^2} \tag{2.8}$$

C_1 is an arbitrary constant. Adding together (2.7) and (2.8) we obtain the displacement field

$$U_r=A\frac{r^3}{10}+\frac{C}{3}r+\frac{C_1}{r^2} \tag{2.9}$$

The deformation field for (2.9) is written as

$$\varepsilon_{rr}=\frac{3A}{10}r^2+\frac{C}{3}-2\frac{C_1}{r^3}$$

$$\varepsilon_{\vartheta\vartheta}=\varepsilon_{\varphi\varphi}=\frac{A}{10}r^2+\frac{C}{3}+\frac{C_1}{r^3} \tag{2.10}$$

The equalities $\varepsilon_{\vartheta\vartheta}=\varepsilon_{\varphi\varphi}$ and $\sigma_{\vartheta\vartheta}=\sigma_{\varphi\varphi}$ follow and the dilatation is

$$\upsilon = \varepsilon_{rr} + \varepsilon_{\vartheta\vartheta} + \varepsilon_{\varphi\varphi} = \frac{A}{2} r^2 + C \tag{2.11}$$

The arbitrary constants C and C_1 are derived from the boundary condition.

On the Earth surface the pressure is equal to the atmospheric pressure p_0, therefore, on the Earth Surface, when $r = r_0$, normal stresses must be equal to the atmospheric pressure

$$\sigma_{rr}(R) = \sigma_{\vartheta\vartheta}(R) = p_0 \tag{2.12}$$

Two conditions are involved in (2.12) which determine unambiguously the two arbitrary constants in (2.10). There are no other arbitrary parameters in (2.10), which means that the pressure on the inner boundary of the sphere on the nucleus surface at $r = r_0$ has been defined, which coincides with the postulates of the hydrostatic hypothesis. The elastic energy density is

$$w = \frac{\lambda}{2} \upsilon^2 + \mu \left(\varepsilon_{rr}^2 + \varepsilon_{\vartheta\vartheta}^2 + \varepsilon_{\varphi\varphi}^2 \right) \tag{2.13}$$

or, in the representation of the deformation resolved into pure shear and pure compression [2] :

$$w = \frac{K}{2} \upsilon^2 + \mu \left[\left(\varepsilon_{rr} - \frac{\upsilon}{3} \right)^2 + \left(\varepsilon_{\vartheta\vartheta} - \frac{\upsilon}{3} \right)^2 + \left(\varepsilon_{\varphi\varphi} - \frac{\upsilon}{3} \right)^2 \right] \tag{2.14}$$

Here K is the modulus of cubic compressibility, λ and μ are the Lamé coefficients. The first component in (2.14) is the elastic energy density caused only by the change of the volume, while the second component is the elastic energy density caused only by the shear.

The stress field is

$$\sigma_{ii} = \frac{\partial w}{\partial \varepsilon_{ii}} = \lambda \upsilon \delta_{ii} + 2\mu \varepsilon_{ii} \tag{2.15}$$

The deformation tensor is written for the main axes and, therefore the nondiagonal components are zero. The Equation (2.15) can also be written in this way

$$\sigma_{ii} = K\upsilon\delta_{ii} + 2\mu\left(\varepsilon_{ii} - \frac{\upsilon}{3}\delta_{ii}\right) \tag{2.16}$$

Similarly to (2.14) the first component in (2.16) is the contribution of the volume change into stress, whereas the second component is the shear deformation contribution. From (2.17) it follows that at the shear modulus $\mu = 0$, that is for a liquid,

$$\sigma_{ii} = K\upsilon\delta_{ii} = p \tag{2.17}$$

The value of $K\upsilon$ is the hydrostatic pressure in a solid. The Poisson coefficient is

$$\nu = \frac{1}{2}\frac{3K - 2\mu}{3K + \mu}$$

For a liquid $\nu = 1/2$, then, according to (2.4),

$$A = 4\pi\gamma\frac{\rho^2}{K}$$

For a liquid sphere in the expression for a shear (2.9) we assume $C_1 = 0$ to avoid divergence at the origin of coordinates. The constant C written as C_0 for a liquid to differentiate it from the same constant in a solid (2.10) is specified by the pressure on the external boundary of a liquid sphere. Then from (2.10), (2.11) and (2.16) we obtain for a liquid

$$p = K\left(\frac{A}{2}r^2 + C_0\right) \tag{2.18}$$

The above formula yields the dependence of pressure on the depth in a liquid at its constant density. From the equality $\sigma_{rr}(R) = p_0$ we derive an arbitrary constant

$$C_0 = \frac{p_0}{K} - \frac{AR^2}{2}$$

If a gravitating sphere has a solid shell with a liquid nucleus inside, then when determining the stress field of the lithosphere, unlike a liquid sphere, we should consider that $C_1 \neq 0$. The deformation field in the solid shell is given by (2.10) and, according to (2.12), the arbitrary constants and the lithosphere stresses including those on its internal boundary are determined at $r = r_0$. The above solution is not implemented in nature. However, it is needed as a starting point for finding a realizable solution.

It is worth noting some specific features of the solution (2.9). There are no stresses in the absence of gravitation. Application of gravitation results in an elastic field (2.10). If gravitation is removed, then the elastic field disappears, that is, there is no plastic flow at loading, which disagrees with materials science data. At any value of loading the processes of plasticity and mass transfer take place in a solid. The assumption of the absence of plastic deformation holds true only for short time intervals. Taking into account the long time of the lithosphere existence, mass transfer caused by diffusion, a magma flow, etc., makes a large contribution to the stress field. As a result, the elastic field changes so that its elastic energy decreases. During the geological period, the lithosphere elastic field energy reaches its minimum irrespective of the initial stress distribution. The elastic filed with minimum energy is the GEF.

The mass transfer effect can be considered as an additional dilatation to (2.10)

$$\hat{\upsilon}(r) = \nabla \cdot \hat{U}$$

which must be treated as a displacement source and one should use the method for the solution of the equation of elastic equilibrium (2.3) with displacement sources [3] . Here $\hat{U}$ is the field of a displacement source caused by the mass transfer and producing dilatation, which is similar to expansion of a solid at heating. Another example is swelling of a material at radioactive irradiation when point defects are introduced, which increases its volume. The unknown vector $\hat{U}$ is found from the condition of the minimum of elastic energy. The problem is spherically symmetric, therefore, $\hat{U}$ depends only on the distance to the Earth center and it has only one non-zero radial component. According to (2.6), we can define the deformation produced by the vector $\hat{U}$

$$\hat{\varepsilon}_{rr} = \frac{\partial \hat{U}_r}{\partial r}$$

$$\hat{\varepsilon}_{\vartheta\vartheta} = \hat{\varepsilon}_{\varphi\varphi} = \frac{\hat{U}_r}{r} \tag{2.19}$$

The other tensor components are zero. The vector $U + \hat{U}$ has only a radial component $U_r + \hat{U}_r$ depending only on r. The equation of elastic equilibrium for the above vector is written as

$$\frac{\mathrm{d}}{\mathrm{d}r}\left[\frac{1}{r^2}\frac{\mathrm{d}}{\mathrm{d}r}\left(r^2 U_r\right)\right] + \frac{\mathrm{d}}{\mathrm{d}r}\left[\frac{1}{r^2}\frac{\mathrm{d}}{\mathrm{d}r}\left(r^2 \hat{U}_r\right)\right] = Ar + F(r) \tag{2.20}$$

Here $F(r)$ is the displacement source function. Due to linearity, Equation (2.20) is divided into two Equations (2.5) and

$$\frac{\mathrm{d}}{\mathrm{d}r}\left[\frac{1}{r^2}\frac{\mathrm{d}}{\mathrm{d}r}\left(r^2 \hat{U}_r\right)\right] = F(r) \tag{2.21}$$

Integrating we get

$$\hat{U}_r = \frac{1}{r^2}\int r^2\left[\int F(r)\mathrm{d}r\right]\mathrm{d}r$$

The dilatation source function $F(r)$ is determined by the value of the volume change due to the mass transfer. Since (2.20) falls apart into two linear equations, then the linear combination of solutions (2.5) and (2.21) will be a common solution

$$T_r = U_r - C_3\hat{U}_r$$

The deformation tensor is written as

$$t_{rr} = \frac{3A}{10}r^2 + \frac{C}{3} - 2\frac{C_1}{r^3} - C_3\hat{\varepsilon}_{rr}$$

$$t_{\vartheta\vartheta} = t_{\varphi\varphi} = \frac{A}{10}r^2 + \frac{C}{3} + \frac{C_1}{r^3} - C_3\hat{\varepsilon}_{\vartheta\vartheta}$$

$$\tau = t_{rr} + t_{\vartheta\vartheta} + t_{\varphi\varphi} \tag{2.22}$$

Here C_3 is an arbitrary constant. The dilatation caused by the mass transfer is written in the spherical coordinates as

$$\hat{\upsilon} = \frac{1}{r^2}\frac{\mathrm{d}}{\mathrm{d}r}\left(r^2\hat{U}_r + 2\frac{\hat{U}_r}{r}\right) \tag{2.23}$$

The following value plays the role of pressure

$$p = K\tau \tag{2.24}$$

The elastic energy density is written as

$$w = \frac{\lambda}{2}\left(\frac{Ar^2}{2} + C - \hat{\upsilon}\right)^2 + \mu\left(\frac{3A}{10}r^2 + \frac{C}{3} - 2\frac{C_1}{r^3} - \frac{\hat{\upsilon}}{3}\right)^2$$

$$+ 2\mu\left(\frac{A}{10}r^2 + \frac{C}{3} + \frac{C_1}{r^3} - \frac{\hat{\upsilon}}{3}\right)^2 \tag{2.25}$$

Integrating (2.25) over the entire lithosphere we obtain the energy of elastic stresses

$$W = \int_0^{2\pi} \mathrm{d}\varphi \int_0^{\pi} \mathrm{d}\vartheta \int_{r_0}^{R} wr^2 \sin\vartheta \mathrm{d}r = 4\pi \int_{r_0}^{R} wr^2 \mathrm{d}r \tag{2.26}$$

The unknown function $\hat{\upsilon}$ should be chosen so that the energy W is minimum. From (2.26) according to the variation analysis theory, the Euler equation for determination of the function expressing the minimum energy W is written as [4]

$$\frac{\partial\left(wr^2\right)}{\partial\hat{U}_r}-\frac{\partial^2\left(wr^2\right)}{\partial\hat{U}'_r\partial r}-\frac{\partial^2\left(wr^2\right)}{\partial\hat{U}'_r\partial\hat{U}_r}\hat{U}'_r-\frac{\partial^2\left(wr^2\right)}{\left(\partial\hat{U}'_r\right)^2}\frac{\mathrm{d}\hat{U}'_r}{\mathrm{d}r}=0 \tag{2.27}$$

Taking into account (2.23) we have the differentiation rule

$$\frac{\partial}{\partial\hat{U}_r}=\frac{\partial}{\partial\hat{\upsilon}}\frac{\partial\hat{\upsilon}}{\partial\hat{U}_r}=\frac{2}{r}\frac{\partial}{\partial\hat{\upsilon}}$$
$$\frac{\partial}{\partial\hat{U}'_r}=\frac{\partial}{\partial\hat{\upsilon}}\frac{\partial\hat{\upsilon}}{\partial\hat{U}'_r}=\frac{\partial}{\partial\hat{\upsilon}} \tag{2.28}$$

Here we designate

$$\hat{U}'_r=\frac{\mathrm{d}\hat{U}_r}{\mathrm{d}r}$$

Then from (2.27) we obtain

$$\frac{2}{r}\frac{\partial\left(wr^2\right)}{\partial\hat{\upsilon}}-\frac{\partial^2\left(wr^2\right)}{\partial\hat{\upsilon}\partial r}-\frac{2}{r}\frac{\partial^2\left(wr^2\right)}{\partial\hat{\upsilon}^2}\frac{\mathrm{d}\hat{U}_r}{\mathrm{d}r}-\frac{\partial^2\left(wr^2\right)}{\partial\hat{\upsilon}^2}\frac{\mathrm{d}^2\hat{U}_r}{\mathrm{d}r^2}=0 \tag{2.29}$$

Differentiating, we derive a differential equation to determine the unknown function $\hat{\boldsymbol{U}}$ [4]

$$\frac{\mathrm{d}^2\hat{U}_r}{\mathrm{d}r^2}+\frac{1}{r}\frac{\mathrm{d}\hat{U}_r}{\mathrm{d}r}-\frac{A}{2}r-\frac{2}{r^2}=0$$

or, taking into account (2.19)

$$\frac{\mathrm{d}\hat{\varepsilon}_{rr}}{\mathrm{d}r}+\frac{1}{r}\hat{\varepsilon}_{rr}-\frac{A}{2}r-\frac{2}{r^2}=0 \tag{2.30}$$

The solution (2.30) is [5] :

$$\hat{\varepsilon}_{rr}=\frac{2}{r}\ln\frac{r}{R}+Ar^2+\frac{C_2}{r}$$

Here C_2 is an arbitrary constant. Integrating $\hat{\varepsilon}_{rr}$ we find the displacement

$$\hat{U}_r = \ln^2 \frac{r}{R} + C_2 \ln \frac{r}{R} + \frac{A}{6} r^3$$

A complete deformation field is written in the form of (2.22). The stress field is written according to (2.15) and has the form

$$\tilde{\sigma}_{ii} = K\tau\delta_{ii} + 2\mu\left(t_{ii} - \frac{\tau}{3}\delta_{ii}\right)$$

The elastic field which provides the lithosphere minimum elastic energy under gravitation is denoted by the tilde. The hydrostatic pressure is (2.24). On the Earth surface at $r = R$ the boundary condition is similar to (2.12)

$$\tilde{\sigma}_{rr} = \tilde{\sigma}_{\vartheta\vartheta} = \tilde{\sigma}_{\varphi\varphi} = p_0 \tag{2.31}$$

On the Earth surface normal stress must be equal to the atmospheric pressure or that of the water near the bottom of a reservoir. As a result, we obtain a set of two linear Equations (2.31) relative to C_1 and C_2. By virtue of the equality $\tilde{\sigma}_{\vartheta\vartheta} = \tilde{\sigma}_{\varphi\varphi}$, one of the equalities $\tilde{\sigma}_{\vartheta\vartheta} = p_0$ or $\tilde{\sigma}_{\varphi\varphi} = p_0$ in (2.31) should be rejected. It is also assumed that the pressure P on the lithosphere internal boundary at $r = r_0$ is known

$$\tilde{\sigma}_{rr}(r_0) = P \tag{2.32}$$

As is shown below, the boundary condition does not significantly affect the solution near the Earth surface. Solving (2.30) we get the values C, C_1 and C_2 in the form

$$C=\frac{1}{\varsigma}\Big\{(2\nu-1)\big[2AKR^{5}(5C_{3}-3)+20R^{3}p_{0}\big]$$
$$+5r_{0}^{2}\bigg[\bigg(-12C_{3}K\nu\ln\frac{r_{0}}{R}+2AK\nu R^{3}(8C_{3}-3)+12R\nu p_{0}+12C_{3}K(\nu-1)\bigg)\ln\frac{r_{0}}{R}$$
$$+AKR^{3}(3-8C_{3})(\nu-1)-6Rp_{0}(\nu-1)\bigg]-10r_{0}^{3}P(\nu+1)$$
$$+r_{0}^{5}AK\big[10C_{3}(2\nu-3)-3(\nu-3)\big]\Big\}$$
$$C_{1}=\frac{R^{3}}{3\varsigma}\bigg\{r_{0}^{2}\bigg[2K\bigg(30\nu C_{3}\ln\frac{r_{0}}{R}+A\nu R^{3}(6-25C_{3})-30C_{3}(\nu-1)\bigg)\ln\frac{r_{0}}{R}$$
$$+AKR^{3}(25C_{3}-6)(\nu-1)\bigg]+r_{0}^{3}(\nu+1)\big[AKR^{2}(3-5C_{3})+10(P-p_{0})\big]$$
$$+AKr_{0}^{5}\big[10C_{3}(3-2\nu)+3(\nu-3)\big]\bigg\}$$
$$C_{2}=\frac{R}{\varsigma C_{3}}\bigg\{2AKR^{5}(2\nu-1)(6-25C_{3})+180r_{0}^{2}KC_{3}\bigg[(\nu-1)-\nu\ln\frac{r_{0}}{R}\bigg]\ln\frac{r_{0}}{R}$$
$$+5r_{0}^{3}(\nu+1)\big[AKR^{2}(8C_{3}-3)-6(P-p_{0})\big]+r_{0}^{5}AK\big[10C_{3}(2\nu-3)-3(\nu-3)\big]\bigg\} \quad (2.33)$$

Here we designate

$$\varsigma=10K\big[6\nu Rr_{0}^{2}\ln(r_{0}/R)+2R^{3}(2\nu-1)+3Rr_{0}^{2}(1-\nu)-r_{0}^{3}(\nu+1)\big]$$

The formulae (2.15), (2.22), and (2.33) yield a solution of the problem of stresses in the lithosphere possessing minimum elastic energy at an arbitrary value C_3. Undoubtedly, one could substitute (2.22) and (2.33) into the expression for stress (2.15) and write it in the explicit form, but then the equations become too long and, therefore, inconvenient. Let us substitute the values of the Earth parameters [6] into the above formulae: $\nu=0.25$, $R=6371$ km, $r_0=2900$ km, $K=10^{11}$ Pa, $\rho=4000$ kg/m^3, $\gamma=6.672\times10^{-11}$ N·m/kg^2, $p_0=-1.03\times10^{5}$ Pa. Here the pressure has a negative value due to the fact that in the elasticity theory the deformation of compression is a negative value while tension is a positive one. As density we take a value $\rho=4000$ kg/m^3 resulting from averaging the data on the lithosphere density from [6] . Figure 1 shows the plots of the stresses $\tilde{\sigma}_{rr}$ and $\tilde{\sigma}_{99}$, the pressure $\tilde{p}$ derived from (2.24) as well as the plot of the hydrostatic pressure taken from [7] . In the plots the sign of tilde is absent. If the value of pressure at the nucleus boundary is taken to be equal to $P = 10^{11}$ Pa, then the energy value (2.27) is $W = 4.2\times10^{30}$ J, and for the hydrostatic stress it is $W = 3.2\times10^{31}$ J. The energy of the nonhydrostatic field stresses is by an order of magnitude less. Since the pressure P at the lithosphere internal boundary is unknown, here you can see calculation for several values of P. The pressure P can also be determined from the condition of minimum energy. Calculation of the minimum energy by P shows that minimum energy will be on the lithosphere internal boundary at a positive value, that is, in tension. However, a liquid nucleus cannot undergo tensile stresses, at least its surface pressure can be zero. Figure 2 shows a plot of the stress field under internal boundary pressure of the lithosphere at $P = 0$ in (2.23). As one can see in the plot, the radial stress is considerably less than for the case in Figure 1, and the elastic energy for this case is $W = 1.78\times10^{30}$ J which is less than for the variant in Figure 1. For example, in Figure 3there are plots of stresses for $P = 5\times10^{9}$ Pa. The elastic energy corresponding for this case is $W = 1.82\times10^{30}$ J. Comparing the plots in Figure 2 and Figure 3 one can see that the stresses nearly at the whole depth of the lithosphere weakly depend on the value of P, whereas in the subsurface layer, which is several hundred kilometers thick, they coincide.

Redistribution of stresses from hydrostatic to the GEF can be illustrated by the following example. If a spring is fixed in a stress condition, then, in terms of the elasticity theory, it is in the equilibrium position. However, as is known from materials science, under stress the process of plastic relaxation is taking place in the spring decreasing its stress field energy. It is a well-known fact that springs under stress weaken with time. If one waits long enough, then the elastic deformation will completely disappear and the spring energy will become zero. The same process occurs in the lithosphere. At first the stress distribution can be arbitrary. But if the energy is not minimum, then in the lithosphere, just like in a loaded spring, the processes of plastic relaxation will start which decrease its elastic energy and in the long run, will reduce it to minimum—the GEF.

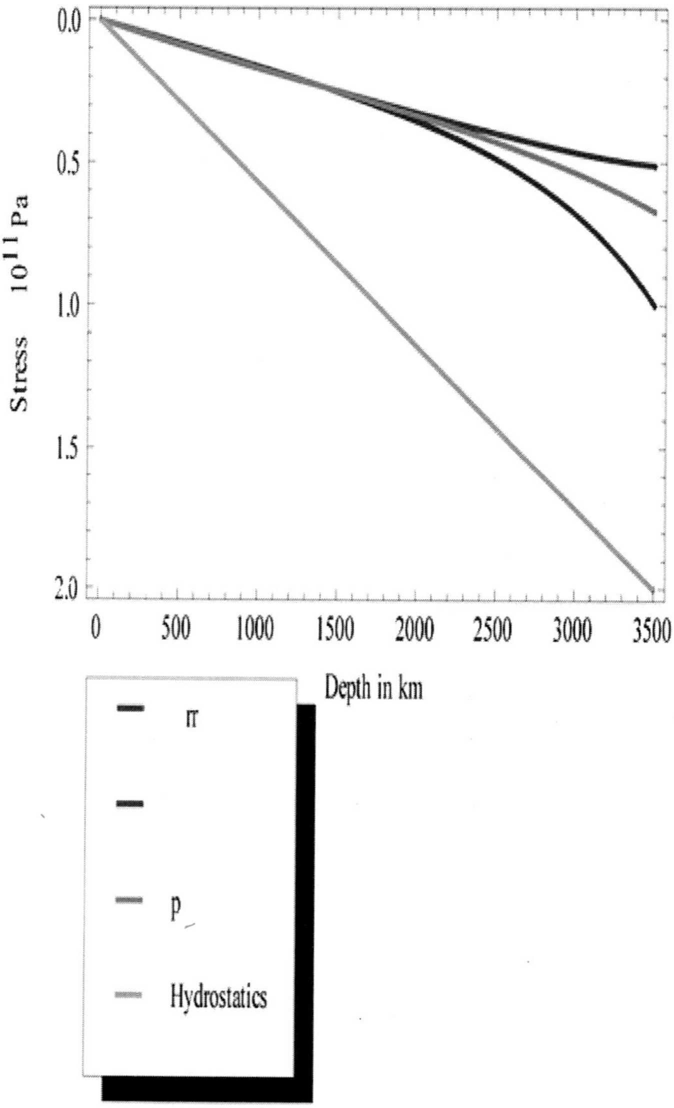

Figure 1. The stresses and their pressure are plotted when the value of pressure at the nucleus boundary is taken to be equal to 10^{11} Pa.

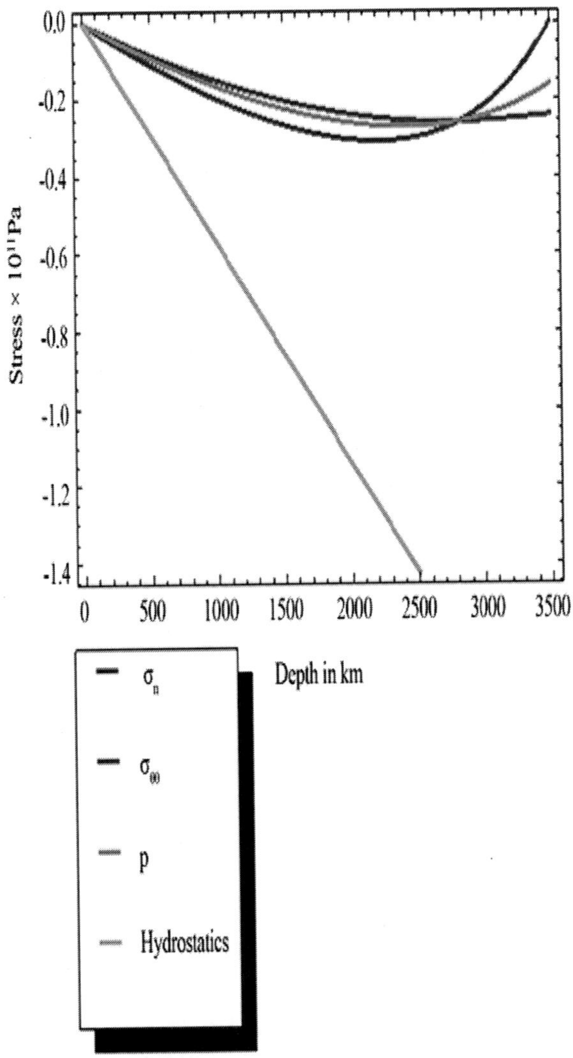

Figure 2. The stresses and their pressure are plotted when the value of pressure at the nucleus boundary is taken to be equal to 0 Pa.

Thus, the lithosphere proves to be a bearing system, which, similar to building structures, bears its load, its weight, without transferring it to the nucleus and can even ensure that the pressure on the nucleus surface is equal to zero.

3. SUBSURFACE REGION

By Equations (2.22), (2.24) and (2.33), one calculates the pressure $\tilde{p}$ in the crust at small depths depending on the lithosphere surface pressure. The results are shown in the Table 1. Both in the atmosphere and the ocean, a stressed state is purely hydrostatic unlike the crust, and therefore, in (2.31) p_0 denotes the pressure under the ocean bottom at various depths.

In the calculation in (2.33) the density value $\rho = 3000 \text{ kg/m}^3$ is the characteristic density value of the Earth crust. The Table 1 below shows the pressure values calculated according to the above formulae taking into account the pressure produced by the water in the ocean. The first column indicates the depth in kilometers starting from sea level. The second column shows the pressure in the crust at atmospheric pressure p_0 on a free surface. The third column shows the pressure in the ocean at a depth of 5 km and the pressure in the crust under the ocean depth. In this case, the water pressure at the ocean bottom will be $500p_0$. In the fourth column, it is the same only in the ocean at a depth of 10 km, where the pressure at the ocean bottom will be $1000p_0$.

Cells related to the ocean are highlighted in blue. Here one can clearly see the role of the solid character of the lithosphere. In spite of the fact that the density of solid rocks is three times higher than that of the water, at a depth of 5 km the pressure in the ocean is by 13% higher than that in the crust. At a depth of 10 km the discrepancy is even larger. This example shows the principal difference of the stressed state of the lithosphere from hydrostatics.

The existence of land is caused by this effect. If the lithosphere had a hydrostatic pressure distribution, then the pressure would be greater under the continents, since their density is three times as much as that of water, and in the course of time, they would sink below sea level. At a hydrostatic stress distribution, the uniform distribution of water on the planet surface is more favorable in terms of energy. The ability of the lithosphere to bear a load, i.e., its weight basically changes the situation.

It can be of interest to carry out energy research into the relation of land and ocean areas. It seems likely that this relation is also determined by minimum elastic energy. But this is a question of a separate investigation.

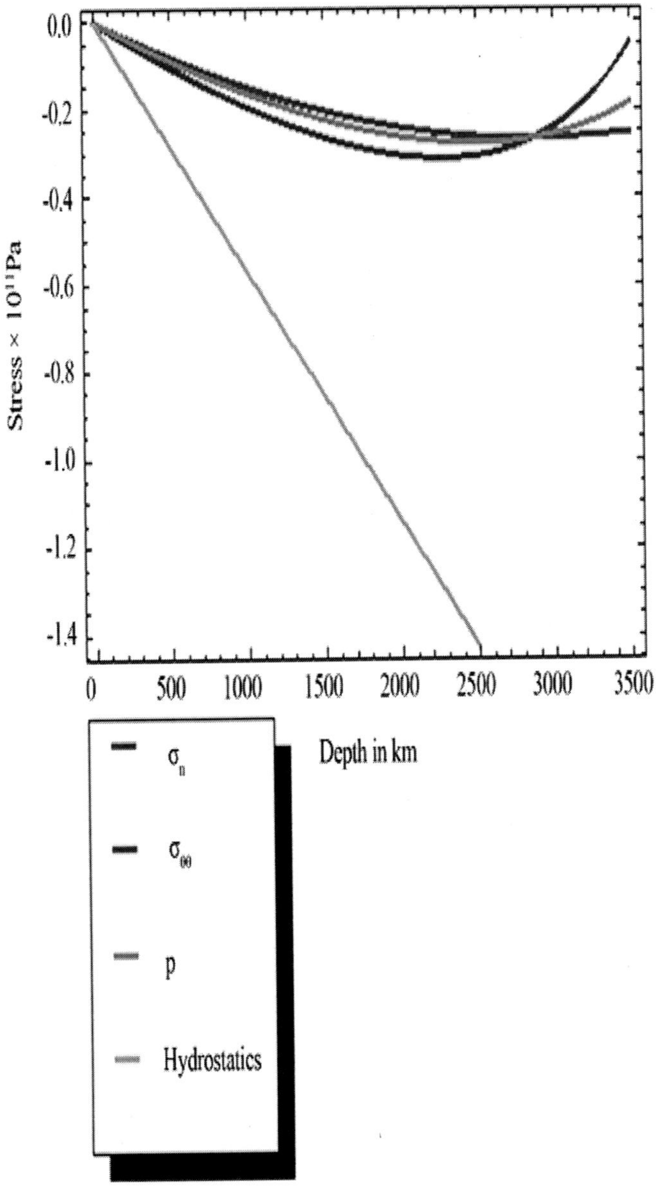

Figure 3. The stresses and their pressure are plotted when the value of pressure at the nucleus boundary is taken to be equal to 5×10^{11} Pa.

Table 1. The pressure on land and under ocean bottom.

Depth in km	Pressure in Pa · 10^5 on land	Pressure in Pa · 10^5 under ocean bottom at a depth of 5 km	Pressure in Pa · 10^5 under ocean bottom at a depth of 10 km
0	1	1	1
5	434	500	500
10	867	933	1000
15	1300	1360	1430
20	1730	1800	1870
25	2170	2230	2300

In modern geophysics when considering stresses in the lithosphere, they postulate them as hydrostatic; there are no shear stresses. The distribution of pressure in a liquid with the same density as in the lithosphere is taken as the equilibrium state [1] , which is considered unquestionable and is not even discussed. The solid properties in the lithosphere do not matter in this hypothesis. On the whole, when dealing with physical problems one must proceed from general physical laws. One of them is the principle of minimum energy in equilibrium. Hydrostatic stress distribution does not ensure the minimum. Minimum elastic energy occurs at the GEF. Even near the Earth surface, the pressure in the crust is greatly different from the hydrostatic pressure in the ocean. It is related to the fact that hydrostatic distribution has no shear stresses, and energy is the quadratic stress function. If shear stresses are introduced, one can reduce normal stresses and decrease energy. In this work the coordinate system is chosen in the main axes, in which only stress tensor components along the main diagonal are different from zero, but, unlike the hydrostatic case, they are not equal. In the nonhydrostatic case, in the coordinate system rotated relative to the main axes, there are nondiagonal components different from zero, whereas in the hydrostatic case, in any coordinate system nondiagonal components, there are components of zero and diagonal ones are equal.

Alexander Ivanchin The nonhydrostatic character of the pressure in the crust can be used for engineering purposes. For example, one can artificially decrease the nonhydrostatic character of the stresses in a local site of the crust, in particular, above an oil field, which will raise the pressure in an oil stratum and, as a result, increase its oil recovery. Other applications are also possible.

REFERENCES

1. Turcotte, D.L. and Schubert, D. (2002) Geadynamics. Cambridge University Press, Cambridge.
2. Landau, L.D. and Lifshits, E.M. (1986) Theory of Elasticity. 3rd Edition, Elsevier Butterworth-Heimenann, Oxford.
3. Ivanchin, A. (2010) Potential. Solution of Poisson's Equation, Equation of Continuity and Elasticity.
4. Elsgolz, L.Y. (1969) Differential Equations and Calculus of Variations. Nauka, Moscow.
5. Zaitsev, V.F. and Polyanin, A.D. (2003) Handbook of Exact Solutions for Ordinary Differential Equations. Chapman & Hall/CRC, Boca Raton.
6. Kikoin, I.K. (1976) Tablicy fizicheskih velichin (Tables of Physical Values). Atomizdat, Moscow. (in Russian)
7. Vikulin, A. (2009) Physics of the Earth and Geodynamics. V. Bering State University, Petropavlovsk-Kamchatskii.

CHAPTER 15

X-Ray Computed Tomography Analysis of Sajau Coal, Berau Basin, Indonesia: 3D Imaging of Cleat and Microcleat Characteristics

Ahmad Helman Hamdani

Faculty of Geology, University of Padjadjaran, Jalan Raya Bandung Sumedang Km 21, Jatinangor, Bandung 45363, Indonesia

ABSTRACT

The Pliocene Sajau coals of the Berau Basin area have a moderately to highly developed cleat system. Mostly the cleat fractures are well developed in both bright and dull bands, and these cleats are generally inclined or perpendicular to the bedding planes of the seam. The presence of cleat networks/fractures in coal seam is the important point in coalbed methane prospect. The 3D X-ray computed tomography (CT) technique was performed to identify cleats characteristics in the Sajau coal seams, such as the direction of coal cleats, geometry of cleat, and cleats mineralization. By CT scan imaging technique two different types of natural fractures observed in Sajau coals have been identified, that is, face cleats and butt cleats. This technique also identified the direction of face cleats and butt cleats as shown in the resulting 3D images. Based on the images, face cleats show a NNE-SSW direction while butt cleats have a NW-SE direction. The crosscutting relationship indicated that NNE-SSW cleats were formed earlier than NW-SE cleats. The procedure also identified the types of minerals that filled the cleats apertures. Based on their density, the minerals are categorized as follows: very high density minerals (pyrite), high density minerals (anastase), and low density minerals (kaolinite, calcite) were identified filling the cleats aperture.

1. INTRODUCTION

Coal is an important source of energy in many parts of the world. Coal is an important source and reservoir rock from which natural gas is formed. Coal is a sedimentary rock with complex polymeric material and dual porosity. The permeability and producibility of coalbed methane reservoir are influenced by distribution of pore and fractures/cleats.

The CBM accumulation, retention, and recovery involve mechanisms that are significantly different from those involved in the concentration of conventional gas reserves. Thus, the pore structure and cleats distribution are very important for gas flow pathway. There are three known gas flow mechanisms in coal [1, 2]: (1) desorption of gas from coal surface inside the microspores; (2) diffusion of gas through microspores as governed by Fick's law; and (3) Darcy flow through the cleat system, natural fracture network in the coal to the wellbore. However, it should be remembered that various types of minerals can fill cracks and when cracks are filled with minerals, CBM permeability and productivity will be reduced [3–5]. Cleats are one of the most important internal structures of coal. Cleats play an important role in CBM as they are responsible for determining cleats permeability networks, including cleat mineralization, morphology, and density, and how cleats are related to macroscale fracture permeability [6, 7].

Maceral type, rank, and mineral contents influence cleats intensity in coals [8–10]. By nature, coals break easily into small pieces from core and outcrop samples. Therefore, to understand the coal cleat pattern at millimeter scale in coal samples, we need an effective and quick investigation technique that will not destroy samples but can give maximum results. An investigative technique that is increasingly used today is the CT scan tomography [11, 12]. The ability to perform 3D imaging without destroying samples is one of the advantages of the CT scan method. However, the CT scan method has a limitation as it produces noise and discretization that affect the imaging artifact and beam hardening and may thus cause CT number value to be higher than it should be [13, 14].

Although in the beginning X-ray CT was mainly used in medicine to image bone structures as introduced by [15], it became soon clear that X-ray CT had a large potential for other applications. Since 1974, the CT scan was developed in many areas of geology, such as paleontology, sedimentology, petroleum geology, and coal geology [16–20]. Computed tomography (CT) is a recent, nondestructive state-of-the-art imaging technique which uses X-ray power to analyze the internal structure of different materials [21, 22]. It works on the basis of density contrast, taking the advantage of different levels of X-ray adsorption. Therefore, by this technique, different materials, interior structure of coals in coals can be identified in 3D by X-ray computed tomography (X-ray CT). In CBM reservoir evaluation, the combined use of X-ray CT techniques has helped to improve understanding about the pores, geometry, and direction of cleats and to examine the evolution of the 3D fracture network during stressing through failure in coals.

Berau Basin is one of the sedimentary basins in Indonesia that have huge coal reserves. Based on a study by [23] the basin has CBM reserves of about 8.4 Tcf. However, there are still insufficient studies of pores, fracture/cleats, and mineralization in coals. In addition, CT characteristics of coals of low rank coal have also yet to be reported. In this research, twelve coal core samples from the potential coal seam in Sajau CBM Prospect, Berau Basin, were investigated to describe cleats characteristics (including cleats direction and cleats mineralization). The aim of this study is to demonstrate capabilities of CT for nondestructive coal characterization, especially for cleat characteristics. The emphasis of this study is to show a new method for the quantification and 3D visualization of cleats orientation and mineralization.

2. EXPERIMENTAL METHODS

2.1. Coal Sample Characteristics

To achieve the aim of this research, 12 (twelve) fresh oriented coal blocks were directly collected from shallow well drilling in the Kasai, Mangkupadi, and Tanah Kuning coal mining areas in Sajau CBM Prospect, Berau Basin (Figure 1). All core samples were CT-scanned prior to coal quality and mercury injection capillary permeability (MICP) routine tests.

The coals selected for the research (Table 1) had a similar rank degree, that is, lignite. The samples also showed a very wide range of porosity, permeability, and mineralogy [24, 25]. The samples were dominated by huminite macerals range from 76.1 to 86.5 vol.% (average 83.81 vol.%), followed by liptinite macerals which have an average of 10.70 vol.% and inertinite macerals (average 5.49%). The samples' average huminite reflectance was 0.32%, which indicated that the coal was in an immature stage. The X-ray CT directly scanned all cores to determine the cleats network. The coal reservoir (including porosity and permeability) was measured by MICP and the results showed significant differences between these samples; the porosity Sajau coal was of 3.02–12.04% with the average 11.64% and standard deviation 2.86%, while the coal permeability ranged from 0.41 md to 128.24 md with an average of about 51.29 md and the standard deviation was 38.69 md. The significant differences in the permeability value indicate that the coals were heterogeneous.

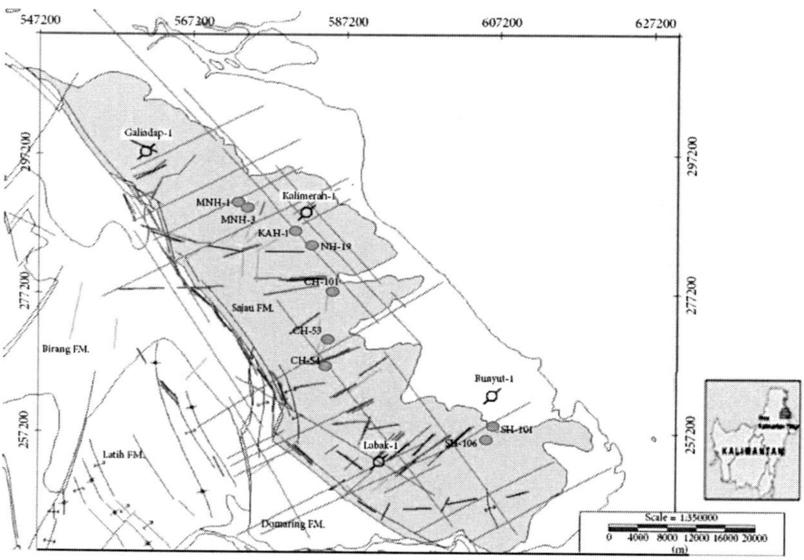

Figure 1: The geological map of Tanjungredeb sheet and samples location in Berau Basin.

Table 1: Sajau coal characteristics, density, porosity, and permeability.

Mining area	Coal maceral			Ro	ρ	Ø	K
Sample ID	H (%)	L (%)	I (%)	(%)	(%)	(%)	(md)
Tanah Kuning							
MNH-1A	84.6	12.8	2.6	0.34	1.27	8.20	57.24
MNH-3C	81.2	11.6	7.2	0.31	1.08	6.04	15.11
MNH-3D	76.1	11.8	12.1	0.32	1.06	4.10	14.02
MNH-3E	84.9	9.2	5.9	0.24	1.02	5.82	28.62
KAH-1F	78.6	12.8	8.6	0.26	1.20	6.92	1.06
NG-19G	84.1	9.8	6.1	0.26	1.12	5.08	0.41
NG-19I	88.2	8.6	3.2	0.26	1.10	5.18	1.23
Mangkupadi							
CH-54A	84.2	11.7	4.1	0.34	1.24	5.20	1.42
CH-53A	86.7	9.6	3.7	0.36	1.21	3.24	14.62
CH-101A	84.4	10.2	5.4	0.36	1.24	3.02	4.18
Kasai							
SH-101A	86.5	9.4	4.1	0.38	1.27	9.92	67.21
SH-106A	86.2	10.9	2.9	0.36	1.27	12.92	128.24

Note. ρ: coal density (g/cm^3), Ø: coal porosity by MICP (%), K: permeability by MICP (md), Ro: huminite reflectance (%), H: huminite (%), L: liptinite (%), and I: inertinite (%).

2.2. CT Imaging Scans

The X-ray CT is a nondestructive technique which detects the attenuation of X-rays passing through an object and is dependent on bulk density and effective atomic number. The X-ray attenuation coefficient data over the plane were stored in slice. The X-ray attenuation is mapped in Hounsfield units and converted to CT numbers, in each voxel of a CT image of the object, whereTheis the calculated X-ray attenuation coefficient andis the attenuation of water. In this study we used a water phantom for CT number calibration; the CT number is 0 for water (1 gr/cc) and −1000 for air [26, 27].

The CT scans were performed on a Brightspeed Type equipment manufactured by GE Corporation, USA. The X-ray source was a 225 kV Fein Focus focal spot, which allowed resolution to fall down to 10 μm for an object of 4.8 mm. The detector system was a Toshiba 3D image intensifier from which data are captured and digitized by a CCD digital camera with a spatial resolution. The CT slice thicknesses (0.3 mm) and slice increments (0.9 mm) were kept as thin as possible from slice to improve both resolution and contrast. The Mimics software was used for slice visualization of coal. The CT scan produced images of 512 × 512 pixels with a pixel resolution of 50 × 50 μm^2 of 4.4 lp/mm.

The cylindrical coal sample was placed perpendicular to the sample pads. Further scanning is in azimuth radiation throughout 0° and 90°. Then, a series of scans were performed several times (slice). At each sample per meter is research sample scanning with a space of 300 μm. The CT scan process produced a series of 2D images called "slices," which resulted from scanning with the azimuth angles 0° and 90°; this is due to the image that will be in line with what would be seen if the object is sliced along the plane of the scan. Each slice would record all X-ray attenuation coefficients along the plane of the scan. Data damping was converted to CT number, of which range was specified by the computer system. Generally, CT numbers were listed as values of gray on gray imaging. A series of CT calibrations were used to maximize the signal-to-noise ratio and minimize or eliminate imaging artifacts that might affect interpretation.

The slice set images from CT scans of cylindrical coal samples were made from three directions, , , and . The scans in cross section of rock core diameter were theplane, while scanning in the direction of the axis of the core rock is the result of scanning the -axis. Combined slices of CT scan on theandplanes were saved as a bitmap image and then converted to SegY format. To build the 3D coal image from X-ray CT scan, the CT image had to be converted to the 3D cleat aperture distribution [28, 29] developed. Furthermore, all slices scanned images were analyzed using the 3D visualization and modeling (Figure 2).

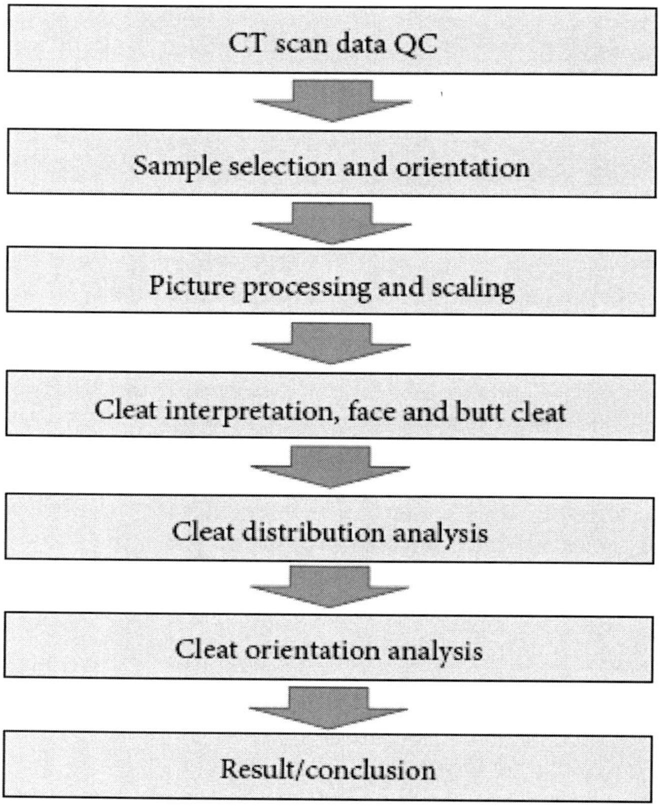

Figure 2: Workflow CT scan for 3D reconstruction.

3. RESULTS AND DISCUSSION

The CT scan of the oriented core rock samples performed on the scan axes,, andidentified two sets of cleats, namely, face and butt cleats. Figure 3 shows the results of the scan axes,, andof various rock samples. The face cleat is characterized by an elongated shape and firm, continuous line that has a regular orientation direction, unlike the butt cleat, which has a form of short lines, always abutting with face cleat. Some authors mention that the angle of intersection between the butt cleats and face cleats is 90° [30]. However, in this study the separation angle is within the range of 81°–87°.

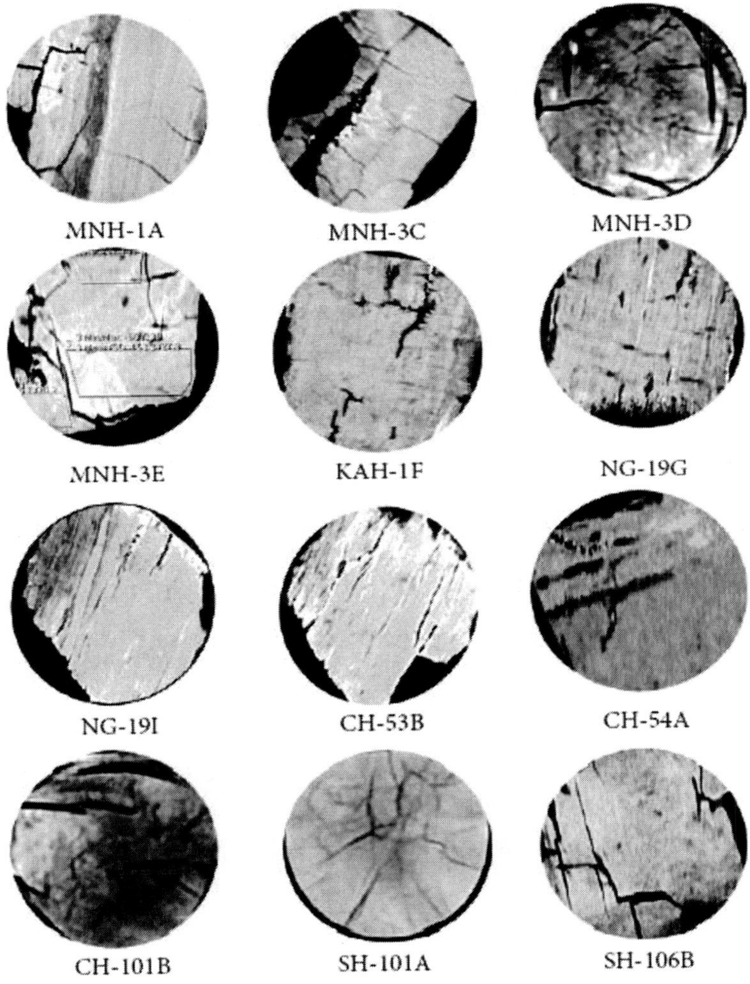

Figure 3: The result of CT scan of slice coal on 12 representative coal samples.

Based on the analysis of 3D cleat orientation work flow (Figure 2), the direction of the orientation of the face cleat and butt cleats could be determined. The various coal samples showed at least two known sets of cleat orientation: NNE-SSW and NW-SE. The face cleats showed a NNE-SSW direction while the butt cleat had a NW-SE direction. The direction orientation of cleats on core rock showed similarities to that which was measured on the outcrop sample. Measurement directions of cleats on various outcrops within the range of N 18°–264° E belonged to face cleats and those within the range of N 62°–337° E belonged to butt cleats. Figures 4and 5 show the scan results and interpretation of the direction of orientation of the cleats. From the same images it can also be seen that the face cleats in the direction NNE-SSW are abutting the cleats with NE-SW direction. This signifies that NNE-SSW cleats were formed first, followed by the cleats with the NW-SE direction. Then it can be stated that, chronologically, the formation of the cleats followed the NNE-SSW and NW-SE directions.

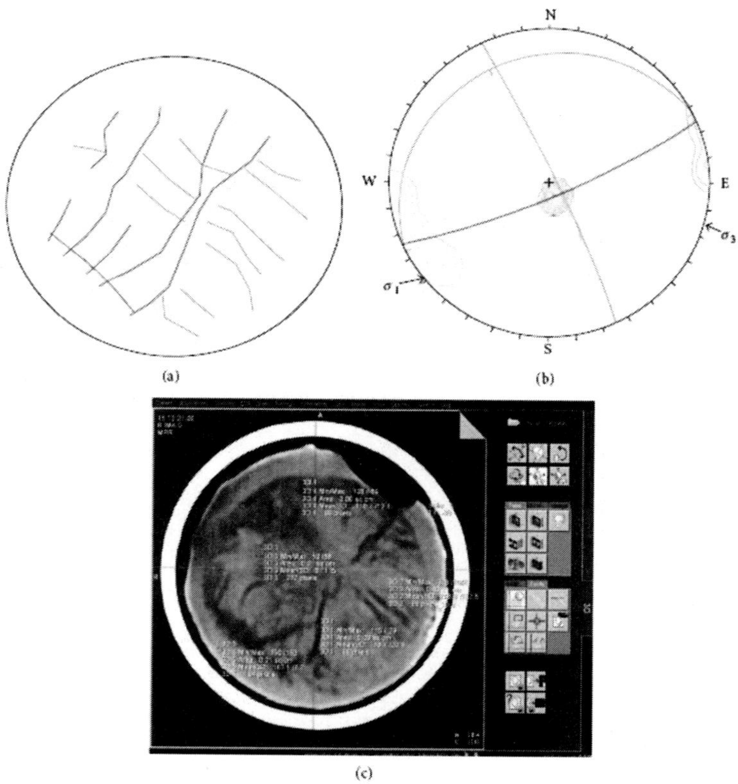

Figure 4: Slice image of SH-101A (-axis direction), illustrating the CT cleat image (c), cleat interpretation and orientation (a), and Schmidt lower hemisphere equal projection net of cleats orientation relationships (b), showing thathas NE-SW direction and its scale bar is 10 mm. Two cleat orientations have been identified, that is, NNE-SSW direction (red color) and NW-SE direction (blue color). The cleats with NNE-SSW direction are abutting in the NE-SW cleats.

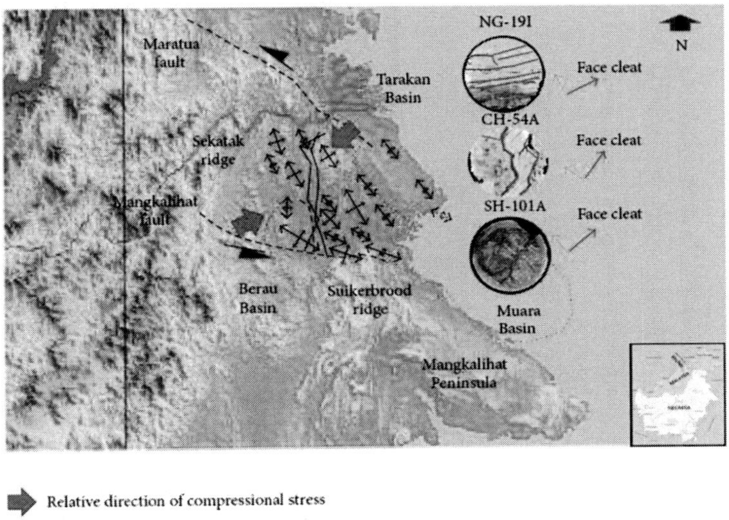

Figure 5: Structural geology (folds, faults, and cleats) arrangements in Berau Basin and CT scan image of representative coal samples.

Tectonically, the structural geology in Berau Basin is influenced by deep-seated NE-SW sinistral Mangkalihat and Maratua. These faults play an important role not only in the arrangement of NW-SE and NNW-SSE fold but also in the cleats direction. The main stressof cleats in Berau Basin is NW-SE (see Figure 4). Based on the reconstruction of the structural elements, the relative compression stress of structural geology (fold and fault) also has approximately NE-SW direction (Figure 5). Therefore, the compressional stressof fold, faults, and cleats has a similar direction. Thus, images produced from the CT scan of rock samples can be used to determine the direction (trend) of the main tectonics of regional patterns.

The 3D reconstruction from slices of core samples of MNH-1A also indicates that the most dominant directions of orientation of face cleats and butt cleats were NNE-SSW and NW-SE, respectively, as shown in Figure 6. The MNH-1A shows that two set of cleats are recognized: the face cleats and butt cleats. The angles of the face cleats were approximately in the range of 35°–45° on - plane, all of which were well developed and well connected. The butt cleats were rare and mostly found on the edge of the samples, and they developed on the - plane, with the angles ranging between 22° and 28°. All cleats had an aperture of approximately 0.2 mm and a cleat spacing of 0.2–0.82 mm.

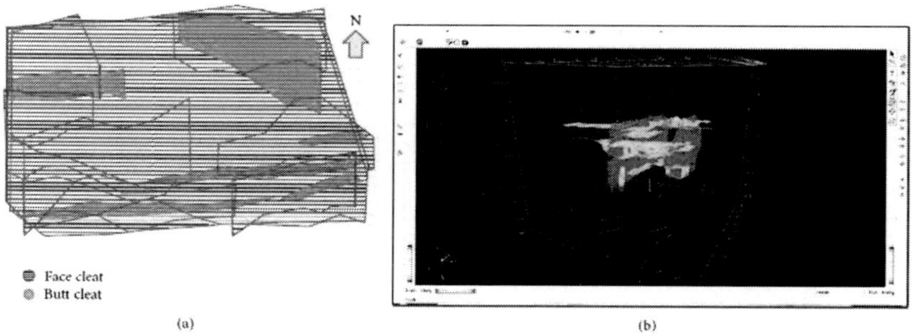

Figure 6: Illustrating 3D reconstruction of the cleats direction (a) from slice image of MNH-1A (-axis direction); 3D rendering CT scan of MNH-1A (b). Face cleats have NNE-SSW direction (red color and gradational color), and butt cleats have NW-SE direction (green color and solid color).

By using CT scan measurements, the study has been able to identify at least the following three types of minerals density that fills cleats aperture: very high, high, and low density minerals. Figure 6 shows the CT scan images of the very high density mineral with CT numbers between 450 and 658 HU, bright white color, probably pyrite; high density mineral images were shown in white color and have CT numbers 250–410 HU, most likely anastase, while low density minerals probably kaolinite and calcite are represented by the grey color, CT numbers being in the range of 180–230 HU. In addition to filling cleats aperture, high density minerals are also scattered in the coal matrix, that is, pyrite in coal matrix as seen in SH-106A (Figure 7). A mineral with a low density (kaolinite and calcite) was frequently detected, surrounding high density minerals (pyrite and anastase), such as rims along cleats. In the northern area (NG-19 and KAH-1) the majority of the cleats network was filled with minerals; these minerals could cause permeability level to be low (0.41–1.23 md). Meanwhile, in the southern part (SH-101A), only a few cleats were filled (= 128 md).

From the discussion above, the 3D reconstruction of the coal samples can provide cleat network imaging and cleats orientation and mineralization, which will give a different effect on gas permeability. This is very important in evaluating the potential of Sajau coal as a CBM reservoir. Furthermore, the CT scan technique will not damage the sample and does not require special preparation as other types of analysis do. This feature is an advantage when compared with 2D microscopic examination techniques.

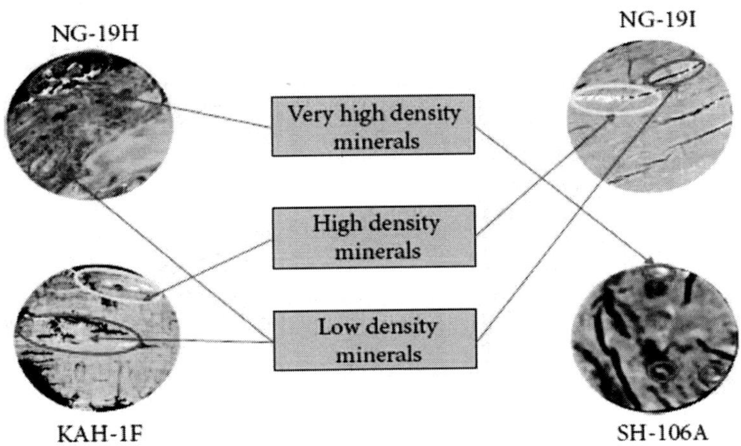

Figure 7: X-section slice images of cleat mineral and their infills as indicated by the X-ray CT technique of coal samples; very high density minerals (bright white, red oval), high density minerals (white, yellow oval), and low density minerals (grey, blue oval).

The 3D modeling from CT image also has some associated technical limitations such as image threshold, beam hardening, ring artifacts, partial-volume effects, and lower plane resolution CT. These limitations may induce the uncertainty of the results. Despite these weaknesses, the results obtained from the analysis are still effective for studies of coal cleats characteristics. For future practical applications it is necessary to develop an image processing method that connects the CT data processing between CT images with coal deformation. With such a method, the size of the cleats and deformation density can be calculated from the CT images.

4. CONCLUSIONS

A total of 12 coals of low rank coal of Sajau Formation in Berau Basin were scanned for 3600 CT slices. These slices were used to reconstruct the 3D imaging cleats orientation. By means of 3D X-ray tomography, face cleats and butt cleats, as well as their orientation and microcleat generation, were successfully determined based on crosscutting relationships.

Based on the CT scan technique used, the Sajau coals were categorized as a moderately to highly developed cleat system. The cleat fractures were parallel, inclined, or perpendicular to the bedding planes of the seam, and the open spaces were filled with mineral matter, such as kaolinite, calcite, anastase, and pyrite. Crosscutting relationship indicated by X-ray CT scan analysis clearly showed that the NNE-SSW cleats were formed earlier than NW-SE cleats.

CONFLICT OF INTERESTS

The author declares no conflict of interests.

ACKNOWLEDGMENTS

Mr. Kosasih and Mr. Alit from Laboratory of Coalbed Methane, Lemigas, Jakarta, Indonesia, are acknowledged for making X-ray CT available for this research. Mr. Wegi Oktariadi, from PT. Patriot Wiraperkasa, a coal mine company operating in Berau Regency, is also acknowledged for giving permission to use the borehole cores for CT scan analysis. The author would like to thank anonymous reviewers, who thoroughly reviewed this paper, and their critical comments and valuable suggestions were very helpful in preparing this paper.

REFERENCES

1. C. Ö. Karacan and E. Okandan, "Adsorption and gas transport in coal microstructure: Investigation and evaluation by quantitative X-ray CT imaging," Fuel, vol. 80, no. 4, pp. 509–520, 2001.
2. T. Ertekin, "Coalbed methane recovery modeling: what we know and what we need to learn," Turkish Journal of Oil and Gas, vol. 1, no. 1, pp. 7–18, 1995.
3. P. Gamson, B. Beamish, and D. Johnson, "Effect of coal microstructure and secondary mineralization on methane recovery," Geological Special Publication, vol. 199, pp. 165–179, 1998.
4. C. R. Ward, "Analysis and significance of mineral matter in coal seams," International Journal of Coal Geology, vol. 50, no. 1–4, pp. 135–168, 2002.
5. J. C. Close, "Natural fractures in coal," in Hydrocarbons from Coal, B. E. Law and D. D. Rice, Eds., Studies in Geology Number 38, pp. 119–132, American Association of Petroleum Geologists, Tulsa, Okla, USA, 1993.
6. K. E. Nick, M. W. Conway, and K. S. Fowler, "The relation of diagenetic clays and sulfates to the treatment of coalbed methane reservoirs," in Proceedings of the SPE Annual Technical Conference and Exhibition, Paper SPE 30736, p. 11, Society of Petroleum Engineers, Dallas, Tex, USA, October 1995.
7. C. I. Pattison, C. R. Fieling, R. H. McWatters, and L. H. Hamilton, "Nature and origin of fractures in Permian coals from the Bowen basin, Queensland, Australia," in Coalbed Methane and Coal Geology, R. Gayer and I. Harris, Eds., vol. 109 of Geological Society Special Publication, pp. 131–150, The Geological Society, London, UK, 1996.

8. P. F. Kendall and H. Briggs, "The formation of rock joints and the cleat of coal," Proceedings of the Royal Society of Edinburgh, vol. 53, pp. 164–187, 1933.
9. B. E. Law, "The relation between coal rank and cleat spacing: implications for the prediction of permeability in coal," in Proceedings of the International Coalbed Methane Symposium II, pp. 435–442, 1993.
10. E. Stach, M. T. H. Mackowsky, M. Teichmueller, G. H. Taylor, and D. R. Chandra, Coal Petrology, Gebruder Borntraeger, Stuttgart, Germany, 1982.
11. K.-H. A. A. Wolf, F. van Bergen, R. Ephraim, and H. Pagnier, "Determination of the cleat angle distribution of the RECOPOL coal seams, using CT-scans and image analysis on drilling cuttings and coal blocks," International Journal of Coal Geology, vol. 73, no. 3-4, pp. 259–272, 2008. r · ·
12. V. Cnudde and M. N. Boone, "High-resolution X-ray computed tomography in geosciences: a review of the current technology and applications," Earth-Science Reviews, vol. 123, pp. 1–17, 2013. r · ·
13. C. D. Montemagno and L. J. Pyrak-Nolte, "Fracture network versus single fractures: measurement of fracture geometry with X-ray tomography," Physics and Chemistry of the Earth Part A: Solid Earth and Geodesy, vol. 24, no. 7, pp. 575–579, 1999. r · ·
14. J. C. Akin and A. R. Kovscek, "Computerised tomography in petroleum engineering research," in Applications of Computerized X-ray Tomography in Geology and Related Domains, P. Jacobs, F. Mees, R. Swennen, and M. van Geet, Eds., vol. 215, pp. 23–28, Geological Society, London, UK, 2003.
15. G. N. Hounsfield, "Computerized transverse axial scanning (tomography). Part 1. Description of system," British Journal of Radiology, vol. 46, no. 552, pp. 1016–1022, 1973. r · ·
16. S. Fourie, "The cranial morphologhy of Thrinaxondon liohinus Seeley," Annals of the South African Museum, vol. 65, pp. 337–400, 1974.
17. G. C. Conroy and M. W. Vannier, "Noninvasive three-dimensional computer imaging of matrix-filled fossil skulls by high-resolution computed tomography," Science, vol. 226, no. 4673, pp. 456–458, 1984.r · ·
18. S. H. Anderson, R. L. Peyton, and C. J. Gantzer, "Evaluation of constructed and natural soil macropores using X-ray computed tomography," Geoderma, vol. 46, no. 1–3, pp. 13–29, 1990. r · ·
19. S. L. Wellington and H. J. Vinegar, "X-ray computerized tomography," Journal of Power Technologies, 1987.
20. Y. Zeng, C. J. Gantzer, R. L. Payton, and S. H. Anderson, "Fractal dimension and lacunarity of bulk density determined with X-ray computed tomography," Soil Science Society of America Journal, vol. 60, no. 6, pp. 1718–1724, 1996. r · ·
21. M. Van Geet, R. Swennen, and M. Wevers, "Towards 3D petrography: application of microfocus computer tomography in geological science," Computers and Geosciences, vol. 27, no. 9, pp. 1091–1099, 2001. r · ·
22. F. Mees, R. Swennen, M. van Geet, and P. Jacobs, Applications of X-Ray Computed Tomography in the Geosciences, vol. 215, Geological Society of London, 2003.
23. Advanced Resources International, "Indonesia CBM data package, advanced resources international," Inc. Migas and ADB Report, ARI, 2003,
24. A. H. Hamdani, E. Sunardi, and A. S. Yoga, "Petrographical characteristics and environmental characteristics and depositional environment of Sajau coal formation in Berau basin, east Kalimantan," in Proceedings of the 10th International Conference on Fluvial Sedimentology (ICFS 10 '13), p. 19, University of Leeds, Leeds, UK, July 2013.
25. A. H. Hamdani, E. Sunardi, and A. S. Yoga, "The coalbed methane potential from Sajau coal in eastern part of Berau basin, east Kalimantan," International Journal of Scientific Engineering and Research, vol. 3, no. 3, p. 4, 2014.
26. R. A. Ketcham and W. D. Carlson, "Acquisition, optimization and interpretation of x-ray computed tomographic imagery: applications to the geosciences," Computers & Geosciences, vol. 27, no. 4, pp. 381–400, 2001. r · ·

27. P. K. Hunt, P. Engler, and C. Bajscrowicz, "Computed tomography as a core analysis tool: applications and artifact reduction techniques," in Proceedings of the SPE Annual Technical Conference and Exhibition, SPE Paper 16952, Dallas, Tex, USA, 1987.
28. R. A. Ketcham, D. T. Slottke, and J. M. Sharp, "Three-dimensional measurement of fractures in heterogeneous materials using high-resolution X-ray computed tomography," Geosphere, vol. 6, no. 5, pp. 499–514, 2010.
29. L. A. Lubis, T. Harith, Z. Zahir, M. Noh, and K. Ariffin, "Workflow to reconstruct 3D pore space from 2D CT-scan image of Berea Sandstone," in Proceedings of the 2011 National Geoscience Conference, Johor Bahru, Malaysia, June 2011.
30. M. Widera, "What are cleats? Preliminary studies from the Konin lignite mine, Miocene of central Poland," Geologos, vol. 20, no. 1, pp. 3–12, 2014.

CHAPTER 16

On the Prime Geodesic Theorem for Non-Compact Riemann Surfaces

Muharem Avdispahić, Dženan Gušić

Department of Mathematics, Faculty of Sciences and Mathematics, University of Sarajevo, Sarajevo, Bosnia and Herzegovina

ABSTRACT

We use B. Randol's method to improve the error term in the prime geodesic theorem for a noncompact Riemann surface having at least one cusp. The case considered is a general one, corresponding to a Fuchsian group of the first kind and a multiplier system with a weight on it.

Keywords: Selberg Trace Formula, Selberg Zeta Function, Prime Geodesic Theorem

1. INTRODUCTION

The Selberg trace formula, introduced by A. Selberg in 1956, describes the spectrum of the hyperbolic Laplacian in terms of geometric data involving the lengths of geodesics on a Riemann surface. Motivated by analogy between this trace formula and the explicit formulas of number theory relating the zeroes of the Riemann zeta function to prime numbers, Selberg [1] introduced a zeta function whose analytic properties are encoded in the

Selberg trace formula. By focusing on the Selberg zeta function, H. Huber ([2] , p. 386; [3] , p. 464), proved an analogue of the prime number theorem for compact Riemann surfaces with the error term $O\left(x^{\frac{3}{4}}(\log x)^{-\frac{1}{2}}\right)$ that agrees with Selberg's one.

Using basically the same method as in [4] , D. Hejhal ([5] , p. 475), established also the prime geodesic theorem for non-compact Riemann surfaces with the remainder $O\left(x^{\frac{3}{4}}(\log x)^{-\frac{1}{2}}\right)$. However, in the compact case there exist several different proofs (see,

B. Randol [6] , p. 245; P. Buser [7] , p. 257, Th. 9.6.1; M. Avdispahić and L. Smajlović [8] , Th. 3.1) that give the remainder $O\left(x^{\frac{3}{4}}(\log x)^{-1}\right)$. Thanks to new integral repre-

sentations of the logarithmicerivative of the Selberg zeta function (cf. [9] , p. 185; [10] , p. 128), M. Avdispahić and L. Smajlović ([11] , p. 13) were in position to improve $O\left(x^{\frac{3}{4}}(\log x)^{-\frac{1}{2}}\right)$ error term in a non-compact, finite volume case up to $O\left(x^{\frac{3}{4}}(\log x)^{-1}\right)$.

Whereas the authors in [8] and [11] approached the prime number theorem in various settings via explicit formulas for the Jorgenson-Lang fundamental class of functions, our main goal is to obtain this improvement for non-compact Riemann surfaces with cusps following a more direct method of B. Randol [6] .

2. PRELIMINARIES

Let X be a non-compact Riemann surface regarded as a quotient $\Gamma \backslash \mathcal{H}$ of the upper half-plane $\mathcal{H}$ by a finitely-generated Fuchsian group $\Gamma \subseteq \mathrm{PSL}(2,\mathbb{R})$ of the first kind, containing $n_1 \geq 1$ cusps. Let $\mathfrak{I}$ denote the fundamental region of Γ. We shall assume that the fundamental region $\mathfrak{I}$ of Γ has a finite non-Euclidean area $|\mathfrak{I}|$. We put

$$\bar{\Gamma} = \left\{ \begin{pmatrix} a & b \\ c & d \end{pmatrix} \in SL(2,\mathbb{R}) : \frac{az+b}{cz+d} \in \Gamma \right\}$$

and denote by v the multiplier system of the weight $m \in \mathbb{R}$ for $\bar{\Gamma}$. Let ψ be an irreducible $r \times r$ unitary representation on Γ and $W(T) = \psi(T)v(T)$, $T \in \bar{\Gamma}$. For an r dimensional vector space V over $\mathbb{C}$ we consider an essentially self-adjoint operator

$$\Delta_m = y^2\left(\frac{\partial}{\partial x^2} + \frac{\partial}{\partial y^2}\right) - imy\frac{\partial}{\partial x}$$

on the space $\mathcal{D}_m$ of all twice continuously differentiable functions $f : \mathcal{H} \to V$, such that f and $\Delta_m(f)$ are square integrable on $\mathfrak{I}$, and satisfy the equality

$$f(Sz) = \frac{(cz+d)^m}{|cz+d|^m} W(S) f(z), \text{ for all } z \in \mathcal{H} \text{ and } \mathcal{S} = \begin{pmatrix} a & b \\ c & d \end{pmatrix} \in \bar{\Gamma}.$$

The operator $-\Delta_m$ has the unique self-adjoint extension $-\tilde{\Delta}_m$ to the space $\tilde{\mathcal{D}}_m$, a dense subspace of $L^2(\Gamma \backslash \mathcal{H})$. Let T_j, $j = 1, \cdots, n_1$ be the set of parabolic transformations corresponding to n_1 cusps of Γ. $W(T_j)$ does not depend on the choice of a representative of the parabolic class $\{T_j\}$ and can be considered as a matrix from $\mathbb{C}^{r \times r}$. By m_j we will denote the multiplicity of 1 as an eigen-value of the matrix

$W(T_j)$, and $n_1^* = \sum_{j=1}^{n_1} m_j$ will be the degree of singularity of W. We mention that operator $-\tilde{\Delta}_m$ has both the discrete and continuous spectrum in the case $n_1^* \geq 1$, and only the discrete spectrum in the case $n_1^* = 0$. The discrete spectrum will be denoted as $\{\lambda_n\}_{n>0}$ ($0 = \lambda_0 < \lambda_1 < \cdots < \lambda_n \to \infty$). The continuous spectrum is expressed through zeros (or equivalently poles) of the hyperbolic scattering determinant (see, [12]).

3. SELBERG ZETA FUNCTION

Let denotes the set of Γ-conjugacy classes of a primitive hyperbolic element P_0 in Γ, and Γ_h denotes the set of Γ-conjugacy classes of a hyperbolic element P in Γ that satisfy property $\mathrm{Tr}(P) > 2$. Assume that $|m| \leq 1$. We define the Selberg zeta function associated to the pair (Γ, W) by

$$Z_{\Gamma,W}(s) = \prod_{P_0 \in P\Gamma_h} \prod_{k=0}^{\infty} \det\left(I_r - W(P_0) N(P_0)^{-s-k}\right).$$

$Z_{\Gamma,W}(s)$ is absolutely convergent for $\mathrm{Re}(s) > 1$. Analytic considerations given in ([5] , pp. 499-501) yield that the Selberg zeta function in this setting satisfies the functional equation

$$Z_{\Gamma,W}(s)\Psi(s) = Z_{\Gamma,W}(1-s)$$

with the fudge factor

$$\Psi(s) = \phi(s) \cdot \eta\left(\frac{1}{2}\right)\left(\int_{\frac{1}{2}}^{s} \frac{\eta'}{\eta}(u)\,\mathrm{d}u\right). \tag{1}$$

Here, ϕ denotes the hyperbolic scattering determinant. It can be represented in the form

$$\phi(s)=\left(\frac{\sqrt{\pi}\Gamma(s)\Gamma\left(s-\frac{1}{2}\right)}{\Gamma\left(s+\frac{m}{2}\right)\Gamma\left(s-\frac{m}{2}\right)}\right)^{n_1^*}\sum_{n=1}^{\infty}\frac{a_n}{g_n^{2s}},$$

where the coefficients a_n and g_n depend on the group Γ (see, [5] , p. 437). Here, n_1^* denotes the degree of singularity of W (see Section 2). An explicit expression for the fudge factor η in the Equation (1) is given in ([5] , p. 501, Equation (5.10)).

The logarithmic derivative of the Selberg zeta function $Z_{\Gamma,W}(s)$ is given by

$$\frac{Z'_{\Gamma,W}(s)}{Z_{\Gamma,W}(s)}=\sum_{P\in\Gamma_h}\frac{\Lambda(P)}{N(P)^s}\operatorname{Tr}(W(P)),$$

where $N(P)$ denotes the norm of the class P and $\Lambda(P)=\dfrac{\log N(P_0)}{1-N(P)^{-1}}$ for a primi-

tive element P_0 such that $P=P_0^n$ for some $n\in\mathbb{N}$. We will omit the indices in $Z_{\Gamma,W}$ in the sequel.

4. COUNTING FUNCTIONS $\psi_n(x,W)$

Lemma 1. For $\operatorname{Re}(s)>1$,

$$\frac{Z'(s)}{Z(s)} = \sum_{P\in\Gamma_h} \Lambda_1(P)\operatorname{Tr}(W(P))N(P)^{-s} + \frac{Z'(s+1)}{Z(s+1)},$$

where $\Lambda_1(P) = \log N(P_0)$ for a primitive element P_0 such that $P = P_0^n$ for some $n \in \mathbb{N}$.

Proof.

$$\frac{Z'(s)}{Z(s)} = \sum_{P\in\Gamma_h} \Lambda_1(P)\operatorname{Tr}(W(P))\left(1-N(P)^{-1}\right)^{-1} N(P)^{-s}$$

$$= \sum_{P\in\Gamma_h} \Lambda_1(P)\operatorname{Tr}(W(P))N(P)^{-s} + \sum_{P\in\Gamma_h} \Lambda_1(P)\operatorname{Tr}(W(P))\left(N(P)-1\right)^{-1} N(P)^{-s}$$

$$= \sum_{P\in\Gamma_h} \Lambda_1(P)\operatorname{Tr}(W(P))N(P)^{-s} + \sum_{P\in\Gamma_h} \Lambda_1(P)\operatorname{Tr}(W(P))\left(1-N(P)^{-1}\right)^{-1} N(P)^{-(s+1)}.$$

We shall spend the rest of this section to derive a representation of $\psi_2(x,W)$ in the form (11) bellow. We choose not to write it in a separate statement because of the length of expressions involved. However, it will serve as a base for the proof of the prime geodesic theorem in Section 5.

Let us recall the following theorem given in ([13] , p. 51, Th. 40).

Theorem 1. If the Dirichlet's series $f(s) = \Sigma a_n e^{-\lambda_n} = \Sigma a_n l_n^{-s}$ is summable (l,k) for $s = \beta$ and $c > 0$, $c > \beta$, then

$$\omega^{-k} \sum_{l_n<\omega} a_n(\omega - l_n)^k = \frac{1}{2\pi i}\int_{c-i\infty}^{c+i\infty} f(s)\frac{\Gamma(k+1)\Gamma(s)}{\Gamma(k+1+s)}\omega^s \mathrm{d}s. \tag{2}$$

By Lemma 1,

$$\frac{Z'(s)}{Z(s)} - \frac{Z'(s+1)}{Z(s+1)} = \sum_{P\in\Gamma_h} \Lambda_1(P)\operatorname{Tr}(W(P))N(P)^{-s}.$$

We have,

$$\frac{1}{2\pi i}\int_{c-i\infty}^{c+i\infty}\left(\frac{Z'(s)}{Z(s)}-\frac{Z'(s+1)}{Z(s+1)}\right)s^{-1}(s+1)^{-1}\cdots(s+k)^{-1}x^{s}\mathrm{d}s$$

$$=\frac{1}{k!}\frac{1}{2\pi i}\int_{c-i\infty}^{c+i\infty}\left(\sum_{P\in\Gamma_h}\Lambda_1(P)\mathrm{Tr}(W(P))\left(\frac{N(P)}{x}\right)^{-s}\right)\frac{\Gamma(k+1)\Gamma(s)}{\Gamma(k+1+s)}\mathrm{d}s.$$

Therefore, substituting ω = 1, $f(s)=\sum_{P\in\Gamma_h}\Lambda_1(P)\mathrm{Tr}(W(P))\left(\frac{N(P)}{x}\right)^{-s}$, and hence $a_n=\Lambda_1(P)\mathrm{Tr}(W(P))$, $l_n=\frac{N(P)}{x}$ in (2), we get

$$\frac{1}{2\pi i}\int_{c-i\infty}^{c+i\infty}\left(\frac{Z'(s)}{Z(s)}-\frac{Z'(s+1)}{Z(s+1)}\right)s^{-1}(s+1)^{-1}\cdots(s+k)^{-1}x^{s}\mathrm{d}s$$

$$=\frac{1}{k!}\sum_{\frac{N(P)}{x}\le 1}\Lambda_1(P)\mathrm{Tr}(W(P))\left(1-\frac{N(P)}{x}\right)^{-s}$$

$$=\frac{1}{k!}\sum_{N(P)\le x}\Lambda_1(P)\mathrm{Tr}(W(P))\left(1-\frac{N(P)}{x}\right)^{-s}.$$

Then,

$$\frac{1}{2\pi i}\int_{c-i\infty}^{c+i\infty}\frac{Z'(s)}{Z(s)}s^{-1}(s+1)^{-1}\cdots(s+k)^{-1}x^{s}\mathrm{d}s$$

$$=\frac{1}{k!}\sum_{N(P)\le x}\Lambda_1(P)\mathrm{Tr}(W(P))\left(1-\frac{N(P)}{x}\right)^{k}$$

$$+\frac{1}{2\pi i}\int_{c-i\infty}^{c+i\infty}\frac{Z'(s+1)}{Z(s+1)}s^{-1}(s+1)^{-1}\cdots(s+k)^{-1}x^{s}\mathrm{d}s. \quad (3)$$

Now, put

$$\psi_0(x,W)=\sum_{N(P)\le x}\Lambda_1(P)\mathrm{Tr}(W(P))$$

and

$$\psi_j(x,W) = \int_0^x \psi_{j-1}(t,W)\,dt$$

for $j = 1,2,\cdots$. Using ([14] , p. 12, Th. 1.3.5), it is easy to get that

$$\psi_j(x,W) = \frac{1}{j!} \sum_{N(P)\le x} \Lambda_1(P)\operatorname{Tr}(W(P))(x - N(P))^j . \tag{4}$$

For $0 < \lambda_n < \frac{1}{4}$, let $s_n = \frac{1}{2} - ir_n = \frac{1}{2} - i\sqrt{\lambda_n - \frac{1}{4}}$, $n = 1,2,\cdots,K$, be the zeros of $Z(s)$ in $\left(\frac{1}{2},1\right)$. Let ρ_k, $k = 0,1,\cdots,M_e$ denote all zeros of the hyperbolic scattering determinant in $\left(\frac{1}{2},1\right]$.

Assume $T > 2$, $T \pm l \ne r_n$, $l \in \mathbb{N}$, where $s_n = \frac{1}{2} + ir_n$ and $\tilde{s}_n = \frac{1}{2} - ir_r$ for $r_n = -i\sqrt{\frac{1}{4} - \lambda_n}$, $\lambda_n > \frac{1}{4}$. Following ([5] , p. 468), we may also assume $T \pm l \ne \gamma$, $l \in \mathbb{N}$, where, $1-\rho$, $1-\bar{\rho}$ are the zeros of the Selberg zeta function $Z(s)$ for each zero $\rho = \frac{1}{2} + \eta + i\gamma$, $\eta \ge 0$, $\gamma > 0$, of the hyperbolic scattering determinant f. Let $A_0 \ge \frac{3}{2}$ be a large constant such that $A_0 \notin \mathbb{Z}$, $A_0 + \frac{1}{2} \notin \mathbb{Z}$, $A_0 \pm \frac{m}{2} \notin \mathbb{Z}$. We put $A = A_0 + 1$.

Without loss of generality we may assume that $c = 1 + \varepsilon$, $\varepsilon > 0$, $x \ge 2$. Let $R(A,T) = [-A,c] \times [-T,T]$. By the Cauchy residue theorem one has

$$\frac{1}{2\pi i}\int_{c-iT}^{c+iT}\frac{Z'(s)}{Z(s)}s^{-1}(s+1)^{-1}\cdots(s+k)^{-1}x^{s}\mathrm{d}s$$

$$=\frac{1}{2\pi i}\left(\int_{-A+iT}^{-A_0+iT}+\int_{-A_0+iT}^{\frac{1}{2}-\varepsilon+iT}+\int_{\frac{1}{2}-\varepsilon+iT}^{\frac{1}{2}+\varepsilon+iT}+\int_{\frac{1}{2}+\varepsilon+iT}^{c+iT}+\int_{c-iT}^{\frac{1}{2}+\varepsilon-iT}+\int_{\frac{1}{2}+\varepsilon-iT}^{\frac{1}{2}-\varepsilon-iT}+\int_{\frac{1}{2}-\varepsilon-iT}^{-A_0-iT}+\int_{-A_0-iT}^{-A-iT}\right)$$

$$+\frac{1}{2\pi i}\int_{-A-iT}^{-A+iT}+\sum_{z\in R(A,T)}\mathrm{Res}_{s=z}\left(\frac{Z'(s)}{Z(s)}s^{-1}(s+1)^{-1}\cdots(s+k)^{-1}x^{s}\right) \quad (5)$$

and

$$\frac{1}{2\pi i}\int_{c-iT}^{c+iT}\frac{Z'(s+1)}{Z(s+1)}s^{-1}(s+1)^{-1}\cdots(s+k)^{-1}x^{s}\mathrm{d}s$$

$$=\frac{1}{2\pi i}\left(\int_{-A+iT}^{-A_0+iT}+\int_{-A_0+iT}^{\frac{1}{2}-\varepsilon+iT}+\int_{\frac{1}{2}-\varepsilon+iT}^{\frac{1}{2}+\varepsilon+iT}+\int_{\frac{1}{2}+\varepsilon+iT}^{c+iT}+\int_{c-iT}^{\frac{1}{2}+\varepsilon-iT}+\int_{\frac{1}{2}+\varepsilon-iT}^{\frac{1}{2}-\varepsilon-iT}+\int_{\frac{1}{2}-\varepsilon-iT}^{-A_0-iT}+\int_{-A_0-iT}^{-A-iT}\right)$$

$$+\frac{1}{2\pi i}\int_{-A-iT}^{-A+iT}+\sum_{z\in R(A,T)}\mathrm{Res}_{s=z}\left(\frac{Z'(s+1)}{Z(s+1)}s^{-1}(s+1)^{-1}\cdots(s+k)^{-1}x^{s}\right). \quad (6)$$

Arguing as in [5] (p. 474) and [4] (pp. 105-108), we easily find that the sum of the first eight integrals on the right hand side of (5) is $O\left(\frac{x^{1+\varepsilon}}{\varepsilon T^{k}}\right)$.

Similarly, taking into account that $\frac{Z'(s)}{Z(s)}$ is bounded for $\mathrm{Re}(s)>1$, we obtain that the sum of the first eight integrals on the right hand side of (6) is $O\left(\frac{x^{1+\varepsilon}}{\varepsilon T^{k}}\right)$. Following [5] (p. 474) and [4] (p. 85, Prop. 5.7), we obtain that the ninth resp. the third integral on the right hand side of (5) resp. (6) are $O\left(x^{-A}\right)$. Now, if we take $k=2$, (5) and (6) will give us

$$\frac{1}{2\pi i}\int_{c-iT}^{c+iT}\frac{Z'(s)}{Z(s)}\frac{x^s}{s(s+1)(s+2)}ds$$

$$=O\left(x^{-A}\right)+O\left(\frac{x^{1+\varepsilon}}{\varepsilon T^2}\right)+\sum_{z\in R(A,T)}\mathrm{Res}_{s=z}\left(\frac{Z'(s)}{Z(s)}\frac{x^s}{s(s+1)(s+2)}\right) \tag{7}$$

and

$$\frac{1}{2\pi i}\int_{c-iT}^{c+iT}\frac{Z'(s+1)}{Z(s+1)}\frac{x^s}{s(s+1)(s+2)}ds$$

$$=O\left(x^{-A}\right)+O\left(\frac{x^{1+\varepsilon}}{\varepsilon T^2}\right)+\sum_{z\in R(A,T)}\mathrm{Res}_{s=z}\left(\frac{Z'(s+1)}{Z(s+1)}\frac{x^s}{s(s+1)(s+2)}\right). \tag{8}$$

Bearing in mind location of the poles of $\frac{Z'(s)}{Z(s)}$ given in ([5] , p. 439, Th. 2.16; or [5] , p. 498, Th. 5.3) and the fact that $|m|\leq 1$, we may assume without loss of generality that

$$-1-\frac{m}{2},-1+\frac{m}{2}\in\left(-\frac{5}{4},-\frac{1}{2}\right],-3-\frac{m}{2},-3+\frac{m}{2}\notin\left[-A,-\frac{5}{2}\right].$$

Calculating residues and passing to the limit $T\to+\infty, A\to+\infty$ in (7) and (8) we get

$$
\begin{aligned}
&\frac{1}{2\pi i}\int_{c-i\infty}^{c+i\infty}\frac{Z'(s)}{Z(s)}\frac{x^{s}}{s(s+1)(s+2)}\mathrm{d}s=\frac{1}{6}x+\Bigg(\sum_{n=1}^{K}\frac{x^{s_n}}{s_n(s_n+1)(s_n+2)}+\sum_{n=1}^{K}\frac{x^{\tilde{s}_n}}{\tilde{s}_n(\tilde{s}_n+1)(\tilde{s}_n+2)}+\sum_{k=0}^{M_e}\frac{x^{1-\rho_k}}{(1-\rho_k)(2-\rho_k)(3-\rho_k)}\\
&+A_0\frac{x^{-\frac{m}{2}}}{-\frac{m}{2}\left(-\frac{m}{2}+1\right)\left(-\frac{m}{2}+2\right)}+A_1\frac{x^{-1-\frac{m}{2}}}{\left(-1-\frac{m}{2}\right)\left(-\frac{m}{2}\right)\left(-\frac{m}{2}+1\right)}+B_0\frac{x^{\frac{m}{2}}}{\frac{m}{2}\left(\frac{m}{2}+1\right)\left(\frac{m}{2}+2\right)}\\
&+B_1\frac{x^{-1+\frac{m}{2}}}{\left(-1+\frac{m}{2}\right)\left(\frac{m}{2}\right)\left(\frac{m}{2}+1\right)}+\frac{8}{15}\mathrm{Tr}\left(I_r-\Phi\left(\frac{1}{2}\right)\right)x^{\frac{1}{2}}-\frac{8}{3}n_1^{*}x^{-\frac{1}{2}}-g_1n_1^{*}x^{-1}-n_1^{*}x^{-1}\log x\\
&+\left(-\frac{3}{2}+h_1\right)\frac{n_1^{*}}{2}+\frac{n_1^{*}}{2}\log x\Bigg)+\Bigg(\sum_{r_n>0}\frac{x^{s_n}}{s_n(s_n+1)(s_n+2)}+\sum_{r_n>0}\frac{x^{\tilde{s}_n}}{\tilde{s}_n(\tilde{s}_n+1)(\tilde{s}_n+2)}\\
&+\sum_{\gamma>0}\frac{x^{\frac{1}{2}-\eta-i\gamma}}{\left(\frac{1}{2}-\eta-i\gamma\right)\left(\frac{3}{2}-\eta-i\gamma\right)\left(\frac{5}{2}-\eta-i\gamma\right)}+\sum_{\gamma>0}\frac{x^{\frac{1}{2}-\eta+i\gamma}}{\left(\frac{1}{2}-\eta+i\gamma\right)\left(\frac{3}{2}-\eta+i\gamma\right)\left(\frac{5}{2}-\eta+i\gamma\right)}\Bigg)\\
&+\Bigg(A_2\frac{x^{-2-\frac{m}{2}}}{\left(-2-\frac{m}{2}\right)\left(-1-\frac{m}{2}\right)\left(-\frac{m}{2}\right)}+B_2\frac{x^{-2+\frac{m}{2}}}{\left(-2+\frac{m}{2}\right)\left(-1+\frac{m}{2}\right)\frac{m}{2}}+\frac{8}{3}n_1^{*}x^{-\frac{3}{2}}-\frac{8}{15}n_1^{*}x^{-\frac{5}{2}}+\left(\frac{3}{2}+f_1\right)\frac{n_1^{*}}{2}x^{-2}+\frac{n_1^{*}}{2}x^{-2}\log x\Bigg)
\end{aligned}
\tag{9}
$$

and

$$
\begin{aligned}
&\frac{1}{2\pi i}\int_{c-i\infty}^{c+i\infty}\frac{Z'(s+1)}{Z(s+1)}\frac{x^{s}}{s(s+1)(s+2)}\mathrm{d}s=\Bigg(\sum_{n=1}^{K}\frac{x^{s_n-1}}{(s_n-1)s_n(s_n+1)}+\sum_{n=1}^{K}\frac{x^{\tilde{s}_n-1}}{(\tilde{s}_n-1)\tilde{s}_n(\tilde{s}_n+1)}+A_0\frac{x^{-\frac{m}{2}-1}}{\left(-\frac{m}{2}-1\right)\left(-\frac{m}{2}\right)\left(-\frac{m}{2}+1\right)}\\
&+B_0\frac{x^{-1+\frac{m}{2}}}{\left(-1+\frac{m}{2}\right)\frac{m}{2}\left(\frac{m}{2}+1\right)}-\frac{8}{3}\mathrm{Tr}\left(I_r-\Phi\left(\frac{1}{2}\right)\right)x^{-\frac{1}{2}}+\sum_{k=0}^{M_e}\frac{x^{-\rho_k}}{(-\rho_k)(-\rho_k+1)(-\rho_k+2)}-g_1'n_1^{*}x^{-1}-n_1^{*}x^{-1}\log x\\
&+\left(-\frac{3}{2}+h_1'\right)\frac{1}{2}+\frac{1}{2}\log x\Bigg)+\Bigg(\sum_{r_n>0}\frac{x^{s_n-1}}{(s_n-1)s_n(s_n+1)}+\sum_{r_n>0}\frac{x^{\tilde{s}_n-1}}{(\tilde{s}_n-1)\tilde{s}_n(\tilde{s}_n+1)}+\sum_{\gamma>0}\frac{x^{-\frac{1}{2}-\eta-i\gamma}}{\left(-\frac{1}{2}-\eta-i\gamma\right)\left(\frac{1}{2}-\eta-i\gamma\right)\left(\frac{3}{2}-\eta-i\gamma\right)}\\
&+\sum_{\gamma>0}\frac{x^{-\frac{1}{2}-\eta+i\gamma}}{\left(-\frac{1}{2}-\eta+i\gamma\right)\left(\frac{1}{2}-\eta+i\gamma\right)\left(\frac{3}{2}-\eta+i\gamma\right)}\Bigg)+\Bigg(A_1\frac{x^{-2-\frac{m}{2}}}{\left(-2-\frac{m}{2}\right)\left(-1-\frac{m}{2}\right)\left(-\frac{m}{2}\right)}\\
&+B_1\frac{x^{-2+\frac{m}{2}}}{\left(-2+\frac{m}{2}\right)\left(-1+\frac{m}{2}\right)\frac{m}{2}}+\left(\frac{3}{2}+f_1'\right)\frac{n_1^{*}}{2}x^{-2}+\frac{n_1^{*}}{2}x^{-2}\log x+\frac{8}{3}n_1^{*}x^{-\frac{3}{2}}-\frac{8}{15}n_1^{*}x^{-\frac{5}{2}}\Bigg).
\end{aligned}
\tag{10}
$$

The implied constants on the right sides of (9) and (10) depend solely on Γ, m and W. With $k=2$ in (3), $j=2$ in (4), Equations (4), (3), (9) and (10) yield

$$
\begin{aligned}
\psi_2(x,W) &= \frac{1}{2!}\sum_{N(P)\le x}\Lambda_1(P)\,\mathrm{Tr}(W(P))(x-N(P))^2 \\
&= \frac{1}{2\pi i}\left(\int_{c-i\infty}^{c+i\infty}\frac{Z'(s)}{Z(s)}\frac{x^{s+2}}{s(s+1)(s+2)}\mathrm{d}s-\int_{c-i\infty}^{c+i\infty}\frac{Z'(s+1)}{Z(s+1)}\frac{x^{s+2}}{s(s+1)(s+2)}\mathrm{d}s\right) \\
&= \frac{1}{6}x^3+\left(\sum_{n=1}^{K}\frac{x^{s_n+2}}{s_n(s_n+1)(s_n+2)}+\sum_{n=1}^{K}\frac{x^{\tilde{s}_n+2}}{\tilde{s}_n(\tilde{s}_n+1)(\tilde{s}_n+2)}+\sum_{k=0}^{M_g}\frac{x^{3-\rho_k}}{(1-\rho_k)(2-\rho_k)(3-\rho_k)}\right. \\
&\quad +A_0\frac{x^{-\frac{m}{2}+2}}{-\frac{m}{2}\left(-\frac{m}{2}+1\right)\left(-\frac{m}{2}+2\right)}+A_1\frac{x^{1-\frac{m}{2}}}{\left(-1-\frac{m}{2}\right)\left(-\frac{m}{2}\right)\left(-\frac{m}{2}+1\right)}+B_0\frac{x^{\frac{m}{2}+2}}{\frac{m}{2}\left(\frac{m}{2}+1\right)\left(\frac{m}{2}+2\right)} \\
&\quad +B_1\frac{x^{1+\frac{m}{2}}}{\left(-1+\frac{m}{2}\right)\left(\frac{m}{2}\right)\left(\frac{m}{2}+1\right)}+\frac{8}{15}\mathrm{Tr}\left(I_r-\Phi\left(\frac{1}{2}\right)\right)x^{\frac{5}{2}}-\frac{8}{3}n_1^*x^{\frac{3}{2}}-g_1n_1^*x+\left(-\frac{3}{2}+h_1\right)\frac{n_1^*}{2}x^2 \\
&\quad +\frac{n_1^*}{2}x^2\log x-\sum_{n=1}^{K}\frac{x^{s_n+1}}{(s_n-1)s_n(s_n+1)}-\sum_{n=1}^{K}\frac{x^{\tilde{s}_n+1}}{(\tilde{s}_n-1)\tilde{s}_n(\tilde{s}_n+1)}-A_0\frac{x^{-\frac{m}{2}+1}}{\left(-\frac{m}{2}-1\right)\left(-\frac{m}{2}\right)\left(-\frac{m}{2}+1\right)} \\
&\quad -B_0\frac{x^{1+\frac{m}{2}}}{\left(-1+\frac{m}{2}\right)\frac{m}{2}\left(\frac{m}{2}+1\right)}+\frac{8}{3}\mathrm{Tr}\left(I_r-\Phi\left(\frac{1}{2}\right)\right)x^{\frac{3}{2}}-\frac{1}{2}x^2\log x-\sum_{k=0}^{M_g}\frac{x^{-\rho_k+2}}{(-\rho_k)(-\rho_k+1)(-\rho_k+2)} \\
&\quad \left.+g_1'n_1^*x-\left(-\frac{3}{2}+h_1'\right)\frac{1}{2}x^2\right)+\left(\sum_{r_n>0}\frac{x^{s_n+2}}{s_n(s_n+1)(s_n+2)}+\sum_{r_n>0}\frac{x^{\tilde{s}_n+2}}{\tilde{s}_n(\tilde{s}_n+1)(\tilde{s}_n+2)}\right. \\
&\quad +\sum_{\gamma>0}\frac{x^{\frac{5}{2}-\eta-i\gamma}}{\left(\frac{1}{2}-\eta-i\gamma\right)\left(\frac{3}{2}-\eta-i\gamma\right)\left(\frac{5}{2}-\eta-i\gamma\right)}+\sum_{\gamma>0}\frac{x^{\frac{5}{2}-\eta+i\gamma}}{\left(\frac{1}{2}-\eta+i\gamma\right)\left(\frac{3}{2}-\eta+i\gamma\right)\left(\frac{5}{2}-\eta+i\gamma\right)} \\
&\quad -\sum_{r_n>0}\frac{x^{s_n+1}}{(s_n-1)s_n(s_n+1)}-\sum_{r_n>0}\frac{x^{\tilde{s}_n+1}}{(\tilde{s}_n-1)\tilde{s}_n(\tilde{s}_n+1)}-\sum_{\gamma>0}\frac{x^{\frac{3}{2}-\eta-i\gamma}}{\left(-\frac{1}{2}-\eta-i\gamma\right)\left(\frac{1}{2}-\eta-i\gamma\right)\left(\frac{3}{2}-\eta-i\gamma\right)} \\
&\quad \left.-\sum_{\gamma>0}\frac{x^{\frac{3}{2}-\eta+i\gamma}}{\left(-\frac{1}{2}-\eta+i\gamma\right)\left(\frac{1}{2}-\eta+i\gamma\right)\left(\frac{3}{2}-\eta+i\gamma\right)}\right)+\left(A_2\frac{x^{-\frac{m}{2}}}{\left(-2-\frac{m}{2}\right)\left(-1-\frac{m}{2}\right)\left(-\frac{m}{2}\right)}\right. \\
&\quad +B_2\frac{x^{\frac{m}{2}}}{\left(-2+\frac{m}{2}\right)\left(-1+\frac{m}{2}\right)\frac{m}{2}}+\left(\frac{3}{2}+f_1\right)\frac{n_1^*}{2}+\frac{n_1^*}{2}\log x-A_1\frac{x^{-\frac{m}{2}}}{\left(-2-\frac{m}{2}\right)\left(-1-\frac{m}{2}\right)\left(-\frac{m}{2}\right)} \\
&\quad \left.-B_1\frac{x^{\frac{m}{2}}}{\left(-2+\frac{m}{2}\right)\left(-1+\frac{m}{2}\right)\frac{m}{2}}-\left(\frac{3}{2}+f_1'\right)\frac{n_1^*}{2}-\frac{n_1^*}{2}\log x\right) \\
&= \frac{1}{6}x^3+\sum_{\mathrm{I}}+\sum_{\mathrm{II}}+\sum_{\mathrm{III}},
\end{aligned}
\tag{11}
$$

where the first sum ranges over the finite set of poles s of

$$\frac{Z'(s)}{Z(s)}\frac{x^s}{s(s+1)(s+2)}, \frac{Z'(s+1)}{Z(s+1)}\frac{x^s}{s(s+1)(s+2)}$$

with $\mathrm{Re}(s) > -\frac{5}{4}$, $\mathrm{Im}(s) = 0$, the second sum ranges over the set of poles s of the same functions with $\mathrm{Im}(s) > 0$, and the third sum ranges over the finite set of their poles s with $\mathrm{Re}(s) < -\frac{5}{4}$.

5. PRIME GEODESIC THEOREM

In our setting, the prime geodesic counting function is defined by

$$\pi_0(x,W) = \sum_{N(P_0)\le x} \mathrm{Tr}(W(P_0)), x \ge 1,$$

where the sum on the right is taken over all primitive hyperbolic classes $P_0 \in \mathrm{P}\Gamma_{\mathrm{h}}$ with respect to $\overline{\Gamma}$ (see, [5] , p. 473, [11] , p. 13).

Theorem 2. For $x \ge 2$, the formula

$$\pi_0(x,W) = \sum_{n=0}^{K} li(x^{s_n}) + O\left(x^{\frac{3}{4}}(\log x)^{-1}\right)$$

holds true, where, $s_n = \frac{1}{2} + \sqrt{\frac{1}{4} - \lambda_n}$ for $0 \le \lambda_n < \frac{1}{4}$, and the implied constant depends solely on Γ, m and W.

Proof. Following [6] (p. 245) and [15] (p. 11), for a positive number $d > 0$, we define the second difference operator Δ_2^+ by

$$\Delta_2^+ f(x) = \int_x^{x+d} \int_y^{y+d} f''(t)\, \mathrm{d}t\mathrm{d}y. \tag{12}$$

Here, d is a constant which will be fixed later. By the mean value theorem, we have

$$\Delta_2^+ x^\theta = d^2 \theta(\theta - 1)\tilde{x}^{\theta-2} \tag{13}$$

for some $\tilde{x} \in [x, x+2d]$. It is easy to verify that

$$\Delta_2^+ f(x) = f(x+2d) - 2f(x+d) + f(x). \tag{14}$$

Reasoning as in [5] (p. 475), we may assume without loss of generality that $\psi_0(x, W)$ is non-decreasing. Hence, (12) implies

$$\psi_0(x, W) \le d^{-2} \Delta_2^+ \psi_2(x, W) \le \psi_0(x + 2d, W). \tag{15}$$

Since (14) holds true, one can easily deduce that $\Delta_2^+ Cf(x) = C\Delta_2^+ f(x)$, $\Delta_2^+ C = 0$, $\Delta_2^+ x = 0$, $d^{-2}\Delta_2^+ x^2 = O(1)$, $d^{-2}\Delta_2^+ x^2 \log x = O\left(x^{\frac{1}{4}} \log x\right) = O\left(x^{\frac{3}{4}}\right)$, $d^{-2}\Delta_2^+ \log x = O(1)$, $d^{-2}\Delta_2^+ \left(\frac{1}{6}x^3\right) = x + O(d)$. Thus, (13) and finiteness of the sums contained in $\sum_{\mathrm{I}}$ on the right hand side (11) yield

$$d^{-2}\Delta_2^+ \left(\frac{1}{6}x^3 + \sum_{\mathrm{I}}\right) = x + O(d) + \sum_{n=1}^{K} \frac{x^{s_n}}{s_n} + O\left(x^{\frac{3}{4}}\right). \tag{16}$$

Similarly,

$$d^{-2}\Delta_2^+ \sum_{\mathrm{III}} = O\left(x^{-\frac{3}{2}}\right). \tag{17}$$

In order to estimate $d^{-2}\Delta_2^{+}\sum_{\text{II}}$, we will first consider

$$d^{-2}\Delta_2^{+}\sum_{r_n>0}\frac{x^{s_n+2}}{s_n\left(s_n+1\right)\left(s_n+2\right)}.$$

By (14) it is evident that

$$d^{-2}\Delta_2^{+}\frac{x^{s_n+2}}{s_n\left(s_n+1\right)\left(s_n+2\right)}=O\left(d^{-2}\left|s_n\right|^{-3}x^{\frac{5}{2}}\right). \tag{18}$$

On the other hand, the mean value theorem (13) gives us

$$d^{-2}\Delta_2^{+}\frac{x^{s_n+2}}{s_n\left(s_n+1\right)\left(s_n+2\right)}=O\left(\left|s_n\right|^{-1}x^{\frac{1}{2}}\right). \tag{19}$$

Let $N(t)$ be the number of roots of $Z(s)$ on the critical line $\frac{1}{2}+ix$ in the interval $0<x\le t$. It is known ([5] , p. 477, Th. 3.8) that $N(t)\sim\frac{r|\Im|}{4\pi}t^2$. Taking $M>1$ and following ([3] , pp. 463-464; [6] , p. 246), we use (19) resp. (18) in the sums over s_n, $\frac{1}{2}<\left|s_n\right|<1$, $1\le\left|s_n\right|<M$ resp. sum over s_n, $\left|s_n\right|>M$ (below) to get

$$\left| d^{-2}\Delta_2^{+} \sum_{r_n>0} \frac{x^{s_n+2}}{s_n(s_n+1)(s_n+2)} \right|$$
$$\leq \sum_{\frac{1}{2}<|s_n|<1} \left| d^{-2}\Delta_2^{+} \frac{x^{s_n+2}}{s_n(s_n+1)(s_n+2)} \right| + \sum_{1\leq|s_n|<M} \left| d^{-2}\Delta_2^{+} \frac{x^{s_n+2}}{s_n(s_n+1)(s_n+2)} \right|$$
$$+ \sum_{|s_n|>M} \left| d^{-2}\Delta_2^{+} \frac{x^{s_n+2}}{s_n(s_n+1)(s_n+2)} \right|$$
$$\leq C_1 x^{\frac{1}{2}} \sum_{\frac{1}{2}<|s_n|<1} |s_n|^{-1} + C_1 x^{\frac{1}{2}} \sum_{1\leq|s_n|<M} |s_n|^{-1} + C_2 d^{-2} x^{\frac{5}{2}} \sum_{|s_n|>M} |s_n|^{-3}$$
$$= O\left(x^{\frac{1}{2}}\right) + C_1 x^{\frac{1}{2}} \int_1^M t^{-1} \mathrm{d}N(t) + C_2 d^{-2} x^{\frac{5}{2}} \int_M^{+\infty} t^{-3} \mathrm{d}N(t)$$
$$= O\left(x^{\frac{1}{2}}\right) + O\left(Mx^{\frac{1}{2}}\right) + O\left(d^{-2} x^{\frac{5}{2}} M^{-1}\right).$$

Thus,

$$d^{-2}\Delta_2^{+} \sum_{r_n>0} \frac{x^{s_n+2}}{s_n(s_n+1)(s_n+2)} = O\left(x^{\frac{1}{2}}\right) + O\left(Mx^{\frac{1}{2}}\right) + O\left(d^{-2} x^{\frac{5}{2}} M^{-1}\right). \tag{20}$$

Similarly,

$$d^{-2}\Delta_2^{+} \sum_{r_n>0} \frac{x^{\tilde{s}_n+2}}{\tilde{s}_n(\tilde{s}_n+1)(\tilde{s}_n+2)} = O\left(x^{\frac{1}{2}}\right) + O\left(Mx^{\frac{1}{2}}\right) + O\left(d^{-2} x^{\frac{5}{2}} M^{-1}\right). \tag{21}$$

Observe that $\sum_{|\gamma|\leq t} 1 = O(t^2)$ (see, [5] , p. 437, Prop. 2.13). Thus, application of $d^{-2}\Delta_2^{+}$ to the third and the fourth sum in $\sum_{\text{II}}$ gives us

$$O\left(x^{\frac{1}{2}}\right) + O\left(Mx^{\frac{1}{2}}\right) + O\left(d^{-2} x^{\frac{5}{2}} M^{-1}\right).$$

Let us write

$$\sum_{\text{II}} = \sum_{1} - \sum_{2}, \tag{22}$$

where $\sum_1$ denotes the sum of the first four sums in $\sum_{II}$ and $\sum_2$ denotes the sum of the last four sums in $\sum_{II}$. Now, Equations (11), (16), (17), (20), (21) and (22) give us

$$d^{-2}\Delta_2^{+}\psi_2(x,W) = x + O(d) + \sum_{n=1}^{K}\frac{x^{s_n}}{s_n} + O\left(x^{\frac{3}{4}}\right) + O\left(Mx^{\frac{1}{2}}\right) + O\left(d^{-2}x^{\frac{5}{2}}M^{-1}\right) - d^{-2}\Delta_2^{+}\sum_2. \quad (23)$$

Putting $M = x^{\frac{1}{4}}$, $d = x^{\frac{3}{4}}$, the Equation (23) becomes

$$d^{-2}\Delta_2^{+}\psi_2(x,W) = x + \sum_{n=1}^{K}\frac{x^{s_n}}{s_n} + O\left(x^{\frac{3}{4}}\right) - d^{-2}\Delta_2^{+}\sum_2. \quad (24)$$

Since the left sides of Equations (20), (21) are $O\left(x^{\frac{3}{4}}\right)$ for such choice of M and d, we get $d^{-2}\Delta_2^{+}\sum_1 = O\left(x^{\frac{3}{4}}\right)$. Now, it is obvious that $d^{-2}\Delta_2^{+}\sum_1 = O\left(x^{-\frac{1}{4}}\right)$. Finally, Equation (24) gives us

$$d^{-2}\Delta_2^{+}\psi_2(x,W) = x + \sum_{n=1}^{K}\frac{x^{s_n}}{s_n} + O\left(x^{\frac{3}{4}}\right).$$

Returning to (15), we conclude that inequality

$$\psi_0(x,W) \le x + \sum_{n=1}^{K}\frac{x^{s_n}}{s_n} + O\left(x^{\frac{3}{4}}\right)$$

holds true. Following ([15] , p. 11), we analogously obtain that

$$x+\sum_{n=1}^{K}\frac{x^{s_n}}{s_n}+O\left(x^{\frac{3}{4}}\right)\le\psi_0\left(x,W\right).$$

Hence,

$$\psi_0\left(x,W\right)=x+\sum_{n=1}^{K}\frac{x^{s_n}}{s_n}+O\left(x^{\frac{3}{4}}\right). \tag{25}$$

Arguing as in [5] (p. 475) and [4] (p. 113), one immediately sees that equality (25) proves the theorem.

ACKNOWLEDGEMENTS

We thank the Editor and the referee for their comments.

CITE THIS PAPER

Avdispahić, M. and Gušić, Dž. (2016) On the Prime Geodesic Theorem for Non-Compact Riemann Surfaces. Advances in Pure Mathematics, 6, 903-914.

REFERENCES

1. Selberg, A. (1956) Harmonic Analysis and Discontinuous Groups in Weakly Symmetric Riemannian Spaces with Applications to Dirichlet Series. Journal of the Indian Mathematical Society, 20, 47-87.
2. Huber, H. (1961) Zur analytischen Theorie hyperbolischer Raumformen und Bewegungsgrupen II. Mathematische Annalen, 142, 385-398.
3. Huber, H. (1961) Nachtrag zu. Mathematische Annalen, 143, 463-464.
4. Hejhal, D. (1973) The Selberg Trace Formula for , Vol. I. Lecture Notes in Mathematics, Volume 548. Springer-Verlag, Berlin-Heidelberg.
5. Hejhal, D. (1983) The Selberg Trace Formula for , Vol. II. Lecture Notes in Mathematics, Volume 1001. Springer-Verlag, Berlin-Heidelberg.[Citation Time(s):14]
6. Randol, B. (1977) On the Asymptotic Distribution of Closed Geodesics on Compact Riemann Surfaces. Transactions of the American Mathematical Society, 233, 241-247.
7. Buser, P. (1992) Geometry and Spectra of Compact Riemann Surfaces, Progress in Mathematics, Vol. 106. Birkhäuser, Boston-Basel-Berlin.

8. Avdispahić, M. and Smajlović, L. (2009) On the Prime Number Theorem for a Compact Riemmann Surface. Rocky Mountain Journal of Mathematics, 39, 1837-1845.
9. Avdispahić, M. and Smajlović, L. (2006) An explicit Formula and Its Application to the Selberg Trace Formula. Monatshefte für Mathematik, 147, 183-198.
10. Avdispahić, M. and Smajlović, L. (2008) Euler Constants for a Fuchsian Group of the First Kind. Acta Arithmetica, 131, 125-143.
11. Avdispahić, M. and Smajlović, L. (2016) Selberg Trace Formula as an Explicit Formula and the Prime Geodesic Theorem. (Submitted)
12. Fischer, J. (19760 An Approach to the Selberg Trace Formula via Selberg Zeta-Function. Lecture Notes in Mathematics, Volume 1253. Springer-Verlag, Berlin-Heidelberg.
13. Hardy, G.H. and Riesz. M. (1915) The General Theory of Dirichlet's Series. Cambridge University Press, Cambridge.
14. Jameson, G.J. (2003) The Prime Number Theorem. Cambridge University Press, Cambridge.
15. Park, J. (2010) Ruelle Zeta Function and Prime Geodesic Theorem for Hyperbolic Manifolds with Cusps. In: van Dijk, G. and Wakayama, M., Eds., Casimir Force, Casimir Operators and Riemann Hypothesis, de Gruyter, Berlin, 89-104.

CHAPTER 17

Geophysical Surveys at Khirbat Faynan, an Ancient Mound Site in Southern Jordan

Alexandre Novo,[1] Matthew L. Vincent,[2] and Thomas E. Levy[2]

[1]Geostudi Astier, via A. Nicolodi 48, 57121 Livorno, Italy
[2]Department of Anthropology & California Institute for Telecommunications and Information Technology (Calit2), University of California, San Diego, CA 92093, USA

ABSTRACT

Faynan in Jordan contains the largest copper ore resource zone in the southern Levant (Israel, Jordan, Palestinian territories, Lebanon, Syria, and the Sinai Peninsula). Located 50 km southeast of the Dead Sea, it is home to one of the world's best-preserved ancient mining and metallurgy districts encompassing an area of ca. 400 km^2. During the past three decades, archaeologists have carried out numerous excavations and surveys recording hundreds of mines and sites related to metallurgical activities that span the past 10 millennia. Khirbat Faynan (Biblical Punon), is situated in the main Faynan Valley and is the largest (ca. 15 ha) settlement site in the region and has remained unexcavated until 2011. As Jordan's most southern mound site with indications of widespread ancient architecture, we employed a suite of noninvasive geophysical survey methods to identify areas suitable for excavation. Earlier geophysical surveys were carried out in the Faynan region by our team in the late 1990s when only EMI (electromagnetic induction) proved successful, but with relatively poor resolution. As reported here, by 2011, improvements in data processing software and 3D ERT (electrical resistivity tomography) sampling protocols made it possible to greatly improve the application of noninvasive geophysical surveying in this hyperarid zone.

1. INTRODUCTION

Improvements in processing software and sampling protocols for shallow geophysical prospecting make it a key tool for initiating archaeological exploration. In terms of research design, geophysics have increasingly become an essential part of the methodology for planning archaeological excavation in a number of unexplored regions around the world [1, 2]. As part of a long-term archaeological study of the role of mining and metallurgy in the evolution of societies from the Neolithic period (ca. 8000 BC–6000 BC) to Islamic times (7th c. CE–early 20th c. CE) in Jordan's Faynan copper ore district, the University of California, San Diego (UCSD) Edom Lowlands Regional Archaeology Project (ELRAP) conducted intensive systematic surveys and large-scale excavations at sites spanning all these periods. From 2002–2010, the primary research focus was on the formative Iron Age (ca. 1200–500 BCE) when the first historical state-level societies evolved in this part of the eastern Mediterranean and the ancient Near East's first industrial revolution took place. Large-scale excavations were carried out at Khirbat en-Nahas (ca. 10 hectares), the largest copper ore processing and smelting site in the region (cf. [3]), smaller production sites [4], fortresses [5], mining complexes [6], and other sites making it possible to reconstruct the Iron Age "metallurgical landscape" of this natural resource zone [7]. While it was possible to reconstruct the network of ore exploitation, processing, and manufacture, the location and nature of political and economic center remains a mystery. Based on previous surveys and research in the main Wadi Faynan valley, more researchers believe that a large (ca. 15 hectares) ancient mound site called Khirbat Faynan (known as Punon in the Hebrew Bible and Phaino in late Roman/Byzantine texts) may have been that center during the Iron Age [8]. In 2011, the UCSD ELRAP team initiated the first multiyear large-scale excavation project at this site, the most southern ancient mound site in Jordan. To initially tackle the excavation of this site, a geophysical survey was carried out that helped guide the selection of the first excavation at the site.

Khirbat Faynan (Figure 1) is a multiperiod site located where Wadi Dana and Wadi Ghuwayr meet (becoming Wadi Faynan) in the hyperarid Saharo-Arabian vegetation zone [9]. The site has a significant amount of well-preserved Roman and Byzantine ruins still visible. Previously, the

Deutsches Bergbau Museum under Andreas Hauptman and the Council for British Research in the Levant (CBRL) under Graeme Barker and colleagues surveyed the site [8, 10]. To facilitate a Paleoenvironmental study in the region, a small sounding was made north of the site behind a large wall structure called the "Barrage" [11, 12]. The site's prominent position next to the convergence of the two previously mentioned wadis sets it apart in the landscape, and it is easy to postulate that it controlled the agricultural field systems to the north and west, and access to the road leading from the copper-ore-rich lowlands to the highland plateau to the east. While many have suggested the role of the "Khirbat" in the region, there still lacks archaeological evidence to fully support these postulations. The 2011 ELRAP expedition was codirected by Thomas Levy and Mohammad Najjar to begin to answer questions concerning the socioeconomic function of the site during the Iron Age. With several significant periods of occupation, it will require several seasons of excavation to fully articulate and understand the site. As part of this expedition, a geophysical survey of the site and its immediate surroundings was conducted in order to create a geological profile of the tell (Arabic: ancient mound site) and identify primary candidates for excavation.

Figure 1: Aerial view of Khirbat Faynan looking east showing the excavation trench and excavators. Geophysical surveys were made around the entire site. Scale, note that the trench is 40 m in length from west to east. Arrow indicates the location of 3D ERT survey (photo, UCSD Levantine Archaeology Lab).

Full characterization of archaeological sites by conventional archaeological techniques can take years. Moreover, drilling and excavations, the most often applied methods for archaeological assessment, are ground disturbing. Therefore, nondestructive geophysical techniques have been used to detect and map underground features in a fast and noninvasive manner [13–16]. Among these techniques and based on previous results in the same area [17], we decided to use EMI (electromagnetic induction) and ERT (electrical resistivity tomography). Electromagnetic techniques have been used for archaeological investigations of different sites [18, 19], including places with similar conditions in the Middle East [20, 21]. Similarly, resistivity surveying has commonly used in noninvasive archaeological prospection [22, 23]. ERT is quite popular in archaeological investigations [24–27]. The success of the method depends on the difference between the resistivity properties of the potential archaeological targets (walls, roads, buildings, etc.) and the surrounding environment [25]. In recent years, 3D ERT is being increasingly used because it provides a three-dimensional view of the subsoil resistivity. Besides, this technology is considered well-suited for mapping Near Eastern tell sites [28].

The combination of EMI and ERT methods provides a flexible approach for planning archaeological excavations [29]. Our aim was to survey buried structures, especially Iron Age ones, around the site. This paper shows the preliminary results of our archaeogeophysical investigations at Khirbat Faynan undertaken in 2011. After a brief description of the area and past results of the ongoing project, we describe our survey strategies and data processing methods and offer an overview of our findings to date.

2. METHODOLOGIES

Conditions around the site of Khirbat Faynan make some of the more successful techniques such as GPR (Ground Penetrating Radar) nearly impossible to employ. The ground surface is generally covered with debris from earlier occupation that would have to be cleared before any survey could take place. The process of clearing rocks would not only be expensive in terms of time required to complete such a task, but would also be potentially invasive as it might put at risk exposed architecture which is not always immediately visible among the rock debris. It was decided that the next best options were EMI and ERT, both being well suited for the difficult terrain. These techniques allow for the characterization of the subsoil as well as the ability to detect anomalies that could be associated with buried structures or artifacts.

We first used EMI, requiring no direct contact with the ground surface, which allowed us to survey large areas quickly even with the difficult terrain. We then selected locations to perform ERT dipole-dipole profiles across the site in order to obtain vertical cross sections of the first 5 meters depth. These data were inverted to obtain 2D electrical images. Finally, we selected an appropriate place to perform a high-resolution 3D mesh. The 3D electrical image obtained from these data was used to decide the location of the season's excavation.

2.1. EMI Equipment and Discussion

Among many applications, electromagnetic induction instruments are used for near-surface archaeological investigations, the detection of buried structures, such as building foundations as well as for the detection of highly conductive metallic objects. EMI was, therefore, used in this study to attempt to locate buried remains of past human activity in the lower flat areas of Khirbat.

The system used in this campaign is a Profiler EMP-400 (GSSI, USA) that is a digital, portable, multifrequency electromagnetic induction sensor. The system bandwidth extends from 1 kHz to 16 kHz. The system collects three different frequencies simultaneously. The in-phase and quadrature components of the induced secondary field as well as the apparent conductivity are collected and stored for each reading along with a time stamp. GPS data, in the form of a NMEA string, were also recorded for every line.

The system sensor electronics are controlled via a wireless Bluetooth communications interface. The system user interface and data storage are incorporated into a TDS RECON-400 personal digital assistant (PDA) provided with the system. This instrument is easy to use in the field, and the operator can cover several hectares in a week (Figure 2). Due to extremely rough topography, some areas were not scanned.

We studied an area of approximately 6 ha. The system was calibrated every day before starting the data acquisition. Data were collected in rectangular grids of parallel lines spaced 1 m apart. After marking the grid corners, two measuring tapes were placed in the shorter pair of parallel sides of the grid for measuring the 1 m spacing between the two survey tapes to mark the exact profile location. The surveyor precisely followed a string in order to ensure straight profiles. For data acquisition consistency, the same person collected all EMI datasets. Field system calibration was a crucial requirement for every survey grid because of extreme temperature changes during data acquisition as well as variable topography across the site.

Figure 2: Operator follows the string to ensure straight profiles on the slopes of the site of Khirbat Faynan, Jordan. The instrument (photo, M. L. Vincent, UCSD Levantine Archaeology Lab) is kept at a constant distance from the ground surface.

After collection, data were sent to Geostudi Astier (Livorno, Italy) for processing. First, data trajectories were corrected for spurious GPS points in order to properly align the profiles. Therefore, and second, the data were gridded in MVS/EVS (CTech, USA) using the Kriging algorithm [30]. Finally, the processed results were then sent back to us to be incorporated in the project's geodatabase [31]. The rapid processing allowed for minor adjustments to be made in the survey planning, prioritizing areas with greater potential and leaving areas where there was clearly no subsurface cultural remains.

2.2. ERT Equipment and Discussion

We used a Syscal Pro resistivimeter (Iris Instruments, France) with 48 electrodes spaced by 0.5 meters. Four different sets of 2D ERT lines (24 electrodes at 1 m separation each) were collected over 4 sides (north, south, east, and west) of the Khirbat Faynan mound. The main objective was to obtain vertical sections up to 6 meters along the mound in order to investigate the different layers and distinguish cultural from natural layers.

Additionally, a small 3D ERT survey was collected in order to facilitate excavation. The 3D ERT survey used a set of parallel 2D profiles with a separation of 0.5 m among electrodes. The surveyed area of 24 m × 5 m included 6 ERT profiles. Both dipole-dipole and Wenner arrays were performed.

Data acquisition proved to be particularly challenging due to the extremely dry soil, preventing electrical conductivity, and abundance of rocks under the surface, impeding easy introduction of the metal stakes into the ground. While the later was not easily overcome, the former was improved through a mixture of water, clay, and salt producing a remarkably conductive medium which could be poured over each electrode.

After collection, data were sent to Geostudi Astier (Livorno, Italy) for inversion using ERTLab software (Multi-Phase Technologies LLC, USA and Geostudi Astier srl, Italy) and generation of final maps within the software MVS/EVS (CTech, USA) and ArcMap 10 (ESRI, USA). The main purpose of the 3D ERT survey was to delineate buried stonewalls to guide archaeological excavations.

In our processing, full three-dimensional inversions of apparent resistivities are performed using a specific customized tool, derived from the ERTLab 3D inversion software, which uses a finite-element (FEM) approach to model the subsoil by adopting mesh of hexahedrons to correctly incorporate terrain topography [32]. Throughout the inversion iterations, the effect of non-Gaussian noise is appropriately managed using a robust data weighting algorithm [33, 34].

This past season, new benchmarks were placed around and on top of the tell with an RTK GPS system to ensure precise control of the surveying. Unlike EMI, the ERT has no self-georeferencing system. Therefore, the ERT was geo-referenced using a reflectorless total station, which could measure the precise position of the stakes with or without a prism. Each individual line was shot in its entirety in order to provide redundancy when the lines were then matched up during the postprocessing. Since GIS has been a major component of the ELRAP field recording system since 1999 [35], there was already an infrastructure in place to relate the geophysical data with the archaeological data. As part of this recording system, daily aerial photographs are taken documenting the position of significant architectural features. Each photograph can be correlated with the daily elevations to know exactly what the elevations of the exposed layers are. This enabled us to lay the 3D ERT slices that corresponded to each aerial photograph's elevation in order to study the relationship of the geophysics with the archaeological record.

3. DISCUSSION

No features suggesting buried structures appeared in the lower-frequency EMI data. For this reason, EMI results presented here are for the higher (15 kHz) operating frequency (Figure 3). The most interesting areas, likely related to past human activity, have been highlighted. Some of them show clear linear patterns, other show areas with very high conductivity values (magenta and blue color) which might indicate the presence of accumulated metallurgic production materials. Unfortunately, in some areas, the presence of many pieces of slag covering the surface might have affected the EMI data due to its high sensitivity to metals. Thus, areas with higher conductivity may indicate just places of high concentration of slag covering the surface. However, other areas present subtle features of lower values (green and yellow colors) that may represent walls. As discussed by Witten [17], who worked in similar conditions, buried walls can appear as low values in the EMI data.

Figure 3: EMI data acquired at Khirbat Faynan. While numerous ancient walls can be seen on the surface of this site with Google Earth, the geophysics survey aimed at identifying subsurface features suitable for excavation. Blue color ellipses show interesting areas of low conductivity. Brown color ellipses indicate areas of high conductivity that might be related to past human activity. The site measures approximately 270 × 640 meters.

Past surveys in the area have included EMI and GPR, but ERT had not yet been employed. The 2D lines have enabled us to characterize the mound and we now have a better idea of the overall archaeological record. While the 2D does not give us a precise picture of the archaeology, it does at least suggest possible depths of archaeological depositions (Figure 4). Because of the lack of ERT performed in the area, it was felt that a ground truthing should be conducted in addition to the survey to correlate the results with the archaeological record. Half of square 16–56 was excavated, an area representing over half of the 3D ERT. The 3D volume showed what were most likely buried walls underneath the surface (see Figures 5 and 6). On the other hand, the ERT sections and volumes have shown that the very low-conductivity values, between 0 and 1 mS/m, match well with areas where the resistivity is higher than 1000 Ohm·m. Therefore, the ERT sections were used also to verify and calibrate properly the conductivity value in a condition that is typically not favorable for EM induction instruments.

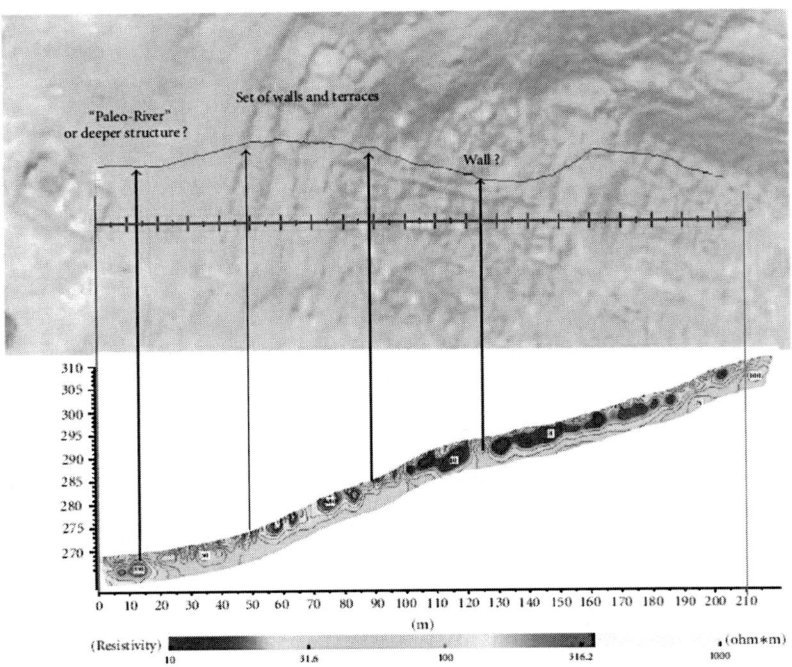

Figure 4: 2D ERT line in the west side of the ancient mound.

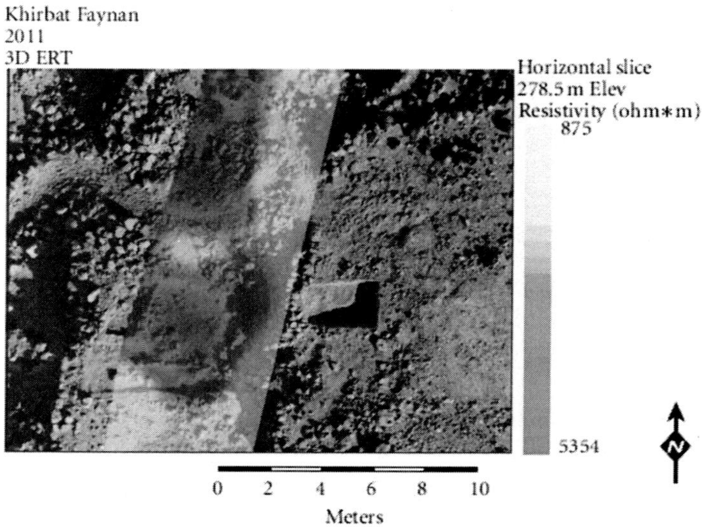

Figure 5: Semitransparent horizontal ERT slice at 1 m depth, overlaid on top of an aerial image of the excavation.

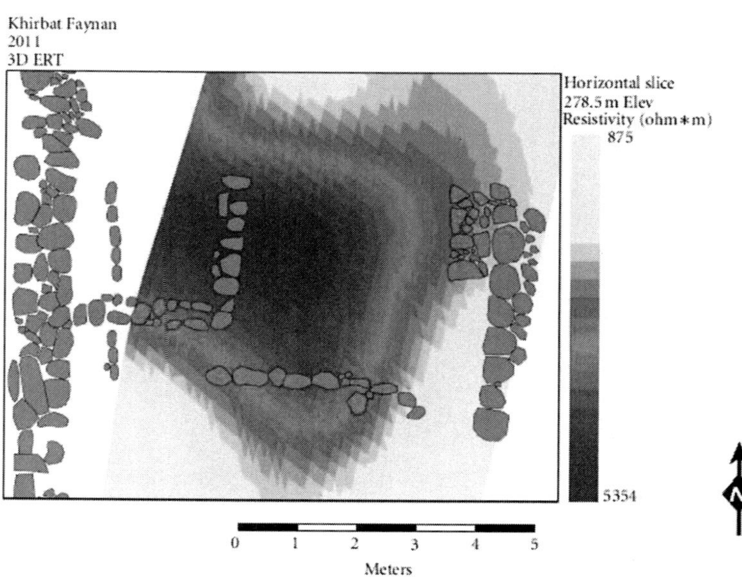

Figure 6: Walls found during the excavation superimposed on a horizontal ERT slide at 1 m depth.

The most significant correlation can be found between walls 1032 and 1057 (Figure 7). Because the lines went from north to south, the east-west walls seem to be the most visible in the data. In other words, every wall crossed is more visible than those running parallel to the survey lines. Wall 1032 was the closer of the two walls to the surface and was also more substantial than wall 1057 which had less preservation, these two factors together likely account for the stronger area of resistivity around wall 1032.

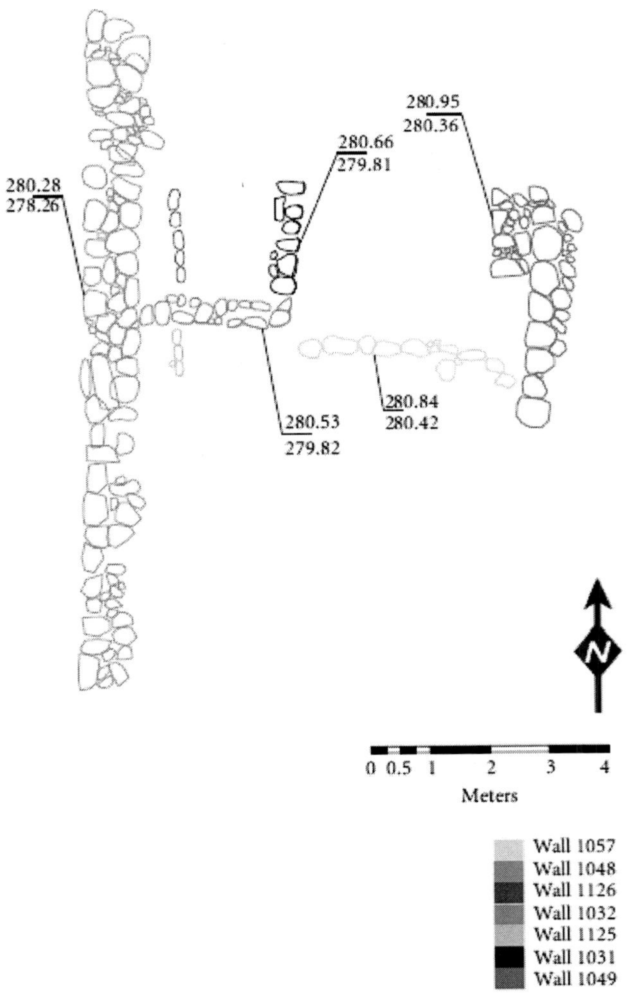

Figure 7: Ground truth, archaeological plan of the excavated walls detected using the 3D ERT geophysics.

The GIS centric system of archaeological recording employed by the ELRAP team [35] allowed us to closely compare the archaeology with the geophysical survey. We were able to drape the appropriate ERT slices with the corresponding exposures at precise elevations. This allowed us to both compare the results of the excavation alongside the geophysics as well as predict what was coming next through the same sort of comparison. The final results were a series of images (see Figures 5 and 6) that show both the archaeology and the geophysics, with the later overlaid over the former. This further highlights the need for accurate georeferencing in geophysical survey, something which was central in our data collection.

4. CONCLUSION

The EMI data give us a general idea of what might be interesting to look at in detail in subsequent seasons in the lower area of the site. Future geophysical surveys should focus on the use of 3D GPR to locate more structures and accurately delineate those suggested by the EMI maps. The use of mobile multicoil EMI systems is also being considered due to their high lateral spatial resolution with accurate vertical measurements [36]. 3D GPR provides high-resolution volumetric data [14, 37, 38]. However, in order to use a GPR system in this area, rocks will have to be cleared from the study surface beforehand. Considering the quantity of rocks strewn about the surface, this will be a difficult task. On the other hand, the high-precision 3D ERT is potentially one of the more useful survey methods for such difficult terrain as a rock-strewn tell. The time required to clean the surface of rocks as well as sort out what was tumble and what was part of existing architecture, would have proved far too costly to make it a viable technique for wide-area surveys. Indeed, the 3D ERT is still expensive in terms of human hours required to employ it; however, the technique is much less disturbing with high potential for excellent results.

The results from this season's work suggest that 3D ERT is a useful method for geophysical survey in difficult conditions that prevent the use of GPR. However, future work should include continued 3D ERT survey to conduct further correlation between the data and the archaeological record. Likewise, continued use of EMI can be helpful in revealing promising areas for more precise geophysical survey. With newer instruments, EMI is continuing to increase in precision and could potentially become one of the more useful tools as precision becomes more like that of ERT or GPR. Until such a time, however, EMI surveys like ours do not appear to offer much beyond that of general characterization of subsurface features. Southern Jordan continues to be an ideal place for testing geophysical survey techniques in difficult and harsh environments. The two techniques employed at Khirbat Faynan both resulted in success, each in their own way. We found that EMI helped to guide us towards the regions that we wanted to investigate further with higher-precision survey tools. This saved us countless hours considering the difficulty of employing the ERT versus that of EMI. Ultimately, the two tools together offered the best geophysical kit for our environment.

ACKNOWLEDGMENTS

The authors are grateful to the Department of Antiquities (DOA) of Jordan for its help and interest in the 2011 ELRAP expedition. In particular, Jehad Haroun, Khalil Hamdan, Qutaiba al-Dasouqi, and especially Abdelrahim Al Dwikat and Zouhair Zoubi who served as DOA representatives to our project. Thanks also to Barabara Porter (director) and Christopher Tuttle (associate director) of the American Center of Oriental Research (ACOR), Amman, for their help and support. Members of most of the Bedouin tribes from the Qurayqira/Faynan villages worked on the project. Funding for this project came from: Research Opportunities (CSRO) Grant program; NSF grant EAR-Geophysics, 09-539; National Geographic Society (Grant no. 8974-11), the Jerome and Miram Katzin Family Foundation, and the Norma and Reuben Kershaw Family Foundation. M. Vincent was supported by the National Science Foundation IGERT (Integrative Graduate Education and Research): Training, Research and Education in Engineering for Cultural Heritage Diagnostics (NSF 0966375), where T. E. Levy serves as co-PI. A special thanks to Ramesh Rao, director, Calit2 San Diego Division for his support of the ELRAP expedition. They deeply appreciate the hard work carried out by the undergraduate students enrolled in the UCSD Middle East Archaeology Field School that forms part of the ELRAP research endeavor. The 2011 ELRAP expedition was directed by T. E. Levy and Mohammand Najjar. They thank M. Najjar for his collegiality and interaction in the field. Thanks also to Gianfranco Morelli, Geostudi Astier, for help with data processing.

REFERENCES

1. R. Slatt, W. C. King, J. Molyneux et al., "Tribute to Dr. Alan Witten," Earth Scientist, pp. 9–23, 2005.
2. A. J. Witten, Handbook of Geophysics and Archaeology, Equinox Pub, 2006.
3. T. E. Levy, T. Higham, C. B. Ramsey et al., "High-precision radiocarbon dating and historical biblical archaeology in southern Jordan," Proceedings of the National Academy of Sciences of the United States of America, vol. 105, no. 43, pp. 16460–16465, 2008.

4. E. Ben-Yosef, T. E. Levy, T. Higham, M. Najjar, and L. Tauxe, "The beginning of Iron Age copper production in the southern Levant: new evidence from Khirbat al-Jariya, Faynan, Jordan," Antiquity, vol. 84, no. 325, pp. 724–746, 2010. ·
5. E. Ben-Yosef, T. E. Levy, and M. Najjar, "Rās al-Miyāh fortresses: new discoveries at one of the gateways to the iron age copper production district of Faynan, Jordan," pp. 832–841, 2009.
6. E. Ben-Yosef, T. E. Levy, and M. Najjar, "New iron age copper-mine fields discovered in southern Jordan," Near Eastern Archaeology, vol. 72, no. 2, pp. 98–101, 2009. ·
7. T. E. Levy, E. Ben-Yosef, and M. Najjar, "New perspectives on iron age copper production and society in the Faynan Region, Jordan," in Eastern Mediterranean Metallurgy and Metalwork in the Second Millenium BC, V. Kassianidou, Ed., Oxbox Books, Oxford, UK, 2012.
8. G. Barker, D. D. Gilbertson, and D. J. Mattingly, Archaeology and Desertification: The Wadi Faynan Landscape Survey, Southern Jordan, Oxbow, 2007.
9. C. E. Cordova, Millennial Landscape Change in Jordan: Geoarchaeology and Cultural Ecology, The University of Arizona Press, 2007.
10. A. Hauptmann, The Archaeometallurgy of Copper: Evidence from Faynan, Jordan, Springer, 2007.
11. J. P. Grattan, D. D. Gilbertson, and C. O. Hunt, "The local and global dimensions of metalliferous pollution derived from a reconstruction of an eight thousand year record of copper smelting and mining at a desert-mountain frontier in southern Jordan," Journal of Archaeological Science, vol. 34, no. 1, pp. 83–110, 2007. r · ·
12. C. O. Hunt, D. D. Gilbertson, and H. A. El-Rishi, "An 8000-year history of landscape, climate, and copper exploitation in the Middle East: the Wadi Faynan and the Wadi Dana National Reserve in southern Jordan," Journal of Archaeological Science, vol. 34, no. 8, pp. 1306–1338, 2007. r · ·
13. B. W. Bevan, "Geophysical exploration for archaeology: an introduction to geophysical exploration," Midwest Archaeological Center Special Report 1, U.S. Department of the Interior, National Park Service, Lincoln, Neb, USA, 1998.
14. L. B. Conyers and D. Goodman, Ground-Penetrating Radar: An Introduction for Archaeologists, AltaMira Press, 1997.
15. C. Gaffney, "Detecting trends in the prediction of the buried past: a review of geophysical techniques in archaeology," Archaeometry, vol. 50, no. 2, pp. 313–336, 2008. r · ·
16. K. L. Kvamme, "Geophysical surveys as landscape archaeology," American Antiquity, vol. 68, no. 3, pp. 435–457, 2003. ·
17. A. J. Witten, T. E. Levy, R. B. Adams, and I. Won, "Geophysical surveys in the Jebel Hamrat Fidan, Jordan," Geoarchaeology, vol. 15, no. 2, pp. 135–150, 2000.
18. D. Simpson, A. Lehouck, M. Van Meirvenne, J. Bourgeois, E. Thoen, and J. Vervloet, "Geoarchaeological prospection of a medieval manor in the Dutch polders using an electromagnetic induction sensor in combination with soil augerings," Geoarchaeology, vol. 23, no. 2, pp. 305–319, 2008. r · ·
19. A. Tabbagh, "Applications and advantages of the Slingram electromagnetic method for archaeological prospecting," Geophysics, vol. 51, no. 3, pp. 576–584, 1986. ·
20. B. Frohlich and W. J. Lancaster, "Electromagnetic surveying in current Middle Eastern archaeology: application and evaluation.," Geophysics, vol. 51, no. 7, pp. 1414–1425, 1986. ·
21. A. Witten, G. Calvert, B. Witten, and T. Levy, "Magnetic and electromagnetic induction studies at archaeological sites in Southwestern Jordan," Journal of Environmental and Engineering Geophysics, vol. 8, no. 3, pp. 209–215, 2003. ·
22. M. Dabas, A. Hesse, and J. Tabbagh, "Experimental resistivity survey at Wroxeter archaeological site with a fast and light recording device," Archaeological Prospection, vol. 7, no. 2, pp. 107–118, 2000.
23. I. Scollar, A. Tabbagh, A. Hesse, and I. Herzog, Archaeological Prospecting and Remote Sensing, Cambridge University Press, Cambridge, UK, 1990.

24. M. Drahor, M. Berge, and Ö. Caner, "Integrated geophysical surveys for the subsurface mapping of buried structures under and surrounding of the Agios Voukolos Church in Izmir, Turkey," Journal of Archaeological Science, vol. 38, no. 9, pp. 2231–2242, 2001.

25. S. Negri, G. Leucci, and F. Mazzone, "High resolution 3D ERT to help GPR data interpretation for researching archaeological items in a geologically complex subsurface," Journal of Applied Geophysics, vol. 65, no. 3-4, pp. 111–120, 2008. r · ·

26. G. N. Tsokas, P. I. Tsourlos, G. Vargemezis, and M. Novack, "Non-destructive electrical resistivity tomography for indoor investigation: the case of Kapnikarea Church in Athens," Archaeological Prospection, vol. 15, no. 1, pp. 47–61, 2008. r · ·

27. P. I. Tsourlos and G. N. Tsokas, "Non-destructive electrical resistivity tomography survey at the South Walls of the Acropolis of Athens," Archaeological Prospection, vol. 18, no. 3, pp. 173–186, 2011. r · ·

28. J. Casana, J. T. Herrmann, and A. Fogel, "Deep subsurface geophysical prospection at Tell Qarqur, Syria," Archaeological Prospection, vol. 15, no. 3, pp. 207–225, 2008. r · ·

29. A. Osella, M. de la Vega, and E. Lascano, "3D electrical imaging of an archaeological site using electrical and electromagnetic methods," Geophysics, vol. 70, no. 4, pp. G101–G107, 2005. r · ·

30. G. Matheron, "Principles of geostatistics," Economic Geology, vol. 58, no. 8, pp. 1246–1266, 1963.

31. T. E. Levy, M. Najjar, A. D. Gidding et al., "The 2011 Edom Lowlands Regional Archaeology Project (ELRAP): Excavations and Surveys in the Faynan Copper Ore District, Jordan," Annual of the Department of Antiquities of Jordan.

32. Z. Bing and S. A. Greenhalgh, "Finite element three-dimensional direct current resistivity modelling: accuracy and efficiency considerations," Geophysical Journal International, vol. 145, no. 3, pp. 679–688, 2001. r · ·

33. D. J. LaBrecque, M. Miletto, W. Daily, A. Ramirez, and E. Owen, "The effects of noise on Occam's inversion of resistivity tomography data," Geophysics, vol. 61, no. 2, pp. 538–548, 1996. ·

34. G. Morelli and D. J. Labrecque, "Advances in ERT inverse modeling," European Journal of Environmental and Engineering Geophysics, vol. 1, no. 2, pp. 171–186, 1996. ·

35. T. Levy, V. Petrovic, T. Wypych et al., On-Site Digital Archaeology 3.0 and Cyber-Archaeology: Into the Future of the Past–New Developments, Delivery and the Creation of a Data Avalanche, Introduction to Cyber-Archaeology, Archaeopress, Oxford, UK, 2010.

36. P. De Smedt, M. Van Meirvenne, E. Meerschman et al., "Reconstructing palaeochannel morphology with a mobile multicoil electromagnetic induction sensor," Geomorphology, vol. 130, no. 3-4, pp. 136–141, 2011. r · ·

37. J. Leckebusch, "Ground-penetrating radar: a modern three-dimensional prospection method," Archaeological Prospection, vol. 10, no. 4, pp. 213–240, 2003. r ·

38. A. Novo, M. Grasmueck, D. Viggiano, and H. Lorenzo, "3D GPR in archaeology: what can be gained from dense data acquisition and processing," in Proceedings of the 12th International Conference on Ground Penetrating Radar, pp. 16–19, 2008.

CHAPTER 18

Physico-Chemical and Petrographic Characterization of Carbonated Rocks of Mintom (South-Cameroon) and Their Potential Uses

Zo'o Zame Philémon[1], Mpakam Hernanie Grelle[2], Samba Assomo Philippe[1]

[1]*Department of Earth Sciences, Faculty of Sciences, University of Yaoundé I, Yaoundé, Cameroon*

[2]*The Higher Institute of the Sahel, University of Maroua, Maroua, Cameroon*

ABSTRACT

The characterization of carbonated rocks of Mintom allowed identifying three features (massive limestone, banded limestone and dolomite). Samples taken in these features were subjected to the physico-chemical analyses and to the mechanical tests. The results obtained show a convergence in the mineralogical composition in particular in the calcareous features (smectite, micas and calcite); the dolomite feature being mainly constituted by dolomite associated with the smectite. The chemical composition allows finding contents raised in CaO, in all the samples. However, in the dolomite and the banded limestone, the concentrations in MgO and in SiO_2 are also important. The mechanical properties showed a good compression resistance for the massive limestone and the dolomite, the mechanical behavior of the banded limestone being low. The exploitation of these results allows recommending carbonated rocks of Mintom in the production of the clinker, agricultural amendments and in the constructions of the civil engineering.

Keywords: Mintom, Carbonated Rocks, Characterization, Physico-Chemical, Petrographic

1. INTRODUCTION

The need for major infrastructure work in Cameroon imposes a cement addition use for various concrete formulations. This ambition faces a very low national production for this type of binder because of a limited number of cement manufactures. The characterization of carbonated rocks of Mintom appears as well as one of the avenues to explore in order to consider the production of clinker from these rocks that may serve as raw material for cement manufacture. It is also important to consider other uses for these materials. Under these conditions, a more holistic characterization appears necessary in order to optimize the exploitation of carbonated rocks of the study area. This publication has two main objectives, firstly, chemical and petrographic characterization of rocks and secondly the identification of potential uses of each material.

2. PRESENTATION OF THE STUDY AREA

The study area is located in the Dja and Lobo Division between the districts of Djoum and Mintom in the administrative region of South Cameroon. It is hosted in the great valley of the River Dja at 592 m altitude [1] . This area is subjected to the equatorial climate with four seasons and extends from 13°10' - 13°30' Long E and from 2°44' - 2°55' Lat N (Figure 1).

3. PREVIOUS WORK ON THE SITE

3.1. Geology

Carbonated rocks of Mintom constitute the upper series of Dja [2] - [4] . A part from the sedimentary formations, the area reveals the presence of metamorphic rocks (gneiss, schist, mica) and plutonic rocks (gabbro, granite). Structurally, the region is crossed by numerous of tectonic structures such as faults, folds and mylonite. The work of [5] show that this basin is limited at North by the Pan African belt and at south by the Congo Craton (Figure 2). The main geological formations (shales, dolomites and limestones) are covered with argillites and laterite. Two main soils run throughout the area including yellow lateritic soils in land and hydromorphic soils in the lowlands [1] [6] [7] .

3.2. Lithostratigraphy

The study of [4] led to subdivide the upper Dja series in four lithological units from bottom to top, there are: The unity of Kol, made with diamictite on its base and pelit on the summit; Metou unit composed of massive dolomite; Mominbolé unit that has an alternation of limestone and clay: it is a pelitic limestone and the Atog Adjap unit composed of clear limestone and yellowish gray band. From the photogeological analysis [2] [4] , it appears that, the different features which form the sedimentary basin of Mintom stick out in a powerful soil layer near the Dja River. The survey S1 of [2] in BI was used to estimate the relative thickness of limestone feature (Figure 3). Thus, the massive limestone has a thickness of 21 m, while the banded limestone features extends over a thickness of 24 m.

3.3. Methodology

The methodology of work is divided into two major activities including field campaigns and laboratory work.

3.4. Field Campaigns

They took place in four stages. The first phase consisted of an exploration of the study area in order to locate outcrops; the second was used to determine the sampling points and the two last stages were devoted to sample in previously defined sampling points. The samples were collected by geological hammer at three points defined on various features. Then, they were packed in sampling plastic bags. Field campaigns have also helped to give the description of the different features encountered.

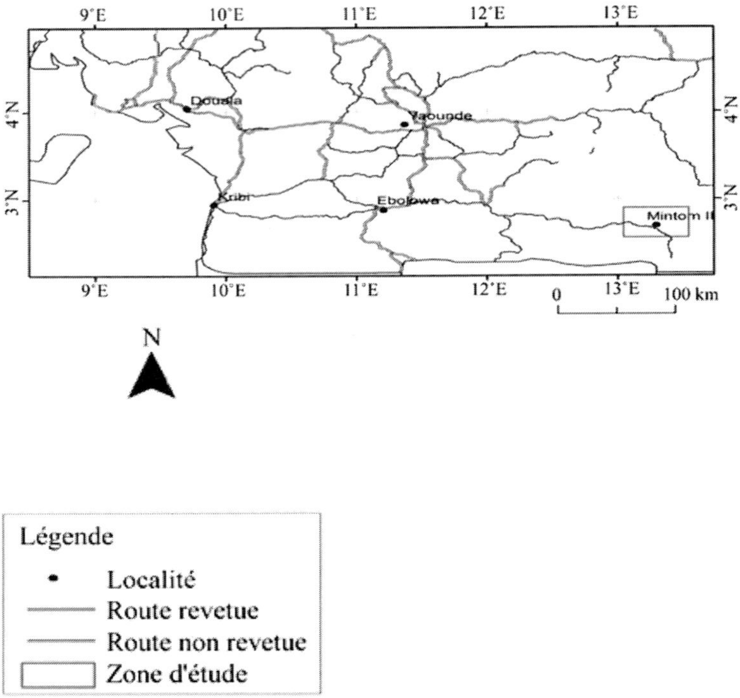

Figure 1. Location of the study area.

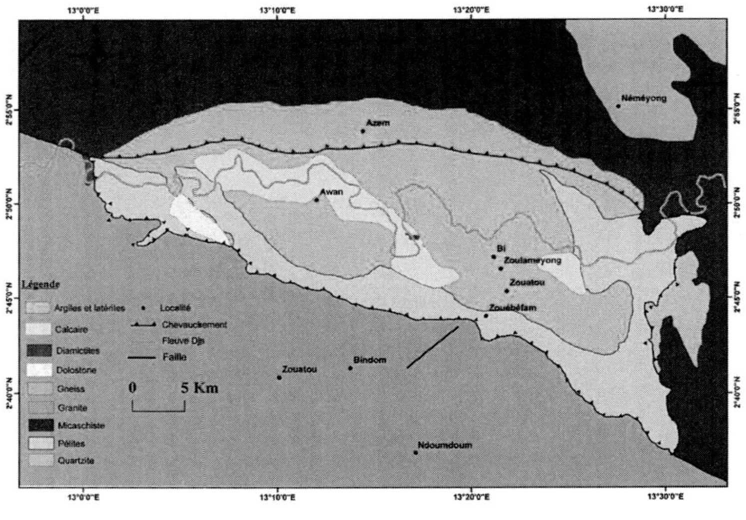

Figure 2. Geology of the area.

3.5. Laboratory Work

Laboratory work concern the analysis of different samples collected in the field. These analyzes were performed in different laboratories. The Table 1 gives the synthesis of laboratory analysis.

4. PRESENTATION OF THE RESULTS

4.1. Outcrops and Geological Structures

Several geological structures and outcrops were encountered (block, slab, escarpment, cleavage, bend, fault) (Figure 4).

4.2. Features

Three major features have been described from outcrops and geological structures (Figure 5). They are:

- Massive limestone feature

It is observed mainly in Atog-Adjap where it is very extensive. It is a hard gray rock and very effervescent in HCL. Its structure shows alternating benches separated by thin layer of dark argillite. In optical microscope, the rock is homogeneous and composed with very small crystals (<20 microns).

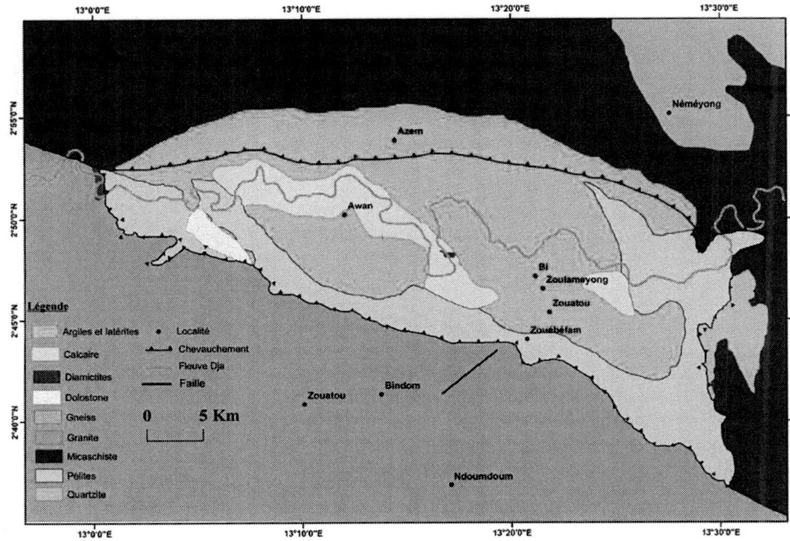

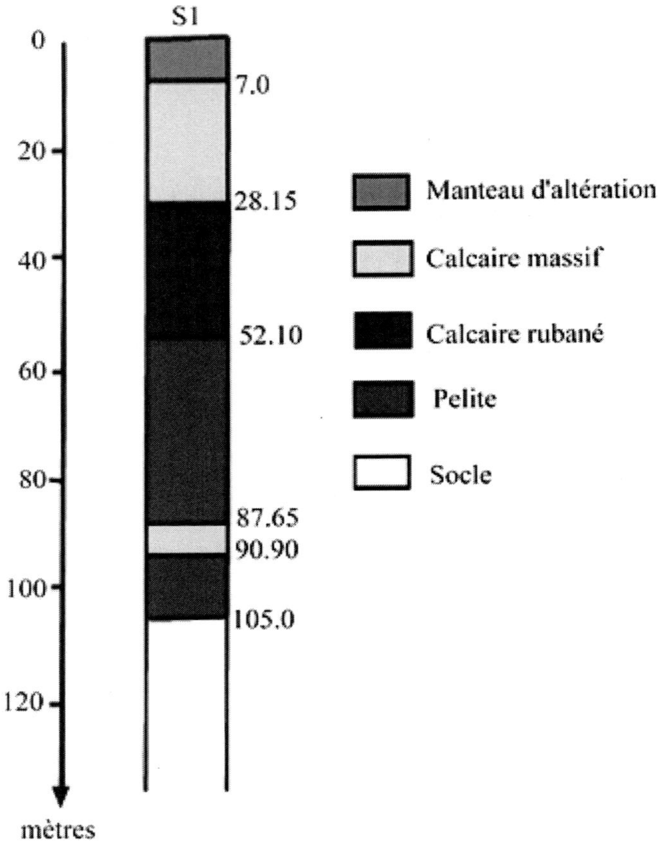

Figure 3. Stratigraphic log of survey S1.

- Banded limestone feature

It is a dark gray rock characterized by alternating light and dark bands. The thickness of a band varies between 10 - 15 cm. In thin blade, the texture is cryptogranular with small calcite crystals (0.03 mm). We also note the presence of veins (e = 60 microns) filled with black politic material.

- Dolomite feature

It is a massive ocher rock, very resistant to hammer and very low effervescence in HCl. In microscope, the bottom is dolomicritic with many microcracks showing dolomite and quartz recrystallization. It is noted the presence of micro-cracks and wall rock piles in pyrite grains.

Table 1. Summary of laboratory works.

Works and analysis performed	Laboratories involved
Cutting rocks, obtaining powders and preparation of thin sections	Minerals Processing Laboratory (MTL) of the Institute of Geological and Mining Research (IRGM) in Cameroon
DRX whole rock	Mission of Promotion of Local Materials (MIPROMALO), Cameroon
Preparation of thin sections Optical microscopy, major element (geochemistry)	Mineral Sciences Laboratory at the National Office of Geological Research, Algeria Office of Geological and Mining Research (BRGM), France
Physico-mechanical analyzes and mechanical tests	National Laboratory of Civil Engineering, Cameroon
Gravimetry, geochemistry of major elements	Laboratory of Chemical Analyzes, Figuil, Cameroon

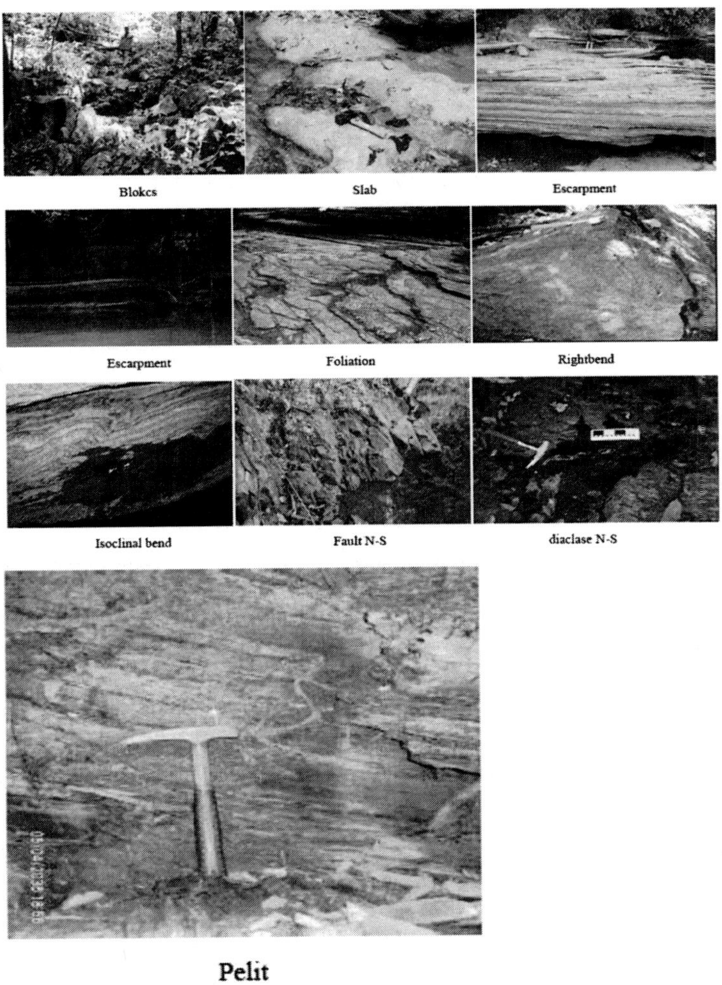

Figure 4. Outcrops and structures of the study area.

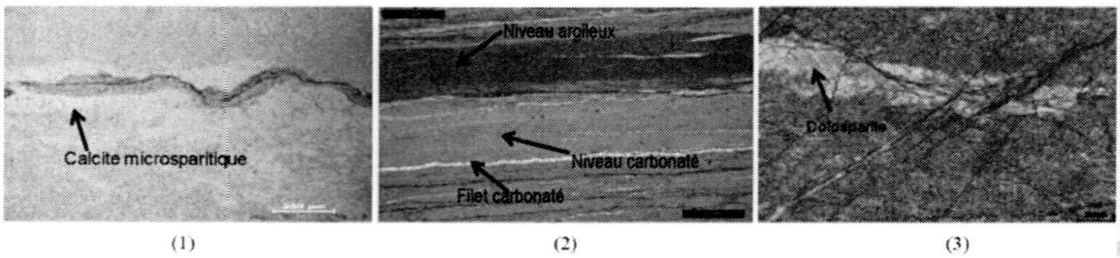

Figure 5. Features encountered. (1) Massive limestone feature; (2) Banded limestone feature; (3) Dolomite feature.

4.3. Mineralogy

The X-ray diffraction analysis sample's powders obtained from different features gave the results that are illustrated by corresponding diffractograms. From these results (Figure 6), it appears that calcite, mica and smectite constitute the major elements of massive limestone. The banded limestone is composed of calcite; mica and smectite while the mineralogical composition of dolomite is made only by dolomite and smectite.

4.4. Geochemistry

Chemical analyzes of samples have identified the different chemical compositions of the different features expressed as percentages of oxides. Table 2 presents the summary of the results. The analysis of these results shows that the massive limestone is rich in CaO and SiO_2, the massive limestone in CaO. For dolomite, the most important elements are the CaO and MgO. These contents can be the basis for various industrial uses.

4.5. Mechanical Properties

The exploitation of mechanical results (Table 3) shows that the massive limestone rock is a low-wear, resistant to both, fragmentation, and compression and relatively dense. The banded limestone meanwhile is a rock with high wear, weakly resistant to both fragmentation, and compressive strength and low density. Concerning dolomite, it is a rock with very low wear, very resistant to both, fragmentation and compression and relatively dense.

4.6. Areas of Use of Rocks

The main uses recognized for carbonated material and the main uses of specifications are presented below:

For massive limestone

It can serve as raw material for the production of clinker ($CaO > 42\%$ $MgO < 2\%$ $Fe_2O_3 > 1\%$ $Al_2O_3 > 2\%$ $SiO_2 > 6\%$); Agriculture's amendments and production of quicklime. Mechanically, its low wear (MDE = 17%), high resistance to fragmentation (LA = 20%), its high compressive strength (MPa = 107) and its high density, predispose it in civil engineering work.

For banded limestone

Their high wear MDE = 48%), low resistance to fragmentation (LA = 27.5%), their low density and compressive strength MPa = 49) showed that the main uses of this material are recruited in the areas of ornamental stones.

For dolomite

Geochemical properties of dolomite show that it can be used in the steel industry or agriculture as amendments. Its mechanical properties notably its low wear (MDE = 11%) its high resistance to fragmentation (LA = 19%), its high compressive strength (MPa = 118) and high density, describe it as good material in civil engineering.

5. DISCUSSION OF RESULTS

The Mintom carbonate rocks were already the subject of previous work [4] [8] and [9] . The [2] and [3] have focused on the establishment of a cement factory in Mintom and the determination of its location. The study of the environment deposits of carbonate rocks and their implications on the theory of the earth Snowball in Neprozoic is tackled by [4] . All these studies agree on the fact that the carbonate rocks of Mintom fall into several features organized in different lithostatic units. However, on contrary, the work of [4] , which identify three different features in the basin notably massive limestone, banded limestone and dolomite, [2] had reported only made of limestone feature. Also, there is a clear difference between the origins of the samples analyzed; the samples analyzed in this study were collected in the feature of rock outcrops while other authors have analyzed samples with a drill core S1. The comparison results show that the samples

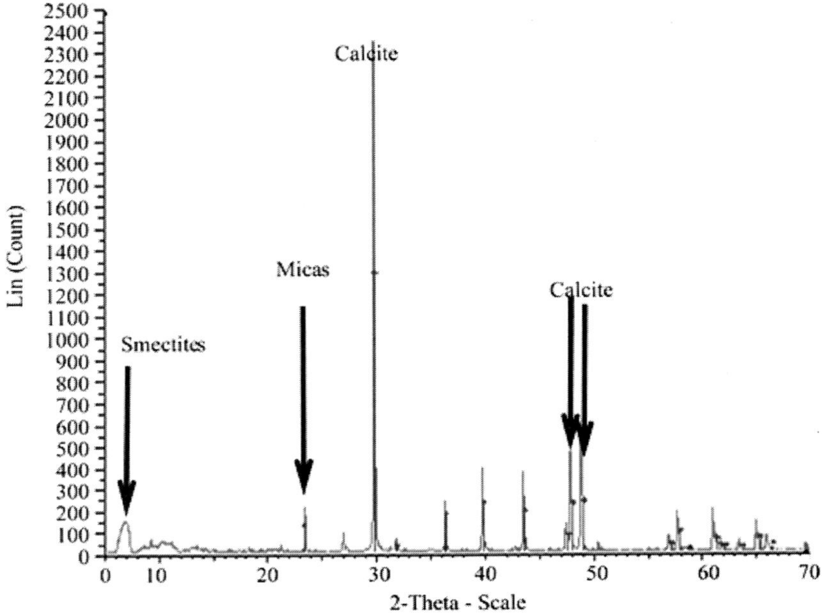
Calcite
Micas
Calcite
Smectites
Lin (Count)
2-Theta - Scale
0
10
20
30
40
50
60
70

(a)

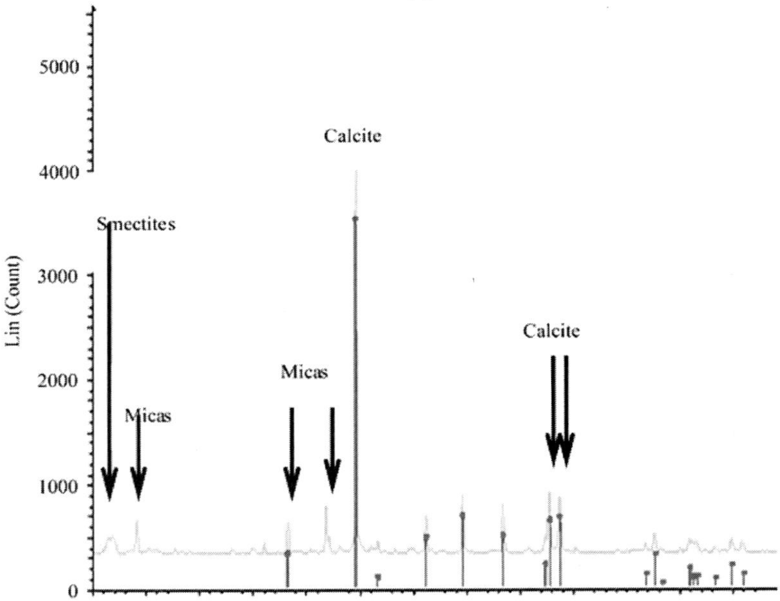
Calcite
Smectites
Micas
Micas
Calcite
Lin (Count)
5000
4000
3000
2000
1000
0

(b)

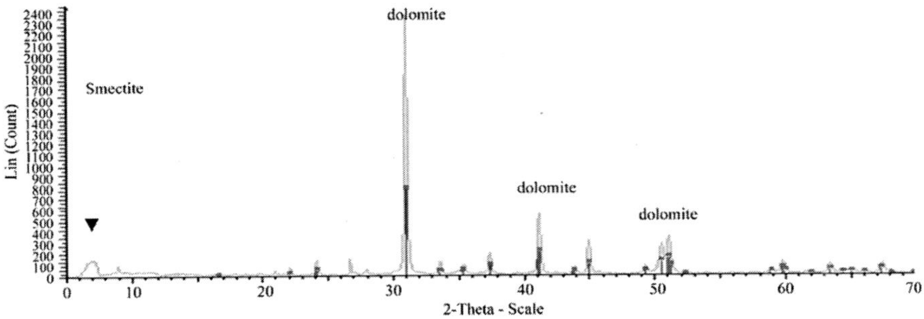

(c)

Figure 6. Mineralogy of features. (a) Massive limestone; (b) Banded limestone; (c) Dolomite.

Table 2. Summary of the results of chemical analysis.

Sample/Feature	SiO_2	Al_2O_3	Fe_2O_3	MnO	MgO	CaO	Na_2O	K_2O	TiO_2	P_2O_5	SO_3	PF	Total
Massive limestone	6.70	1.70	0.74	0.02	1.10	48.30	<Ld	0.63	0.11	<Ld	0.32	39.00	98.30
Banded limestone	23.48	5.43	3.50	0.10	2.53	25.71	0.13	2.35	0.36	0.11	<Ld	29.46	93.16
Dolomite	5.72	1.79	0.93	0.06	18.42	29.29	0.01	0.54	0.11	0.05	0.32	42.80	99.72
L.d	0.2	0.2	0.05	0.02	0.2	0.1	0.2	0.05	0.05	0.05	0.05	0.1	

Table 3. Mechanical properties of the samples collected in different features.

	Nbreof test	MDE in % (wear)	LA in % (fragmentation)	Compressive strength (MPa)	Volumetric weight in (g/cm^3)
Massive limestone	1	16	22	126	2.55
	2	18	18	91	2.52
	3	/	/	104	2.04
	4	/	/	/	2.18
	average	17	20	107	2.32
Banded limestone	1	49	26	45	1.93
	2	47	29	45.6	1.91
	3	/	/	56.2	2.18
	4	/	/	/	2.00
	average	48	27.5	49	2.00
Dolomite	1	10	20	135	2.07
	2	12	18	100	2.20
	3	/	/	120	2.48
	4	/	/	/	2.31
	average	11	19	118	2.30

of the core, especially the upper part with a depth of between 7 m and 28 m have great interest in the production of clinker thank to their very high content CaO. This result is in agreement with the work of the Ministry of Scientific Research and Innovation, which situates the limestone under a layer of laterite and conglomerate at the same depth. All authors also agree on the various highlighted carbonated rocks for direct use in the context of civil engineering constructions or indirect use when chemical composition is favorable to the production of clinker [2] [4] [8] - [11] . However, several security arrangements must be made to face several diseases in the locality of the clinker's manufactory [12] - [15] .

6. CONCLUSION

This work, done in the Mintom sedimentary basin in equatorial South East Cameroon, was to examine all uses of an industrial nature of carbonate rocks of the basin. To do this, petrographic, mineralogical and geochemical characterization prior, associated with physical-mechanical testing was necessary. The main results show that the studied basin contains three features including the massive limestone, banded limestone and dolomite. The massive limestone is hard rock, light gray, to micritic texture composed of calcite, combined with small amounts of mica and smectite. The chemical composition is dominated by the CaO while banded limestone is a dark gray rock with cryptogranular texture composed of calcite and mica. The dolomite is a massive ocher rock with dolomicritic texture and composed essentially of dolomite. Its chemical composition is dominated by CaO and MgO. In mechanical terms, the massive limestone and dolomite rocks are dense, which exhibit good resistance to wear, fragmentation and compression. Conversely, banded limestone is softer and more fragile. The petrological, physical and mechanical properties of materials studied show that they can serve as raw material suitable for the production of clinker, lime and agricultural amendments and civil engineering's work.

CITE THIS PAPER

Zo'o Zame Philémon, Mpakam Hernanie Grelle, Samba Assomo Philippe (2016) Physico-Chemical and Petrographic Characterization of Carbonated Rocks of Mintom (South-Cameroon) and Their Potential Uses. *International Journal of Geosciences*,**07**,775-783.

REFERENCES

1. Bitom, D. (1998) Organization and Evolution of Ferralitic Coverage in the Humid Tropics (Cameroon). Genesis and set Transformation Ferruginous Deep Indurated. Thesis, University of Poitiers, Poitiers, 164 p.
2. Vanhoutte, M. and Salley, P. (1986) Recognition of Limestone Mintom: Mining Project South Cameroon. United Nations Program Report 104691-59.
3. IRGM (2008) Geological Studies, Geophysical, Hydrological and Hydrogeological Limestone of Mintom. IRGM, 162 p.
4. Caron, V., Ekomane, E., Malieux, G., Moussango, P. and Ndjeng, E. (2010) Mintom Formation (New): Sedimentory and Geochemestry of Neoproterozoic, Paralic Succession in South East Cameroon. Journal of African Earth Science, 57, 367-385.
5. Feumoe, S., Ndougsa-Mbarga, T., Dicom-Manguelle, E. and Derek, J. (2012) Delineation of Tectonic Lineaments Using Aeromagnetic Data for the Area South-East Cameroon. Geofizika, 29, 175-192.
6. Nyeck, B. (2005) Organization of Ferralitic Coverage Nsimi-Zoetele Watershed Forest in Southern Cameroon Plateau Central Africa: The Role of the Pore Space and Water Dynamics in Its Transformation. Doctor Thesis, University of Yaoundé I, Yaounde, 434 p.
7. Bekoa, E. (1994) Petrological and Geochemical Study of a Soil Cover on Gneiss in Forest Areas of South Cameroon: Relationship with the Dynamics of the Iron. Doctor Thesis, University of Yaoundé I, Yaounde, 187 p.
8. IRGM (2012) Geological Studies, Geophysical, Hydrological and Hydrogeological Limestone of Mintom. Phases 2, 102 p.
9. Ekomane, E., Bissé, S., Ngueuthcthoua, G. and Ngos, S. (2014) Contribution of Staff, Track and Rare Earth Elements in the Determination of Sandstone and Pelite Source Rocks in Banga (Cameroon). CAM Review, 2, 5-15.
10. Ekodeck, G. (1984) Weathering of Metamorphic Rocks in Southern Cameroon and Their Geotechnical Aspects. Th. Doct. Univ. Scientif. And Medic, Grenoble I, 368 p.
11. Onana, V. (2010) Weathering of Chlorite Series Mbalmayo-Bengbis (Southern Cameroon) and It Geotechnical Implications. PhD Thesis, Faculty of Sciences, University of Yaounde I, Yaounde, 242 p.
12. Zo'oZame, P. (2012) Petrographic Characterization of Carbonated Rocks of Mintom for a Local Industrial Use. Th. Doct. PhD Thesis, University of Yaounde I, Yaounde, 135 p.
13. Galdibé, G. (2008) Petrographic and Structural Study of the Bergui-Career Bidzar. Mem Bachelor Univ Ngaoundéré, 42 p.
14. GENIVAR (2012) Arnaud Mine Project. Summary of the Environmental Impact Study. GENIVAR, 107 p.
15. Toumba, O. and Wakponou, A. (2014) Mining in the District of Figuil (Cameroon): Public Health and Environmental Effects. BELGEO, 4, 14 p.

Index

Index

A

B

C

D

G

H